Colorado and Adjacent Areas

Edited by

David R. Lageson
GSA Field Guide Editor
Department of Earth Sciences
Montana State University
Bozeman, Montana 59717-0348

Alan P. Lester
Co-Editor
Department of Geological Sciences
University of Colorado
Boulder, Colorado 80309-0399

and

Bruce D. Trudgill
Co-Editor
Department of Geological Sciences
University of Colorado
Boulder, Colorado 80309-0399

Field Guide 1
1999

Published by The Geological Society of America, Inc.
3300 Penrose Place, P.O. Box 9140, Boulder, Colorado 80301

Printed in U.S.A.

GSA Field Guide Series Science Editor David Lageson

Library of Congress Cataloging-in-Publication Data

Colorado and adjacent areas / edited by David R. Lageson, Alan Lester, and Bruce Trudgill.
    p. cm. -- (Field guide ; 1)
  Includes bibliographical references.
  ISBN 0-8137-0001-9
    1. Geology--Colorado--Guidebooks. 2. Geology--West (U.S.)--Guidebooks. 3.
Colorado--Guidebooks. 4. West (U.S.)--Guidebooks. I. Lageson, David R. II Lester,
Alan, 1960- III. Trudgill, Bruce, 1964- IV. Field guide (Geological Society of America) ;
1.

QE91 .C685 1999
557.88--dc21
                                                              99-048043

Cover: Andrew's Peak in southwestern Rocky Mountain National Park. Photo by Diane C. Lorenz

10 9 8 7 6 5 4 3 2 1

Contents

Preface... v

Note vi

1. *Bouncing boulders, rising rivers, and sneaky soils: A primer of geologic hazards and engineering geology along Colorado's Front Range* 1
   D. C. Noe, J. M. Soule, J. L. Hynes, and K. A. Berry

2. *Laramide to Holocene structural development of the northern Colorado Front Range* 21
   E. A. Erslev, K. S. Kellogg, B. Bryant, T. K. Ehrlich, S. M. Holdaway, and C. W. Naeser

3. *Laramide faulting and tectonics of the northeastern Front Range of Colorado* 41
   E. A. Erslev and S. M. Holdaway

4. *Hydrogeology and wetlands of the mountains and foothills near Denver, Colorado* 51
   K. E. Kolm and J. C. Emerick

5. *Field trip to Manitou Springs, Colorado, with specific emphasis on the sediments of Cave of the Winds and their relationship to nearby alluvial deposits and spring sediments* 61
   F. G. Luiszer

6. *200,000 years of climate change recorded in eolian sediments of the High Plains of eastern Colorado and western Nebraska* 71
   D. R. Muhs, J. B. Swinehart, D. B. Loope, J. N. Aleinikoff, and J. Been

7. *Walking tour of paleontologist George G. Simpson's boyhood neighborhood* 93
   L. F. Laporte

8. *Active evaporite tectonics and collapse in the Eagle River valley and the southwestern flank of the White River uplift, Colorado* 97
   R. B. Scott, D. J. Lidke, M. R. Hudson, W. J. Perry, Jr., B. Bryant, M. J. Kunk, J. R. Budahn, and F. M. Byers, Jr.

9. *Coal mining in the 21st century: Yampa coal field, northwest Colorado* 115
   M. E. Brownfield, E. A. Johnson, R. H. Affolter, and C. E. Barker

10. *Field guide to the continental Cretaceous-Tertiary boundary in the Raton basin, Colorado and New Mexico* 135
    C. L. Pillmore, D. J. Nichols, and R. F. Fleming

11. *Stratigraphy, sedimentology, and paleontology of the Cambrian-Ordovician of Colorado*.... 157
    P. M. Myrow, J. F. Taylor, J. F. Miller, R. L. Ethington, R. L. Ripperdan, and C. M. Brachle

12. *Field guide for the Heart Mountain detachment and associated structures, northeast Absaroka Range, Wyoming* 177
    D. H. Malone, T. A. Hauge, and E. C. Beutner

Preface

This inaugural volume of the Geological Society of America (GSA) Field Guide series represents an entirely new publication venue for the Society, one that emphasizes the "field roots" of the earth sciences profession. The timing of this new series seems both appropriate and ironic, considering that geology at the close of the twentieth century is immersed in a level of technology that has drawn attention away from its field-oriented beginnings. In an era when universities are altering their traditional field-based curricula in favor of more marketable (computer-based) skills, it seems that going to the field and observing has become a thing of the past. In contrast, we feel that a broad-based field perspective is crucial to understanding most geological problems. Field observations provide for the germination of new ideas and are, ultimately, their testing ground.

The Reverend John Walker, who taught the first systematic course in geology at the University of Edinburgh (1781–1803), had this to say to his students: "The way to knowledge of natural history is to go to the fields, the mountains, the oceans, and observe, collect, identify, experiment, and study." Two hundred years later, there is still much to be said for his approach. It is in this spirit that GSA has created this Field Guide series.

The new Field Guide series will initially showcase field excursions associated with national meetings of the Society, maintaining the high standards of excellence for which other GSA publications are known. The series may be expanded in the near future to showcase selected field trips from sectional meetings as well as special field excursions. Moreover, as professional earth scientists, we hold a responsibility to promote public outreach and education, to address those geological issues relevant to the needs of society, and to disseminate information about the Earth in formats that do the greatest good for the greatest number. Therefore, it is hoped that the Field Guide series will not only serve as a valuable resource for our profession, but also serve as a venue for public education and outreach.

As we prepare to slide into the year 2000, we dedicate this first Field Guide volume to those field-based earth scientists of the twentieth century who walked the outcrops and paid their dues in the field.

DAVID R. LAGESON
GSA Field Guide Editor
Department of Earth Sciences
Montana State University

ALAN LESTER
Co-Editor
Department of Geological Sciences
University of Colorado

BRUCE TRUDGILL
Co-Editor
Department of Geological Sciences
University of Colorado

Note

This volume was compiled from field trips offered as part of the 1999 Geological Society of America Annual Meeting in Denver. The following field trips could not be included in this volume because of unusually tight publication deadlines.

Cretaceous hydrocarbon plays—southern Colorado
P. R. Krutak

Geological reconnaisssance of Dinosaur Ridge and vicinity
N. Cygan, B. Rall, T. Caneeer, and B. Raynolds

Kimberlites of the Colorado-Wyoming state line district
P. Modreski and T. Michalski

Geology tour of Denver buildings and monuments
J. Murphy

Tour of the U.S. Geological Survey mapping and geologic facilities, Denver Federal Center
P. Modreski and J. Kerski

Soil-geomorphic relationships near Rocky Flats, Boulder and Golden, Colorado, with a stop at the pre-Fountain Formation of Wahlstrom
P. Birkeland, R. Shroba, and Penny Patterson

South Park conjunctive use project: A combined look at geology and hydrology in the South Park Basin, Colorado
M. F. McHugh, J. Jehn, and H. Eastman

Geology and paleontology of the gold belt back-country byway: Florissant Fossil Beds and Garden Park fossil area
H. Meyer, T. W. Henry, D. Grenard, and E. Evanoff

Geological Society of America
Field Guide 1
1999

Bouncing boulders, rising rivers, and sneaky soils:
A primer of geologic hazards and engineering geology
along Colorado's Front Range

David C. Noe, James M. Soule, Jeffrey L. Hynes, and Karen A. Berry
Colorado Geological Survey, 1313 Sherman Street, Room 715, Denver, Colorado 80203, United States

INTRODUCTION

The purpose of this one-day field trip is to look at the engineering geology and examples of geologic hazards along the western edge of the Denver Metropolitan Area. We will focus on the foothills of the Colorado Piedmont, an area that marks the transition between the Great Plains and the Front Range of the Rocky Mountains. The trip begins and ends in downtown Denver, at the Colorado Convention Center. We will travel northwest to Boulder, then south through Golden and the unincorporated suburbs of Jefferson County, and then return northwest to downtown Denver (Fig. 1). The trip guide contains narrative descriptions for 21 stops, twelve of which we will "roll by" for the GSA Annual Meeting field trip.

Because of the dynamic nature of engineering geology, it is useful to look at past events as a means of planning for future development projects. In Colorado, the Colorado Geological Survey (CGS) has a statutory requirement to assist local governments with planning issues involving engineering geology. Since the early 1970s, CGS geologists have reviewed plans for hundreds of subdivisions along the Colorado Piedmont. The authors have provided emergency assistance and advice for incidents involving geologic hazards, as well as research and mapping of general geologic hazards (e.g., Soule, 1978) and specific hazards such as heaving bedrock (e.g., Noe and Dodson, 1997) for the CGS. The goal of this field trip is to provide an account of engineering-geologic factors that have affected the Colorado Piedmont in the past and will continue to do so in the future.

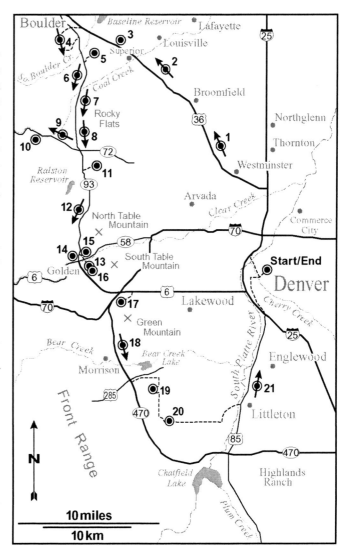

Figure 1. Index map showing the field-trip route in relation to cities and other prominent natural and cultural features in the Denver Metropolitan Area. The locations of actual stops for the GSA Annual Meeting field trip are marked by bull's-eye dots; dots with arrows denote "roll-by" locations.

Noe, D. C., Soule, J. M., Hynes, J. L., and Berry, K. A., 1999, Bouncing boulders, rising rivers, and sneaky soils: A primer of geologic hazards and engineering geology along Colorado's Front Range, *in* Lageson, D. R., Lester, A. P., and Trudgill, B. D., eds., Colorado and Adjacent Areas: Boulder, Colorado, Geological Society of America Field Guide 1.

This field trip guide may be used for self-guided parties. Although it contains general directions and locations of the stops, we recommend bringing along a "close up" atlas and street guide of the Metropolitan Denver Area to aid in locating and driving between the field trip stops. We have found these detailed street maps to be a valuable tool for locating geologic-hazard areas in rapidly developing, suburban areas.

PART 1. DENVER TO BOULDER

Stop 1 (Roll-by)—Geologic Setting
(0.0-12.3 miles; 0.0-19.8 km)

Directions: Leaving downtown Denver, drive north on Speer Boulevard. Exit right onto Interstate 25 (north), then exit onto U.S. Highway 36 to Boulder (west).

The foothills of the Colorado Piedmont are situated between the Denver Basin to the east and the Front Range uplift of the Rocky Mountains to the west. The Denver Basin contains nearly 13,000 feet (4,000 m) of Pennsylvanian to Paleocene sedimentary rocks (Fig. 2; Table 1). The basin is highly asymmetrical in its east-west cross-section (Fig. 3). Most of the Denver Metropolitan Area is underlain by the gently dipping, eastern limb. In contrast, the foothills area is underlain by the steeply dipping, western limb. The Front Range uplift rises 1,500-2,500 feet above the Piedmont along the mountain front. The uplift contains a variety of Precambrian metamorphic and plutonic rocks (see Trimble and Machette, 1979).

The major faults within the Colorado Piedmont are large, en-echelon, high-angle thrust faults associated with the Front Range uplift, between Boulder and Colorado Springs. Of these, we will encounter the Golden Fault at Stop 15. The fault is considered to be inactive by most geologists, although some Quaternary movement has been postulated for an associated fault graben near Golden (Kirkham, 1977). We will encounter a second set of faults in the Boulder-Weld coal field area near Marshall, at Stop 5. These faults have a northeast-southwest trend, with normal displacements on the order of a few feet to over 400 feet. They are interpreted to have formed contemporaneously with the deposition of coals during Cretaceous time, based on coal-thickness relationships (i.e., thicker in grabens, thinner on horsts; see Weimer, 1977).

Unconsolidated, Quaternary deposits in the Denver area include eolian (sand and loess), colluvial, and alluvial-terrace deposits (Fig. 4). The alluvial terraces along the Front Range Piedmont have been studied and subdivided by several authors, most notably Hunt (1954) and Scott (1960). From oldest to youngest, there are four well-defined, older alluvial-terrace deposits on rock-floored pediments: the Nussbaum, Rocky Flats, Verdos, and Slocum Alluviums (Pleistocene). Additionally, there are five younger alluvial-fill terrace deposits: the Louviers and Broadway Alluviums (Pleistocene) and the Piney Creek Alluvium and post-Piney Creek alluvium (Holocene). These deposits are shown in a schematic profile in Figure 5. A volcanic-ash bed in the Verdos Alluvium has been dated at 0.6 m.y. (see Van Horn, 1976).

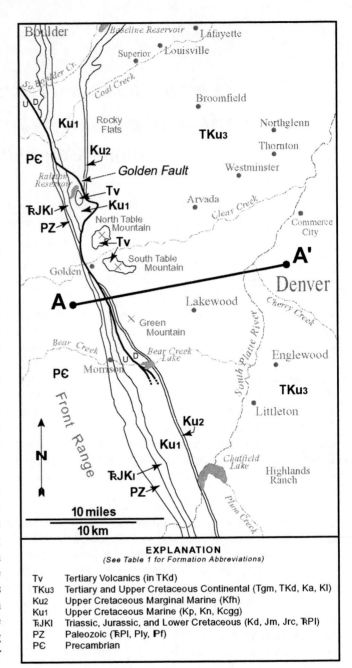

Figure 2. Index map showing the general bedrock geology of the Denver Metropolitan Area (modified from Trimble and Machette, 1979, and Costa and Bilodeau, 1982).

EXPLANATION
(See Table 1 for Formation Abbreviations)

Tv	Tertiary Volcanics (in TKd)
TKu3	Tertiary and Upper Cretaceous Continental (Tgm, TKd, Ka, Kl)
Ku2	Upper Cretaceous Marginal Marine (Kfh)
Ku1	Upper Cretaceous Marine (Kp, Kn, Kcgg)
⊤RJKl	Triassic, Jurassic, and Lower Cretaceous (Kd, Jm, Jrc, ⊤Rpl)
PZ	Paleozoic (⊤Rpl, Ply, Pf)
P€	Precambrian

The general geology of the Colorado Piedmont has been studied and mapped in detail. There have been several engineering-geologic studies published for the area; however, few published case-study investigations involving geologic hazards are available. The U.S. Geological Survey has published 1:24,000 engineering geology maps for the entire field-trip area (Gardner, 1969; Gardner and others, 1971; Miller and Bryant, 1976; McGregor and McDonough, 1980; Simpson and Hart, 1980), and

Table 1. Generalized stratigraphic section for the Colorado Piedmont, Morrison quadrangle (modified from Scott, 1972).

Thickness in
Feet (meters)

Green Mountain Conglomerate (*Paleocene*) – Tgm
    Conglomerate, sandstone, and shale. Contains andesite pebbles..650 (198)
Denver Formation (*Paleocene and Upper Cretaceous*) – TKd
    Brown to olive-gray claystone, siltstone, sandstone, and conglomerate.
    Contains three flows of potassium-rich basalt (shoshonite)
    Rich in andesite pebbles 950 (290)
Arapahoe Formation (*Upper Cretaceous*) – Ka
    White, gray, and yellow sandstone, siltstone, claystone, and conglomerate.
    Conglomerate clasts are sedimentary, igneous, and metamorphic rock 400 (121)
Laramie Formation (*Upper Cretaceous*) – Kl
    Gray siltstone and claystone and yellow and white sandstone.
    Coal in lower part...550 (168)
Fox Hills Sandstone (*Upper Cretaceous*) – Kfh
    Olive to brown silty shale and yellowish-orange sandstone 180 (55)
Pierre Shale (*Upper Cretaceous*) – Kp
    Olive-green shale, some beds of olive to gray sandstone, and limestone concretions..............6,200 (1,890)
Niobrara Formation (*Upper Cretaceous*) – Kn
    Smoky Hill Shale Member:
    Pale to yellowish-brown, thin-bedded, calcareous shale and thin-bedded limestone 140 (43)
    Fort Hayes Limestone Member:
    Yellowish-gray, dense limestone.. 35 (11)
Carlile Shale, Greenhorn Limestone, and Graneros Shale (*Upper Cretaceous*) – Kcgg
    Gray claystone, siltstone, calcarenite, and hard limestone beds 530 (162)
Dakota Group (*Lower Cretaceous*) – Kd
    South Platte Formation:
    Yellowish-gray sandstone and dark gray shale
    Lytle Formation:
    Yellowish-brown sandstone and conglomerate 300 (91)
Morrison Formation (*Upper Jurassic*) – Jm
    Red and green siltstone and claystone.
    Minor beds of brown sandstone and gray limestone 300 (91)
Ralston Creek Formation (*Jurassic*) – Jrc
    Purplish-gray sandstone and siltstone, yellow silty sandstone 90 (27)
Lykins Formation (*Triassic? and Permian*) – TRPl
    Maroon shale, sandy limestone, and maroon and green siltstone...450 (137)
Lyons Formation (*Permian*) – Ply
    Yellowish-orange to yellowish-gray sandstone and conglomerate.. 190 (58)
Fountain Formation (*Permian and Pennsylvanian*) – Pf
    Maroon arkosic sandstone and conglomerate...1,650 (502)
Precambrian rocks – PC
    Igneous and metamorphic rocks.

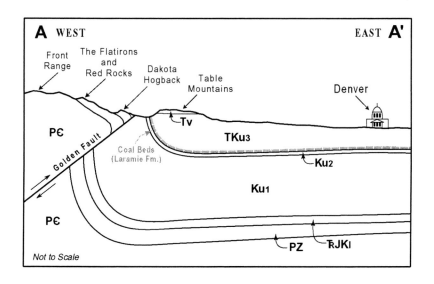

Figure 3. Schematic cross-section through the Golden-Denver area showing the bedrock geology of the Denver Basin and the Colorado Piedmont (modified from Wells, 1967; Van Horn, 1972; and Costa and Bilodeau, 1982). See Figure 2 for cross-section location and explanation of geologic symbols.

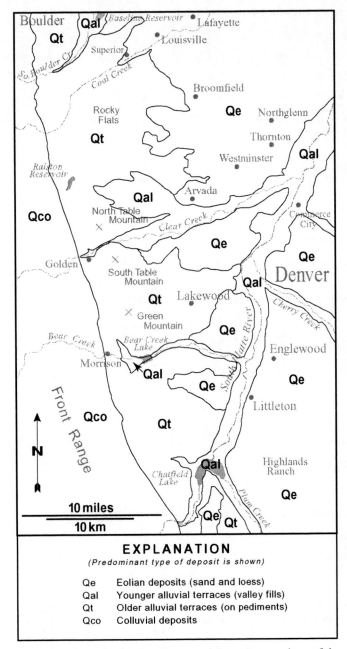

Figure 4. Index map showing the general Quaternary geology of the Denver Metropolitan Area (modified from Trimble and Machette, 1979, and Costa and Bilodeau, 1982).

parts of many of other USGS publications contain descriptions of area-specific engineering-geologic considerations. Many of these maps and publication are out of print. Costa and Bilodeau (1982) have published a synopsis of the engineering geology of Denver, with particular reference to the geologic conditions and engineering of large projects in the downtown area, as part of the Association of Engineering Geologists' "Geology of the World's Cities" series.

Stop 2 (Roll-by)—Development pressure along the Colorado Front Range (12.3-23.2 miles; 19.8-37.4 km)

Directions: Continue along U.S. Highway 36 past the Broomfield exit, toward Boulder (northwest).

The Denver Metropolitan Area and other urban areas along the Colorado Front Range Piedmont were founded and started to grow rapidly during the mid-to-late nineteenth century. Denver, named after a Kansas territorial governor, was founded in 1859. The coming of the railroads greatly accelerated this growth and a secondary service economy ensued. Gen. William Jackson Palmer, a Civil War general, founded the Denver and Rio Grande Western Railroad (now part of the Union Pacific) and the city of Colorado Springs around 1870. Precious metal mining and the near-instant appearance of mining camps in the mountains and coal mining, agriculture, and water development on the Piedmont were responsible for most of this early rapid growth. This continued seemingly unabated until the decline of the metal-mining industry, which accompanied the rapid decline in silver prices around 1893. Many mansions, commercial buildings, and churches were built during the late 1880s. The Colorado State Capitol was started in 1890. Many of these buildings are built from Colorado native stone and are the focus of an interesting tour for a geologist (see Murphy, 1995). Colorado, the Centennial State, was admitted to the Union in 1876.

A century later, this region has seen rapid growth during the regional "energy-boom" years of the mid-1970s to early 1980s, a "bust" from about 1984 to the early 1990s, and another boom accompanying national trends which continues to the present. The Denver Regional Council of Governments estimated the regional population around Denver to be about 2.2 million at the beginning of 1993. The total Front Range regional population is probably now about a million more than that. The latest period of booming growth is characterized by rapid urbanization on the fringes and merging of all of the region's urban centers. This urbanization style is most apparent between Loveland and Fort Collins, the south Denver Metropolitan area (including Highlands Ranch and other communities in northern Douglas County), and Colorado Springs. This is one of the most rapidly growing regions in the nation.

The predominant style of residential growth in this region is now single-family tract-housing development on lots ranging in size from about 3,500 to 8,000 square feet (1/10 to 1/5 acre) with its attendant interspersed infrastructure. In recent years, the tendency has been toward construction on smaller lots with multi-story houses. This has probably slowed the per-capita land consumption by development somewhat, but not its overall rate or amount. A tendency has been for infill of this higher density residential development among older lower density residential and agricultural land uses along the urban fringes.

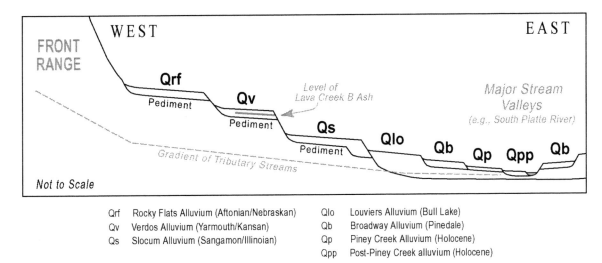

Figure 5. Schematic cross-section showing the Quaternary alluvial-terrace deposits in the Denver area (modified from Scott, 1963, and Madole, 1991).

Stop 3—Boulder Creek Valley Overlook (23.2 miles; 37.4 km)

Directions: About a mile after passing the Louisville (McCaslin Boulevard) exit on U.S. Highway 36, look for the "scenic overlook" sign at the crest of the hill. Exit into the overlook area.

Our first stop is atop Davidson Mesa. It affords an excellent opportunity to overlook the valley of South Boulder Creek, the city of Boulder, the Flatirons (distinctive, reddish slabs of dipping Fountain Formation conglomerate and sandstone), and the Front Range beyond (Fig. 6).

The geologic display board at this site gives a good general stratigraphic portrayal of the area. The mesa we're standing on is a classic example of topographic reversal. The Verdos-age alluvial-terrace gravel deposits that cap the mesa were deposited on this surface about 600 thousand years ago, when this area was a valley. These gravel deposits now act to protect the soft sediments below, while the flanking sediments have been eroded away to become sides of the mesa.

Other topographic highs off to the north have been instrumental in creating a stream piracy episode on South Boulder Creek. The original channel of South Boulder Creek can be seen trending off to the northeast, in a nearby valley just beyond Baseline Reservoir. Farther to the north stands Valmont Butte, an isolated, basalt dike. At the western end of this butte, a north-draining side stream had developed along a north-south, strike valley in the soft shales of the Pierre Shale. This stream eroded headward (southward) along the edge of a more-resistant questa of Fox Hills Sandstone, and it had a significantly steeper gradient than that of South Boulder Creek. Its head eventually intercepted the original channel of South Boulder Creek and "captured" the larger stream, probably during a large flood event. In terms of land use, the piracy hypothesis explains why the current valley of South Boulder Creek is a poor gravel resource, and

why a large, high-quality gravel resource was "unexpectedly" discovered in the older, now-abandoned valley. We'll see a remarkably similar, pre-piracy situation later on the trip (at Leyden Gulch, Stop 11).

Stop 4 (Roll-by)—Table Mesa Area, Boulder (27.1-29.2 miles; 43.6-47.0 km)

Directions: Continue along U.S. Highway 36 towards Boulder (northwest). Take the Table Mesa-Foothills Parkway exit, staying to the right, and then turn left on Table Mesa Drive (west).

For this roll-by, we will look at the engineering geology of an urbanized, mountain-front community. The City of Boulder's Table Mesa area was subdivided and developed during the 1960s and 1970s. During development, many of the small, mountain-front stream drainages in the area were regraded, filled, or rerouted. Unfortunately, the resulting collection system was occasionally overwhelmed by runoff from summer thunderstorms. Driving westward on Table Mesa Drive past Broadway, we will encounter Bear Creek, the major collecting stream for urban runoff from the area. The stream has undergone several episodes of periodic flash flooding because of the increased runoff due to urbanization. A number of flood-control improvements have been made along the drainage in recent years to reduce the effects of aggradation and erosion. These include drop structures, which feature large, native boulders from the Fountain Formation, and an oversized box culvert and bikeway beneath Broadway.

Turning south on Lehigh Street, we will climb up a series of alluvial terraces of Quaternary age as we approach Shanahan Hill. The present valley of Bear Creek contains the Piney Creek and post-Piney Creek alluviums. To the right (west), the scenic National Center for Atmospheric Research (designed by I.M. Pei) is perched atop the area's highest, oldest pediment terrace, under-

Figure 6. Photograph of the valley of South Boulder Creek, showing typical terrain of the Colorado Piedmont. Mountain and local streams have dissected the landscape, leaving flat-topped mesas and broad valleys with gentle to steep side slopes. The Flatirons, a series of rock fins composed of Pennsylvanian Fountain Formation, mark the abrupt transition between the Piedmont and the Front Range uplift in the upper left part of the photograph.

lain by the Rocky Flats Alluvium. We begin a mile-long incline underlain by the Slocum Alluvium. After passing a steeper slope near the crest of the hill, underlain by the Pierre Shale, we finally "top out" Shanahan Hill. There, the road meanders along a terrace underlain by the Verdos Alluvium. Circling to the east on Greenbriar Boulevard, we descend the hill and reverse this sequence.

On Shanahan Hill, most of the Quaternary alluvial units are alluvial fans that were deposited by debris flows and mud flows. These deposits were flushed from the Flatirons above during localized, high-intensity downpour events. They typically consist of a muddy sand matrix with interspersed boulders and cobbles. Note the ubiquitous, large boulders that are now used for yard landscaping. These boulders give Boulder its name, and they present a formidable challenge to the City's home gardeners. The city has been spared from catastrophic debris flows during historic times. Throughout Colorado, however, there is a strong link between wildfires and subsequent debris-flow activity (e.g., Coe, 1987; Soule, 1999; Kirkham et al., in preparation). Boulder's debris-flow safety may depend, to some degree, on careful fire suppression and management in the city-owned forests above town.

The presence of cattails amongst the condominiums, on the southern flank of Shanahan Hill, indicates that there is a shallow water table in the area. This is a result of groundwater "perching" within the Verdos Alluvium, on the top of the relatively impermeable Pierre Shale. Post-development lawn irrigation has added significantly to the overall, ambient level of perched groundwater in this area. Several of the condominiums now experience flooding after multiple-day (monsoon) rainfalls, which raise the water table for days following the storms.

PART 2. BOULDER TO GOLDEN

Stop 5—Coal-Mine Fires and Subsidence, Marshall (32.7 miles; 52.6 km)

Directions: Follow Greenbriar Boulevard past the high school (appropriately named Fairview, with its view of Boul-

der and Longs Peak) to the stoplight. Turn right onto Broadway (south), which becomes State Highway 93. After a mile, turn left at the stoplight and drive eastward through the hamlet of Marshall to the corner of Marshall Road and Cherryvale Road.

Marshall is the discovery point for the Boulder-Weld Coal Field, which stretches some thirty miles to the northeast along the outcrop of the Upper Cretaceous Laramie Formation. The Laramie Formation is a sedimentary sequence of interbedded shale, claystone and sandstone. Coal seams are located throughout the formation. The large swales and ridges in the valley bottom to the south of Marshall Road are related to subsidence caused by both mining and coal fires.

The subsidence pattern in the open-space area to the northeast of the road intersection reflects the room-and-pillar pattern of the workings directly below (Fig. 7). The mine depth here is roughly 30-40 feet, and the extraction height was 5-6 feet. The piston-like nature of these sinkholes is a function of the very brittle, low tensile-strength sandstone beds that formed the roof of the mine. Walking across the area, it is possible to detect steam and sulfur vapors and deposits. These are due to a smoldering coal fire under this site. The ditch on the hillside has undergone significant structural enhancement through this section to maintain its hydraulic integrity and proper grade.

The coal-mine fires and subsidence have imposed considerable land-use constraints on the Marshall area. We will see examples of careful siting of residential buildings on a lot-by-lot basis, based on carefully locating intact coal pillars and the limits of mined areas. In Figure 7, note the location of the new house (N) just outside of the subsidence area. The old, converted schoolhouse (S) has since been relocated to the west in order to increase its setback from the road. It, too, had to be sited carefully to avoid workings of the Fox mine. Boulder County has attained much of the property in the area for use as open space.

Stop 6 (Roll-by)—"Boulder's Stonehenge," a Water Tale (34.5 miles; 55.5 km)

Directions: Backtrack through Marshall, turn left onto Highway 93, and proceed toward Golden (south).

Leaving Marshall, we will climb out of the valley of South Boulder Creek. Notice the stone ruins at the crest of the hill, to the right (west). These stone ruins are the remnants of a twice-burned restaurant called the Matterhorn. The site is locally known as "Boulder's Stonehenge"; Fig. 8). It is located atop a mesa capped by Rocky Flats Alluvium, here underlain by a thick sequence of Pierre Shale.

Both times, the restaurant burned to the ground as the result of a grease fire in the kitchen and the unavailability of on-site water to fight the blaze. The position of the site at the edge of the gravel cap had provided a stunning view; however, it had also doomed the project because no appreciable groundwater could be located or recovered. The shales beneath the site are too tight to yield water, and any perched water entering the granular alluvium from higher on the terrace drains away into the valley near the edge of the alluvial cap because of seepage effects.

Stop 7 (Roll-by)—Mining Operations in Northern Jefferson County (36.3-37.5 miles; 58.4-60.4 km)

Directions: Continue along State Highway 93 toward Golden (south).

As we cross the county line from Boulder County to Jefferson County, you will notice an increase in mining and industrial activities (Fig. 9). On the right (west) side of State Highway 93 is a shale-mining operation. The pit is several hundred feet deep and provides a good exposure of the Pierre Shale, as well an opportunity for fossil collecting (e.g., Cretaceous *Baculites* and other marine fossils). The Pierre Shale is mined and processed as a lightweight aggregate. Across the highway, the shale is "baked" into a hard, light aggregate that is primarily used in the construction of buildings where the weight of concrete is a consideration. The processed fine-sized material is also used as an alternative to sand on area roads during the winter. These materials create less dust than sand, and they help to mitigate some of Denver's PM10 air quality concerns.

Just south of the tall shale processing plant, the Rocky Flats Alluvium is mined for construction aggregate. The clays of the Laramie Formation are also mined for local brick production. A "hogback" of the Fox Hills sandstone was buried during deposition of the Rocky Flats Alluvium. The alluvium that was deposited west of the "buried hogback" tends to contain clay that is difficult to remove from the aggregate. It seems the hogback served as an early sediment basin. The clay-coated aggregate takes a significant amount of washing in order to be used in concrete.

Looking again to the right (west), we see a pronounced, red scar along the base of the flatirons on Eldorado Mountain, about 2 miles (3.2 km) away. This is the Eldorado Canyon Quarry, which produced construction aggregate from the Lyons Sandstone. In the mid-1980s, due to lack of support by Boulder County, an application to expand the quarry was defeated. Boulder County Open Space has since purchased the site and attempted to reclaim it. The long, tan scars across the lower flank of the mountain are abandoned clay mines in the Dakota Group. The refractory-grade, kaolinite-rich clay was used for industrial ceramics.

Stop 8 (Roll-by)—Surficial Geology and Geomorphology of the Rocky Flats Surface (37.3-37.9 miles; 60.0-61.0 km)

Directions: Continue along State Highway 93 toward Golden (south).

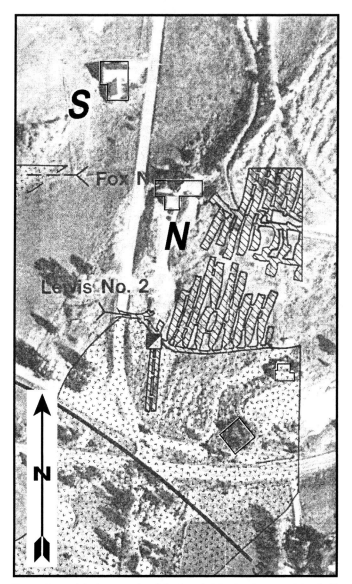

Figure 7. Aerial photograph of the Lewis No. 2 mine near Marshall, showing known (striped) and inferred (stippled) underground-mining areas (courtesy of Colorado Geological Survey, Subsidence Information Center). Room-and-pillar mining patterns are reflected in the subsidence features above these shallow workings.

Figure 8. Photograph of "Boulder's Stonehenge," the ruins of the old Matterhorn restaurant, looking north toward the Flatirons. The lack of a groundwater source spelled this building's doom when it burned on two separate occasions.

Rocky Flats is a gently eastward-sloping, upland surface on the Front Range Piedmont between Coal Creek and Leyden Creek. The surface is dissected by several shallow drainages that radiate across it in an east-to-northeast direction. The depth of dissection stream increases to the east, where the surficial Rocky Flats Alluvium outcrop ends and underlying sandstone and shale bedrock is exposed. The Rocky Flats surface is mantled by this gravelly deposit of variable thickness, which is considered to be the oldest and highest in a sequence of widespread, correlatable Quaternary alluvial-terrace deposits that developed in the Piedmont paleodrainages (Scott, 1960). Presumably, these are high-energy fluvial deposits that formed as alluvial fans near the mountain front and grade eastward into lower-energy stream terraces and overbank-flood deposits, out on the High Plains. The Rocky Flats Alluvium has a strongly developed, old, and unusual soil, and an uncommon plant community. It is considered to be early Pleistocene in age. Its provenance at this locality is considered to be the mountain part of the drainage of ancestral Coal Creek, partially because it contains large numbers of Coal Creek Quartzite boulders that armor the surface (Shroba and Carrara, 1996).

Rocky Flats was used early in this century primarily as grazing land and for access to Coal Creek Canyon. It is not a particularly hospitable place for habitation because of its local weather (which can be severe at times because of high winds and blowing snow), even though it is relatively close to the northwest Denver suburbs, Golden, and Boulder.

The earliest use of the area to directly benefit the local citizenry was in 1936, when the Denver Water Board (now Denver Water) built the South Boulder Diversion Conduit across Rocky Flats. This system of pipes and aqueducts is nearly 10 miles (16.1 km) in length. It originates at the South Boulder [Creek] Diversion Dam and carries raw water to Ralston Reservoir, which is about 2 miles (3.2 km) south of Rocky Flats. It is part of a larger system that carries Fraser River water from the Moffat Tunnel and Gross Reservoir through six siphons, five tunnels and flumes,

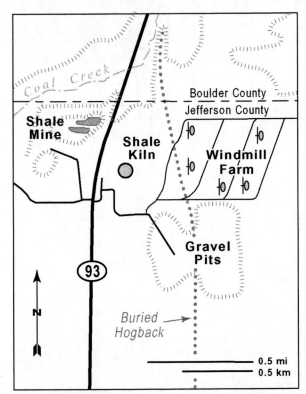

Figure 9. Map of shale and gravel aggregate mining operations at the northern end of Rocky Flats. Note the shale kiln, the windmill farm, and the buried hogback that has played a role in the clay content and quality of the aggregate deposit.

and concrete lined and unlined open channels and canals (Figure 10; Denver Water, 1994). This is a key element of the northern Denver Water collection and storage system.

The Rocky Flats [Nuclear Weapons] Plant was established in 1952 as a manufacturing facility and was renamed the Rocky Flats

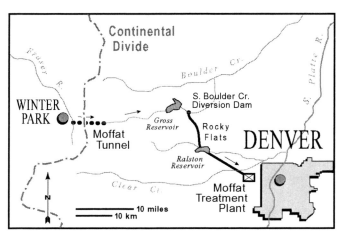

Figure 10. Schematic map of the features of the Denver water system in the vicinity of Rocky Flats (Denver Water, 1994). The South Boulder Diversion Conduit crosses Rocky Flats and links South Boulder Creek and Ralston Reservoir.

Environmental Technology Site in 1996. A contractor to the U.S. Department of Energy (DOE) now manages this site. The site has been the subject of numerous environmental assessments for seismic risk, site remediation, radioactivity hazards, and cleanup over the years. The main facility, located to the left (east) of State Highway 93, is readily visible along the field-trip route. Recently, DOE's National Renewable Energy Laboratory has established the National Wind Technology Center on Rocky Flats. It is located about 1.5 miles (2.4 km) northwest of the Plant at the northwest corner of the Rocky Flats DOE reservation. Sixteen experimental wind turbines are currently in operation at this facility.

A recent human activity at Rocky Flats has been capturing larger resistant stones from the Rocky Flats Alluvium for sale to landscaping companies. Stockpiles of these usually can be seen southwest of the intersection of State Highways 93 and 72.

Stop 9 (Roll-by)—The Moffat Road (39.7-42.2 miles; 63.9-67.9 km)

Directions: Turn right onto State Highway 72, marked by a stop-light, and head toward the mouth of Coal Creek Canyon (west).

Notice the railroad grade that climbs across the rugged face of Eldorado Mountain, ahead to your right, as we turn westward onto State Highway 72 and continue across the Rocky Flats. This is "the Moffat Road," one of Colorado's most famous railroad lines. We will have several opportunities along the trip route to look at the engineering geology of this railroad line in its difficult transition between the Piedmont and the Rocky Mountains.

First, a bit of history (from Bollinger and Bauer, 1962). Denver, a mining town that aspired to be a great city, was in a difficult spot at the turn of the century. It had been bypassed, both to the north and south, by the transcontinental railroads. The numerous railroad lines that radiated westward out of Denver were all narrow gauge lines. These 3-foot-wide rail lines required small engines and small rolling stock to negotiate the narrow, sinuous

canyons and steep grades of the Rocky Mountains. In 1902, David Moffat began constructing a standard gauge line, the Denver, Northwestern & Pacific (later Denver & Salt Lake) Railway Company. The use of larger engines, rolling stock, and track gauge (4-foot 8.5-inches wide) (about 1.43 m) required gentler grades and larger-radius curves. Canyon routes, the traditional choice of the narrow-gauge lines, were out of the question. This presented a tantalizing problem for Moffat's surveyors and engineers as they searched up and down the Front Range for favorable routes and alignments.

Moffat's solution for surmounting the Rocky Mountain Front includes the Big Ten Curve (seen to your left, in Leyden Gulch) and the Tunnel District (seen ahead to your right). The Big Ten is an innovative double-reverse curve that loops onto a flat, Nussbaum-age (pre-Rocky Flats) alluvial terrace. Its name arises from its tight, ten-degree radius of curvature. Note the stationary hopper cars along the upper curve: they are filled with ballast and serve as windbreaks, to keep trains from being blown off the tracks during winter Chinook winds that sometimes exceed 100 mph (161 kph). The Tunnel District consists of 27 tunnels in 13 miles, more than any other railway in the United States over such a short distance. Eight of these tunnels penetrate "Flatirons" of steeply dipping Fountain Formation and Lyons Sandstone along the face of Eldorado Mountain (Fig. 11). Further to the west, the remaining tunnels penetrate Precambrian Boulder Creek Granodiorite and quartz monzanite as the railroad loops along the canyon walls high above South Boulder Creek. The railroad then pierces the Continental Divide through the 6.2-mile (10-km) long Moffat Tunnel.

Moffat's transcontinental link was completed in 1934, several decades after his death. The D&SL railroad was absorbed by the Denver & Rio Grande Western Railroad, and later by the Southern Pacific and Union Pacific Railroads. For a closer look at this engineering marvel, the reader is encouraged to ride the railroad's famous "Ski Train," a one-day excursion that runs on certain summer and winter weekends between Denver and Winter Park.

Stop 10—Water Flooding in the Mountain Canyons, Coal Creek Canyon (42.6 miles; 68.6 km)

Directions: Continue driving westward on State Highway 72 and enter Coal Creek Canyon. After passing underneath the railroad trestle, look for and enter the first large pullout on the left (south). [Note: the GSA bus will proceed further up the canyon to a safe turnaround, and then return to the pullout.]

All larger drainages along the Front Range head in the mountains and fall rapidly from alpine elevations (above 9,500 feet, or 2,900 m) through steep-gradient canyons. These debouch at the mountain front onto the piedmont, where stream gradients lessen substantially. There is typically 6,000 to 8,500 feet (1,830 to 2,590 m) of relief from the highest mountain summits to the highest parts of the piedmont. The resulting orographic and meterological effects on streamflow can be pronounced and extreme.

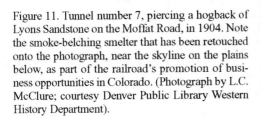

Figure 11. Tunnel number 7, piercing a hogback of Lyons Sandstone on the Moffat Road, in 1904. Note the smoke-belching smelter that has been retouched onto the photograph, near the skyline on the plains below, as part of the railroad's promotion of business opportunities in Colorado. (Photograph by L.C. McClure; courtesy Denver Public Library Western History Department).

At this locality near the mouth of Coal Creek Canyon we can see a likely paleosequence of flood deposits produced by events similar to the Big Thompson Flood. CGS (Soule and Costa, unpublished data) obtained a C-14 date from this locality of 955 +/- 80 ybp from what we interpreted to be the youngest of three flood deposits. Note the sizes of materials in the modern streambed versus those in the bank cuts, and the accordance or discordance of surfaces immediately above the stream. CGS C-14 dates from similar materials at other Front Range localities range from about 10,000 ybp to 300 ybp.

The greatest historical flood along the Colorado Front Range occurred in the drainage of the Big Thompson River, Larimer County, during the evening to nighttime of July 31-August 1, 1976. This flood can be attributed to one thunderstorm event when a large cell remained nearly stationary over the middle part of the drainage basin for about 2.5 hours. This part of the basin extends from about 1 mile (1.6 km) east of Estes Park eastward to the confluence of the Big Thompson River and North Fork Big Thompson River at Drake. Essentially all of the tributary streams in this part of the basin were involved. Below Drake, relatively little rain fell during the event. A peak discharge of the Big Thompson River of about 39,000 cfs was computed at the mountain front, about 4 miles west of Loveland. An estimated 139 persons were killed during this event and property loss of about $40 million resulted. Overbank flooding downstream from the mountain front occurred through the City of Loveland to the River's confluence with the South Platte River. The Colorado Geological Survey documented the geomorphic effects of this event (Soule, et al., 1976; U.S.G.S., 1979). These included debris avalanches and flows, deep scour of ephemeral streambeds, deep sheet ero-

sion on hill slopes, and deposition of prodigious amounts of sediment on gentler slopes above normally active stream channels and on gentle to nearly level slopes near the river on the Piedmont. An example of this compilation is shown in Figure 12.

Flood events of similar or possibly less magnitude have undoubtedly occurred in many Front Range drainages during the Holocene. Soule, Costa, and Jarrett studied paleoflood deposits in fourteen drainages along the Front Range during the late seventies and early eighties. This was done to support a broad research proposal to model Front Range paleo- and modern floods. The study was never completed because of fiscal "malnutrition." Evidence for these occurrences include geomorphology of paleoflood deposits, discharge estimates based on flows necessary to move largest clasts entrained in paleoflood deposits, depth of scour in streambeds, superposition of streamflows in channels, C-14 dating, study of demonstrable periglacial deposits, and meterological computations. Jarrett (1987) offered field evidence for and proposed that the Big Thompson event has a recurrence interval of 10,000 years, and that storms of the magnitude to produce such an event must form below 8,500-foot altitude. Some debate about this continues, however.

Stop 11—Leyden Gulch Area (44.0-49.6 miles; 70.8-79.8 km)

Directions: [Note: this section contains narratives for a roll-by followed by a stop. The roll-by portion begins after exiting Coal Creek Canyon.] Backtrack several miles along State Highway 72 and turn right at the stoplight onto State Highway 93 (south). Proceed to the bottom of a long hill, along a distinctive, fin-like ridge of sandstone to your left. Turn left onto 82nd Avenue (east), an unmarked gravel road that passes through the ridge's water gap. Proceed to the Public Service Company substation, which is enclosed by a chain-link fence. Park in front of the substation.

The valley below and to our right (south) as we backtrack onto the Rocky Flats surface is Leyden Gulch. It is a fairly prominent valley that parallels the pronounced break in slope at the base of the foothills. The upper end of the gulch contains a steep channel that is eroded to within 3/4 of a mile (1.2 km) of the main channel of Coal Creek. It has migrated between 200 and 250 feet (61 and 76 m) northward in the last thirty years. Is this the next piracy episode? What would happen if a flood like the ones just discussed (in Stop 3 and Stop 11) were diverted into Leyden Gulch, through a densely populated area several miles east of here?

Descending into Leyden Gulch, we will travel along the base of a near-vertical exposure of lower Laramie Formation sandstone. Between these vertical sandstone ribs can be seen the collapsed stopes of old, abandoned clay mines. After turning left (east) onto 82nd Avenue, we will pass through the hogback in a water gap cut by Leyden Creek. Some roof-crown arches can still be seen above

the mined-out stopes by looking northward along strike (Fig. 13). In addition to the clay mines, several minor coal mines are located in close proximity to these vertical beds (Fig. 14).

Within 1/4 mile (0.4 km) to the east, these same beds go from vertical to horizontal at a depth of about 500-600 feet (152-183 m). Below this portion of the valley and up the sides lie the Leyden Coal Mines (Fig. 14). These mines produced about 6 million tons of coal in the first half of the 20th Century. Public Service Company of Colorado is currently exploiting the residual void space as a natural gas storage facility. A system of charge wells and withdrawal wells access the mine. Several hundred feet of saturated claystone and shale provide an incredibly effective seal to prevent leakage of the gas to the surface.

These tight claystone and shale beds also provide the seal that allows the landfill to the north to operate with no anticipated groundwater impacts. Several older landfills have been operated and reclaimed along the northern flanks of this valley, to the east of our stop location.

All of these issues come into play as potentially harmful or

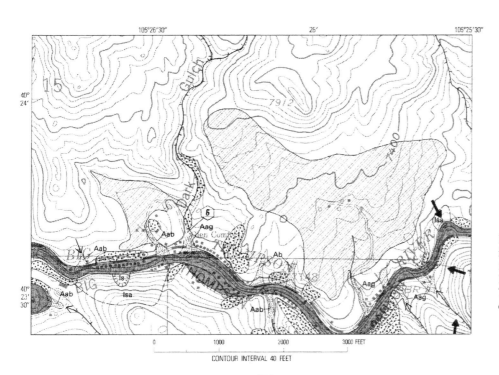

Figure 12. Geologic hazards and flooding in the Glen Comfort area, between mile 53.4 and mile 51.9 on the Big Thompson River, are typical of much of the Big Thompson area (modified from Soule and others, 1976; from U.S.G.S., 1979).

CONTOUR INTERVAL 40 FEET

EXPLANATION

Aag   Debris fans, A—a indicates activity on the debris fan during the 1976 storm; g (pebble to cobble) and b (boulder) indicate the predominant size of material in the debris fans

→   Direction of surface flow

Sheet-erosion areas

Downcut stream channels

Isⅼ   Landslide

Isa   Landslide areas

Rockfall areas

➤   Unstable or potentially unstable slopes

Approximate limit of 1976 flood on the Big Thompson River

6   Site of peak discharge measurement

Figure 13. Photograph of an exposed, vertical stope at the Lindsay Clay Mine near Leyden Gulch. Note the daylight through this remnant of a roof-crown arch. These arches experience ongoing failure as a result of weathering and deterioration of the soil and rock, which reduces their compressive strength.

problematic factors that will require analysis as land-use pressures drive residential development into this valley and along the southern rim.

Stop 12 (Roll-by)—Tertiary Igneous Rocks near Golden (52.1-54.3 miles; 83.9-87.4 km)

Directions: Backtrack to State Highway 93 and turn left (south). Resume driving toward Golden.

Approaching the Golden area, we will pass several intrusive bodies and dikes that are located in the Pierre Shale and lava flows that are interbedded with the Tertiary part of the Denver Formation. The rocks are mafic monzonite and its extrusive equivalent, alternatively called mafic latite by Van Horn (1976) and shoshonite, a potassium-rich basalt, by Scott (1972) and Trimble and Machette (1979). A majority of the intrusives bodies in this area are west of, or associated with, the Golden Fault (Van Horn, 1976).

The Ralston Dike is the largest of the intrusives and is located approximately 1/2 mile (0.8 km) west of State Highway 93, to your left. The dike is about 7,500 feet (2,290 m) long and 2,000 feet (610 m) wide. Its hollowed-out center is used as a water-storage reservoir. The western side of the dike is about 400 feet (122 m) above the reservoir and the eastern side is only about 50 feet (15 m) above the reservoir. The Ralston Dike is being mined for a high-quality aggregate which is used in the manufacture of "Super Pave," a durable pavement that is used in the construction of airport runways and interstate highways. The Ralston Quarry produces about 800,000 tons of aggregate per year.

South of the Ralston Dike and east of State Highway 93 are North and South Table Mountains, ahead and to the left (southeast). Capping the Table Mountains, and interbedded with the Denver Formation, are lavas which flowed from vents, such as the Ralston Dike, about 63 to 64 million years ago. The shoshonite flows can be separated into three separate flows that are about

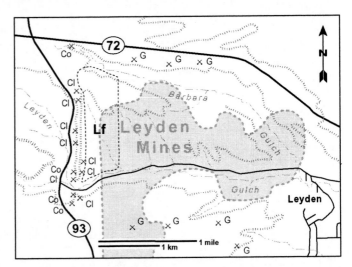

Figure 14. Map showing the extent of coal (Co), clay (Cl), and gravel (G) mining and landfills (L) in Leyden Gulch (modified from Amuedo and Ivey, 1978). Vertical clay-mine stopes in the area are up to 125 feet (38 m) deep, and are still undergoing roof collapse. The extensive, deep Leyden Coal Mines are currently used for natural gas storage.

240 feet (73 m) in total thickness. These lava-capped mesas are another prime example of topographic reversal. The flows occupied valleys when they were deposited, but they have resisted erosion and are now the highest ground.

PART 3. GOLDEN AREA

Stop 13—Lunch (57.9 miles; 92.2 km)

Directions: After entering Golden on State Highway 93, turn left onto Washington Avenue and drive southeast. Parfet Park is located on the left after about a mile (1.6 km), along Clear Creek. There are other parks nearby, along both banks of the creek.

We will stop in or near to the city of Golden for a box lunch. Golden, Colorado's territorial capitol from 1862-1867, is located along Clear Creek. North Table Mountain, South Table Mountain (the profile of which is immortalized on labels of Coors beer, from the famous resident brewery), Lookout Mountain, and Mount Zion (with the white "M" announcing Colorado School of Mines) surround the city and provide a scenic backdrop.

Stop 14—Highway 6-58-93 Junction, Golden (59.0 miles; 95.0 km)

Directions: Backtrack part way along Washington Avenue (northwest), turn left onto State Highway 58 (west), then right onto State Highway 93 (north). Pull off where the shoulder is sufficiently wide.

After circling through the northern part of Golden, we will park and climb a small hillock near the mouth of Clear Creek, to the northeast of the intersection of U.S. Highway 6 and State Highways 58 and 93 converge. Looking to the north and south, we can see the distinctive hogback of the Dakota Sandstone at both ends of the valley. But why is this hogback missing at the valley center? The answer is that the Golden Fault, a large, Laramide thrust fault, weaves along the valley just to the east of the mountain front. Here, the fault has displaced nearly 8,000 feet of the Paleozoic and Mesozoic, sedimentary section (Van Horn, 1976). The trace of the fault can be seen as it cuts up the side of the first large alluvial terrace to the south of Clear Creek. There is a distinct break in the vegetation between the Fountain Formation to the west (covered with mountain mahogany shrubs) and the Pierre Shale to the east (covered with grasses).

Looking directly west, we see a road cut for State Highway 93 (Fig. 15). This portion of the highway was first constructed in 1991. A small landslide formed in the west (opposite) face of the cut shortly thereafter, and the resulting toe bulge closed the southbound lanes of the new highway. Within a year, the landslide had developed a rim of head scarps with up to 12 feet of vertical slippage, and had captured the surface flow of a small stream in Magpie Gulch. Early efforts to drain the landslide were unsuccessful. A full geologic investigation subsequently revealed that the landslide sits directly atop the Golden Fault, with low-permeability Pierre Shale on the eastern side and a wedge of fractured, permeable Fountain Formation on the western side. In 1994, three lines of rock anchors (about 40 anchors in total) were installed across the landslide. The Magpie Gulch stream was piped across the head scarps, longer horizontal drains were installed, and a remote data-logging unit was set up. To date, this mitigative effort appears to have been successful. The combined maintenance and mitigation operations for this incident are reported to have cost about 3 million dollars.

The houses to the north of the landslide were built shortly after the landslide was mitigated and all signs of its existence had been "erased" from view. One can only imagine the concern of these residents had the landslide been active a few years later. This is a good illustration of the often short-term public memory of geologic hazards.

Stop 15—North Table Mountain Rockfall Area, Golden (59.9 miles; 96.4 km)

Directions: Drive north on State Highway 93 to the first stoplight and turn right onto Iowa Street (northeast). Follow Iowa Street across Washington Avenue and turn right onto Ford Street (southeast). Pull into the church parking lot to the left, at the corner of Ford Street and First Street.

We now re-cross the northern part of Golden to get a closer view of the west-facing slope of North Table Mountain (Fig. 15). The rockfall and landslide hazard area around North Table Mountain was mapped by the U.S. Geological Survey during the 1970s (Simpson, 1973a; 1973b). Jefferson County has adopted the mapped hazard area as part of its geologic-hazards overlay, and has considered it to be a "no-build" area. This was challenged in the 1980s by a developer, who staged an actual rock-rolling demonstration on the northwest flank of the mountain in order to prove the fallacy of the outer (distal) hazard-area boundaries. The demonstration was curtailed (and the subdivision application was subsequently denied) after several boulders rolled beyond the outer boundary, but not before one boulder had bounced at least 10 feet (3 m) into the air and knocked a cross-bar off a high-tension power-line tower at the base of the mountain.

In Figure 16, we see that the rockfall hazard area, as mapped by the USGS, exists in both unincorporated Jefferson County and the incorporated city of Golden. Unfortunately, protection of the public from the rockfall hazard does appear to stop at the jurisdictional boundary in this case. The houses we see on the hillside above are located within the city of Golden, which has not adopted Simpson's map. This subdivision was approved and has expanded due the absence of any requirements for home-rule cities to follow the state-mandated, geologic-hazard review process for subdivisions.

Stop 16—Clay Pits and Differential Subsidence, Golden (61.7 miles; 99.3 km)

Directions: Continue southeast on Ford Street, passing the Coors Brewery on your left (note: mileage for side trip to Coors is not included). Shortly after the street splits into a one-way and becomes Jackson Street, turn right onto 19th Street. After passing two stoplights, turn right onto the Colorado School of Mines campus at Elm Street, then immediately to the left onto Campus Road. Drive past the fraternity houses to the married-student housing area.

At this stop, we will look at the challenges of multiple-sequential land use as related to clay mining operations and reclamation along the western side of Golden. Here, an array of open-stope clay pits along the east side of U.S. Highway 6 have been reclaimed for several uses with a variety of problems and solutions, some more successful than others.

Severe differential-settlement problems have been experienced in the student housing at Colorado School of Mines, along Campus Road. The fill used to reclaim this area has settled, per-

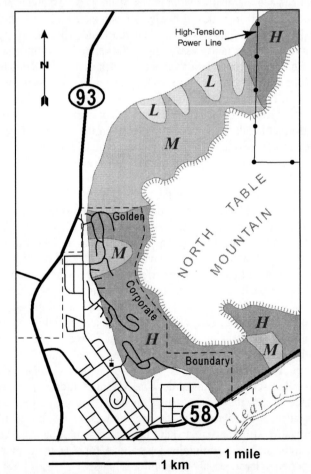

Figure 15. Photograph of the intersections of U.S. Highway 6 and State Highways 58 and 93, looking north from Mt. Zion. In this photograph, we can see the now-mitigated 1991 landslide (L), the North Table Mountain rockfall area (RF), and the new suburban neighborhoods on the outskirts of Golden. Clear Creek crosses the photograph in the foreground.

haps only a few percent, but the bounding sandstone ribs are stable. Major settlement and damage has occurred where structures, flatwork, and roadways straddled these highly variable units.

South of the CSM campus, the stopes were used to effect land reclamation and dispose of fly ash simultaneously. Unresolved water-quality issues led to the termination of this program, and the remaining reclamation is being performed with unregulated materials such as random fill. This may prove to be a future land-use issue for the city of Golden.

PART 4. GOLDEN TO SOUTHERN JEFFERSON COUNTY

Stop 17–Geologic Hazards in the Green Mountain Area (68.3 miles; 110.0 km)

Directions: Backtrack along Campus Road and Elm Street to 19th Street and turn right (southwest). At the stoplight, turn left onto 6th Avenue (U.S. Highway 6). Proceed about 4 miles (6.4 km) and exit onto Indiana Street, turning right (south). Indiana Street curves to the right and becomes Ellsworth Avenue. Stay on Ellsworth Avenue at the Archer Avenue detour by turning right, then circling counter-clockwise by turning left at 1st Avenue, then left on Archer Avenue (east) to the road closure. Walk up Quaker Court, the short connecting street to the right, then turn left on Bayaud Avenue (east). The landslide area is located about halfway down the block.

Leaving Golden, we will travel southward and eastward onto the flanks of Green Mountain, a low summit east of the Rocky Mountain Front in the west-central part of the Denver Metropolitan Area. Green Mountain is capped by the Paleocene Green Mountain Conglomerate, which is about 650 ft thick (Scott, 1972). Its distribution is restricted to this locality and its origin and provenance are debatable, although it may be a latest-Laramide fluvial deposit. It probably originated in penecontemporaneous higher land to the west. It may also be partly

Figure 16. Map showing zones of low (L), moderate (M), and high (H) rockfall-hazard potential along the west flanks of North Table Mountain (modified from Simpson 1973). Jefferson County has adopted the rockfall-hazard area delineated by Simpson as a no-build area. The hazard has been ignored by the city of Golden, however, as witnessed by new suburban development (seen here as curved roads) on the lower slopes of the mountain.

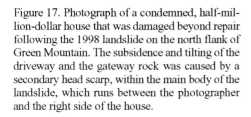
Figure 17. Photograph of a condemned, half-million-dollar house that was damaged beyond repair following the 1998 landslide on the north flank of Green Mountain. The subsidence and tilting of the driveway and the gateway rock was caused by a secondary head scarp, within the main body of the landslide, which runs between the photographer and the right side of the house.

correlative with the nearby volcanic and intrusive rocks of the Table Mountains and the Ralston Dike discussed earlier on this trip. This conglomerate overlies the sandstones and shales of the Denver Formation. Landslides of various ages (including active, modern ones) and colluvium derived from the underlying bedrock or other surficial deposits underlie the side slopes of Green Mountain.

Residential development on Green Mountain began in the 1960s and continues to the present. Much of the remaining, undeveloped land is now in dedicated public open space. Residential development here has been adversely affected by moderate-to-severe drainage problems, erosion, natural-slope and fill instability, and soils-engineering problems. In the early years, ill-advised regrading and fill placements, mostly by one developer on the east slopes of the Mountain, changed the area's natural drainage to the extent that serious local flooding and fill-failure problems occurred. Many of these conditions still exist. These were cause for serious concerns by the City of Lakewood and Jefferson County.

Residential development on Green Mountain continues today with serious geology-related and geotechnical problems. In particular, younger landslides have formed in some of the older landslides on Green Mountain, demonstrating that the older landslides are not necessarily stable. In 1990, Jefferson County allowed development on an area of Green Mountain that had been previously mapped as landslide (earthflow) deposits. In 1998, renewed movement of an older landslide occurred. This resulted in three homes being damaged beyond repair (Fig. 17) and two other homes to be severely damaged. The landslide appears to fill a buried ancient channel or ancient landslide-failure surface that underlies West Bayaud and West Archer Avenues. The soils that underlie the area of renewed slope movement consist mainly of clay and sparse gravel. Because of the lack of gravel, the soils appear to be associated with displaced, greatly weathered claystone bedrock. The clay soils that are now moving

appear to be an earthflow or landslide deposit that originally formed as a bedrock failure.

When Jefferson County approved the subdivision, its approval was based on a geotechnical report submitted by the developer. This report indicated that the subdivision was acceptable from a geological and geotechnical standpoint. The geotechnical engineer recommended several mitigation measures, such as the installation of an underdrain below the sanitary sewer. The county imposed additional restriction on the development. For example, the county required that site-specific slope stability analyses be done for each proposed home site. Many of the recommendations made by the geotechnical engineer and required by the county were not followed as the area was developed. Later, geotechnical studies performed subsequent to the renewed slope movement found that the conditions modeled in the original report did not reflect the deep clay soils located in the ancient channel or failure surface.

Three lines of earth anchors and drainage are currently being constructed to stop or slow movement of the landslide. The anchors will be approximately 110 feet long (34 m) and extend 40 feet (12 m) into the claystone bedrock. After installation of the anchors, horizontal drains will be installed. The drains will be approximately 200 ft long (60 m) and will be placed on three levels along with the anchors.

Stop 18 (Roll-by)—Major Flood-Control Structures in the Denver Metropolitan Area (77.3-78.3 miles; 124.4-126.1 km)

Directions: Backtrack to 6th Avenue (U.S. Highway 6) along Archer Avenue, 1st Avenue, and Ellsworth Avenue. Turn left onto the 6th Avenue freeway (west), then exit right onto Interstate 70 (west) and State Highway C-470 (south). Bear Creek Reservoir lies to the east of C-470 near the Morrison exit. Continue south on C-470 and do not exit.

The master stream in the Denver Metropolitan Area is the

South Platte River. Its four major tributaries are Clear Creek, Bear Creek, Plum Creek, and Cherry Creek. The headwaters of Clear Creek and Bear Creek are located in the mountains. The headwaters of Plum Creek and Cherry Creek are located in the Palmer Divide, south of Denver. All of these streams are subject to the flooding scenarios discussed in a previous section (see Stop 10 narrative). Travelling southward on State Highway C-470, we are afforded views of major flood-control structures that have been emplaced on these major streams.

The South Platte River and Plum Creek (Chatfield Dam and Lake), Cherry Creek (Cherry Creek Dam and Lake), and Bear Creek (Mount Carbon Dam and Bear Creek Lake) have flood-control works built on them as part of the U.S. Army Corps of Engineers' Tri-Lakes Project. These impoundments were completed in 1975, 1950, and 1982, respectively. Prior to this, a small flood-control dam (Kenwood Dam) had been built on Cherry Creek. It was subsequently determined to be grossly undersized and was removed. The present dams are all of the rolled-earth-fill type. Their reservoirs are used for recreation, and are surrounded by parklands. The Colorado Division of Parks and Outdoor Recreation manages Chatfield and Cherry Creek State Parks, and the City of Lakewood manages Bear Creek Lake Park.

The earliest recorded major stream flooding in Denver occurred on Cherry Creek on May 20, 1864, and the latest occurred in May 1973, on the South Platte River and other Piedmont streams. Flows of the South Platte of 55,000 cfs through Denver were documented in June 1965, its greatest historic flood. The 1973 flood produced 18,000 cfs flows of the South Platte. Floods have impacted numerous local communities over the years. Many of the more historic ones are documented in Follansbee and Sawyer (1948). One of the greatest floods occurred in 1921 on the Arkansas River, which caused much damage in Pueblo and its immediate vicinity. An earlier large flood occurred in 1905 on Boulder Creek. Since these flood-control works were installed, the Denver Metropolitan Area has mostly escaped serious damage caused by major flooding of their respective streams. However, some streams such as Boulder Creek, Clear Creek, and the upper reaches of Bear Creek, may be capable of producing flooding in populated areas.

Stop 19—Heaving Bedrock in Southern Jefferson County (80.9-86.1 miles; 130.2-138.6 km)

Directions: [Note: this section contains narratives for a combined stop and roll-by tour. The GSA field trip will visit two areas of historical damage: additional areas may be visited as well.] Continue along State Highway C-470 past the U.S. Highway 285 exit, then exit right (twice) onto Quincy Avenue (north, then east). Descend a long hill and turn right onto Simms Street (south) at the stoplight, then right onto Marlowe Avenue. The first damage-area tour consists of taking a right onto Union Street, left onto Urban Way, and left onto Tanforan Avenue back to Marlowe Avenue (east). Take Marlowe Avenue back to Simms

Street and turn right (south). Proceed for about 2 miles (3.2 km) and turn left onto Coal Mine Avenue (east). At the bottom of the hill, turn left onto Owens Street (north). The second damage-area tour consists of taking a right onto Oak Court, left onto Parfet Street, left onto Walker Drive, and right onto Owens Street again (south).

This field trip mini-tour is designed to show the effects of differentially heaving bedrock on past, present and future development areas of the Upper Cretaceous Pierre Shale outcrop belt. On Quincy Avenue, notice the long, parallel "speed bumps" that have severely deformed the pavement is several areas. After an introductory stop in the Harriman Park neighborhood, we will take a driving tour to view examples of older (mid-1970s) and newer (mid-1990s) damage. If the opportunity exists, we will look at a current development site where overexcavation and fill-replacement technologies are being used.

Differential ground heaving has adversely affected development projects for over 25 years in southern Jefferson County, in an area that is underlain by the Pierre Shale and other Upper Cretaceous formations. This geologic hazard is manifested by the progressive growth of long, somewhat parallel ridges separated by relatively inert swales. Over 2 feet (0.6 m) of differential movement has occurred in some cases within a few years following development. Certain neighborhoods in the Pierre Shale outcrop belt have performed well over the years, while others have sustained tremendous amounts of damage to structures, roads and utilities. One 110-home neighborhood has reportedly incurred more than 5 million dollars worth of damage and mitigation costs after 15 years, and the ground is continuing to heave.

Until about 1990, these problems were often attributed to the geological hazard of swelling soil. However, commonly used mitigative designs for swelling soil, such as drilled-pier foundations, floating-slab floors, and structural floors have been remarkably unsuccessful to date. Subsequent research by the U.S. Geological Survey and the Colorado Geological Survey (Nichols, 1992; Noe and Dodson, 1997) indicates that the differential movements occurs within near-vertical claystone bedrock beneath the ground, resulting from the combined effects of wetting and unloading surface (Fig. 18). The hazard has been called "heaving bedrock," to alert engineers to a need to depart from standard "swelling soil" considerations.

As a result of CGS education efforts, Jefferson County convened a task force of nearly 75 stakeholders in 1994 to innovate minimum standards for site exploration, evaluation, and design to mitigate heaving bedrock. New land-development regulations were enacted for the Dipping Bedrock Area in 1995 (see Noe, 1997). Trenching has been implemented to evaluate the geometric complexities of the dipping bedrock. Overexcavation and fill replacement to at least 10 feet (3 m) beneath foundations are now specified as minimum site-construction standards if the geologic evaluation finds that the bedrock has a potential for differential heave. The new Jefferson County regulations have resulted in a significant reduction in damage to homes and infrastructure.

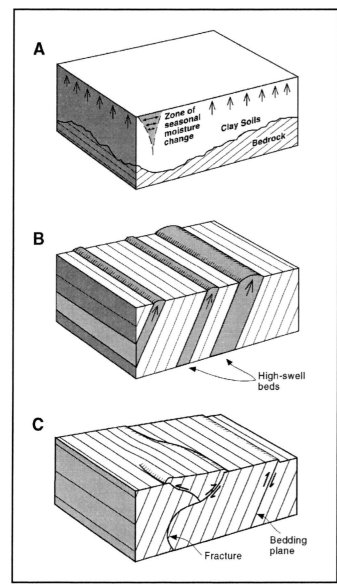

Figure 18. Schematic block diagrams showing basic types of differential ground movement associated with swelling soil and heaving bedrock (modified from Noe and Dodson, 1995): (A) General model for swelling soil, which assumes uniform composition and a zone of seasonal moisture change that becomes more stable with depth. The addition of moisture to the soil results in a somewhat uniform heaving at the ground surface (arrows). (B) Differential heaving of steeply dipping bedrock layers. The wetting of highly expansive layers, flanked by layers having lower swell potential, results in a linear pattern of heaving at the ground surface. (C) Differential heaving of bedding and fracture planes in steeply dipping bedrock. The uneven wetting and subsequent, thrust-like shearing movement along these surfaces results in a linear to curvilinear pattern of heaving at the ground surface.

Stop 20—Road Construction Over the Economy Coal Mine (87.0 miles; 140.0 km)

Directions: From Owens Avenue, turn left onto Coal Mine Avenue (east). The mine crossing is located about 0.3 mile (0.5

km) to the east, just before Kipling Street. It is marked by prominent "call before digging" signs.

The Laramie Formation and the Fox Hills sandstone underlie this site, located on Coal Mine Avenue just west of Kipling Street. The bedrock units strike northwest and dip steeply towards the northeast at approximately 55 degrees. In 1997, the local county government proposed extending a major arterial, Coal Mine Avenue, across the abandoned Economy Coal Mine (Fig. 19). The mine operated from 1932 to 1940, extracting coal from a nine-foot thick coal seam. The coal was mined by the "room and pillar" method. The coal seam was mined from depths of approximately 100 to 300 feet (30 to 91 m) below the ground surface. A 300-foot long (91 m) section of Coal Mine Avenue would be located over the historic workings.

The potential for subsidence exists from failure of both the deep and shallow workings. Broad–based subsidence resulting from collapse of the deeper mine working would have little impact at the surface. However, chimney subsidence from the shallow workings could result in a large sinkhole that would pose a safety hazard. Sinkholes with diameters and depths of 30 feet have formed in adjacent areas and appear without warning. The county looked to engineers to find a practical solution to the problem.

The options that were considered were backfilling the mine, bridging over the mine, grouting the subsurface, installing a geo-grid reinforced mat and constructing a reinforced concrete slab. Backfilling and bridging were quickly eliminated due to cost. Grouting was not practical due to the predominance of claystone bedrock. The consultant to the county recommended that several layers of a high strength geo-grid be placed over the shallow workings. Upon further investigation, it was discovered that this mitigation method would only span a void of 10 feet (3 m) in diameter. The county consulted geologists who verified that historic sinkhole formation in the area had almost always resulted in a void at the surface of 20-30 feet (6-9 m) in diameter. The county concluded that installation of geo-grid would do little to improve public safety.

The coal mine workings were more accurately located with borings, and a reinforced concrete slab that would span a void of 30 feet (9 m) in diameter along any portion of the slab was designed. The slab was installed in 1998 and monitoring of the slab occurs on a yearly basis. The county also overexcavated 5 feet (1.5 m) below the slab to help mitigate heaving bedrock that also underlie the site.

PART 5. SOUTHWEST JEFFERSON COUNTY TO DENVER

Stop 21 (Roll-by)—Trip Summary (87.0-102.5 miles; 140.0-165.0 km)

Directions: Continue eastward on Coal Mine Avenue for about 3.5 miles (5.6 km). Turn left onto Platte Canyon Drive (north), right onto Bowles Avenue (east), and left onto Santa Fe Drive.

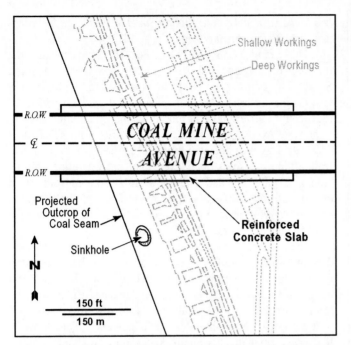

Figure 19. Map showing the alignment of Coal Mine Avenue across the shallow and deep workings of the Economy Coal Mine (modified from NSA Engineering, Inc., 1997). A reinforced concrete slab was installed to bridge the mine, and this road section was completed in 1998.

Follow Santa Fe Drive northward to downtown Denver and the Colorado Convention Center. End of field trip.

As we proceed back to downtown Denver, we can see once again many of the patterns of the historical urban growth and development of this major metropolis that sits on the Piedmont of its famous Rocky Mountains. These growth patterns have tended to exacerbate human land-use problems, especially those caused by or related to water, soils, and bedrock conditions.

In recent years, a tendency has been to build more housing on geologically problematical soils and hillslopes. The increased valuation of regional single-family real estate (estimated earlier this year by the Denver Post to be over $210,000, average) and the relatively high monetary wealth and per capita income in the region has made use of more geologically marginally suitable land economically feasible. These effects have resulted in changes in the amounts and kinds of needed engineering-geologic remediation and geologic-hazards mitigation.

In summary, we have attempted to show the reader how geologic knowledge is critical in many aspects of land-use planning and geologic-hazards mitigation. This is especially true under the growth conditions that are being experienced on the Colorado Piedmont today. Engineering geologists and geotechnical engineers perform increasingly valuable services under these conditions, including the following, crucial activities:

1. Reconnaissance-level engineering-geologic investigations and the characterization of basic surface and subsurface conditions to aid in site layout and to guide subsequent, more-detailed engineering investigations;

2. Detailed, site-specific soils engineering and slope-stability investigations to aid in geologic-hazard mitigation and engineering design;

3. Assessment of the long-term effects of past, present, and future extraction of mineral and mineral-fuel resources on proposed surface-land uses;

4. Creation of sophisticated drainage and erosion-control plans for residential development and its attendant supporting infrastructure; and

5. Rethinking, reevaluating, and continually improving the "standards of practice" for the local engineering, engineering-geologic, and geotechnical consulting communities.

REFERENCES CITED

Amuedo and Ivey, 1978, Coal and clay mine hazard study and estimated unmined coal resources, Jefferson County, Colorado: contract report for Jefferson County and cities of Arvada, Golden, and Lakewood, Colorado, 15 p.

Bollinger, E.T., and Bauer, F., 1962, The Moffat Road: Denver, Sage Books, 359 p.

Coe, J.A., 1987, Characteristics of alluvial fans from tributaries of Clear Creek, Floyd Hill to Georgetown, Colorado: Geological Society of America Abstracts with Programs, v. 29, no.6, p. 316.

Costa, J. E. and Bilodeau, S. W., 1982, Geology of Denver, Colorado, U.S.A.: Bulletin Of The Association Of Engineering Geologists, Vol. XIX No. 3, p. 261-314.

Denver Water, 1994, Features of the Denver Water System: 67 p.

Follansbee, R. and Sawyer, L.R., 1948, Floods in Colorado: U.S. Geological Survey Water Supply Paper 997, 151 p.

Gardner, M.E., 1969, Preliminary report on the engineering geology of the Eldorado Springs quadrangle, Boulder and Jefferson counties, Colorado: U.S. Geological Survey Open File Report 69-102, 9 p.

Gardner, M.E., Simpson, H.E., and Hart, S.S., 1971, Preliminary engineering geology map of the Golden quadrangle, Jefferson County, Colorado: U.S. Geological Survey Map MF-308.

Hunt, C.B., 1954, Pleistocene and Recent deposits in the Denver area, Colorado: U.S. Geological Survey Bulletin 996-C, p. 91-140.

Jarrett, R.D., 1987, Flood hydrology of foothill and mountain streams in Colorado: Ph.D. Dissertation, Colorado State University, Fort Collins, 239 p.

Kirkham, R.M., 1977, Quaternary movements on the Golden Fault, Colorado: Geology, v. 5, p. 689-692.

Kirkham, R.M., Parise, M., and Cannon, S.H., in preparation, Geology of the 1994 South Canyon fire area, and a geomorphic analysis of the September 1, 1994 debris flows, south flank of Storm King Mountain, Glenwood Springs, Colorado: Colorado Geological Survey Special Publication 46.

Madole, R.F., 1991, Colorado Piedmont section, in Morrison, R.B., ed., Quaternary nonglacial geology J12conterminous United States: Boulder, Colorado, Geological Society of America, The Geology of North America, v. K-2, p. 456-462.

McGregor, E.E., and McDonough, J.T., 1980, Bedrock and surficial engineering geology maps of the Littleton quadrangle, Jefferson, Douglas and Arapahoe counties, Colorado: U.S. Geological Survey Open File Report 80-321, 42 p.

Miller, R.D., and Bryant, R., 1976, Engineering geologic map of the Indian Hills quadrangle, Jefferson County, Colorado: U.S. Geological Survey Map I-980.

Murphy, J.A., 1995, Geology Tour of Denver's Buildings and Monuments: Historic Denver, Inc., and the Denver Museum of Natural History. 96 p.

Nichols, T.C., Jr., 1992, Rebound in the Pierre Shale of South Dakota and Colorado - field and laboratory evidence of physical conditions related to processes of shale rebound: U.S. Geological Survey Open-File Report 92-440, 32 p.

Noe, D.C., 1997, Heaving-bedrock hazards, mitigation, and land-use policy, Front Range Piedmont, Colorado: Environmental Geosciences, v. 4, no. 2, p. 48-57 (reprinted as Colorado Geological Survey Special Publication 45, 1997).

Noe, D.C., and Dodson, M.D., 1995, The Dipping Bedrock Overlay District - an area of potential heaving bedrock hazards associated with expansive, steeply dipping bedrock in Douglas County, Colorado: Colorado Geological Survey Open-File Report 95-5, 32 p.

Noe, D.C., and Dodson, M.D., 1997, Heaving-bedrock hazards associated with expansive, steeply dipping bedrock in Douglas County, Colorado: Colorado Geological Survey Special Publication 42, 80 p.

NSA Engineering, Inc., 1997, Subsidence investigation, mitigation, and estimated cost/annual cost/risk evaluation for West Coal Mine Avenue extension: contract report for Jefferson County, 8 p.

Scott, G.R., 1960, Subdivision of the Quaternary alluvium east of the Front Range near Denver, Colorado: Geological Society of America Bulletin, Vol. 71, p. 1541-1543.

Scott, G.R., 1972, Geologic map of the Morrison Quadrangle, Jefferson County, Colorado: U.S. Geological Survey Map I-790-A.

Shroba, R.R. and Carrara, P.E., 1996, Surficial geologic map of the Rocky Flats Environmental Technology Site and vicinity, Jefferson and Boulder Counties, Colorado: U.S. Geological Survey Map I-2526.

Simpson, H.E., 1973a, Map showing landslides in the Golden quadrangle, Jefferson County, Colorado: U.S. Geological Survey Miscellaneous Information Series, Map I-761-B.

Simpson, H.E., 1973b, Map showing areas of potential rockfalls in the Golden quadrangle, Jefferson County, Colorado: U.S. Geological Survey Miscellaneous Information Series, Map I-761-C.

Simpson, H.E., and Hart, S.S., 1980, Preliminary engineering geology map of the Morrison quadrangle, Colorado: U.S. Geological Survey Open File Report 80-654, 104 p.

Soule, J.M., 1978, Geologic hazards study of Douglas County, Colorado: Colorado Geological Survey Open-File Report 78-5, 16 plates.

Soule, J.M., Rogers, W.P., and Shelton, D.C., 1976, Geologic hazards, geomorphic features, and land-use implications in the area of the 1976 Big Thompson flood, Larimer County, Colorado: Colorado Geological Survey Environmental Geology 10.

Soule, J.M., 1999, Active surficial-geologic processes and related geologic hazards in Georgetown, Clear Creek County, Colorado: Colorado Geological Survey Open-File Report 99-13, 6 p.

Trimble, D.E., and Machette, M.N., 1979, Geologic map of the greater Denver area, Front Range Urban Corridor, Colorado: U.S. Geological Survey Miscellaneous Information Series, Map I-856-H.

U.S. Geological Survey, 1979, Storm and flood of July 31-August 1, 1976, in the Big Thompson River and Cache la Poudre River Basins, Larimer and Weld Counties, Colorado: U.S. Geological Survey Professional Paper 1115, 152 p.

Van Horn, R., 1972, Surficial and bedrock geology map of the Golden quadrangle, Jefferson County, Colorado: U.S. Geological Survey Map I-761-A.

Van Horn, R., 1976, Geology of the Golden Quadrangle, Colorado: U.S. Geological Survey Professional Paper 872, 116 p.

Weimer, R.J., 1977, Stratigraphy and tectonics of western coals, in Murray, D.K., ed., Proceedings of the symposium on the geology of Rocky Mountain Coal, 1976: Colorado Geological Survey Resource Series 1, p. 9-27.

Wells, J.D., 1967, Geology of the Eldorado Springs quadrangle, Boulder and Jefferson Counties, Colorado: U.S. Geological Survey Bulletin 1221-D, 85 p.

Printed in U.S.A.

Geological Society of America
Field Guide 1
1999

Laramide to Holocene structural development
of the northern Colorado Front Range

Eric A. Erslev
Department of Earth Resources, Colorado State University, Fort Collins, Colorado 80523, United States
Karl S. Kellogg and Bruce Bryant
U.S. Geological Survey, MS 913, Denver Federal Center, Denver, Colorado 80225, United States
Timothy K. Ehrlich and Steven M. Holdaway*
Department of Earth Resources, Colorado State University, Fort Collins, Colorado 80523, United States
Charles W. Naeser
U.S. Geological Survey, MS 926A, Reston, Virginia 20192, United States

FIELD TRIP OVERVIEW

The field trip will traverse the highly asymmetrical Front Range to examine the west flank's major Laramide thrusts, subsequent volcanic rocks and normal faults, as well as the east flank's higher-angle thrust and reverse faults and their associated fault-propagation folds. Laramide to Holocene tectonics will be debated on the outcrop and during our evening soak at Hot Sulphur Springs.

INTRODUCTION

The Rocky Mountain province of the United States is a classic basement-involved foreland orogen. Deformation during the Late Cretaceous to Eocene Laramide orogeny created an anastomosing system of basement-cored arches that bound the northern and eastern margins of the Colorado Plateau and the elliptical sedimentary basins of the Rockies. The tectonic mechanism for Laramide deformation remains controversial, with proposed mechanisms ranging from subcrustal shear during low-angle subduction (Bird, 1988, 1998; Hamilton, 1988) to detachment of the upper crust during plate collision to the west (Oldow and others, 1990; Erslev, 1993). The Rocky Mountains south of Wyoming have the additional complication of a period of mid-Tertiary igneous activity and sedimentation that coincides with Neogene extension along the Rio Grande rift.

This field trip (Fig. 1) will explore the Laramide to Holocene structural development of the southern Rocky Mountains by examining the geologic record exposed in the northern Front

Range of Colorado. The Front Range starts north of Canon City, Colorado, and trends north-northwest to Golden, Colorado. North of Golden, the range takes a more northerly trend toward the Wyoming border where it bifurcates into the north-trending Laramie Range (Brewer and others, 1982) and the north-northwest-trending Medicine Bow Range.

Specifically, the field trip will visit localities (Figs. 1, 2) whose exposures address the following problems.

1) What were the styles of Laramide basement-involved deformation?

2) Why are the Laramide structures of the eastern, northeastern, and western margins of the Front Range so dissimilar?

3) What was the tectonic regime during mid-Tertiary igneous activity?

4) What is the relationship between Laramide and Tertiary deformation and the regional uplift of the southern Rockies and adjoining High Plains?

REGIONAL SETTING

Laramide to Holocene deformation in the northern Front Range was imposed on a region with a long and complex structural history (Figs. 2 and 3). The oldest rocks in north-central Colorado are supracrustal and plutonic rocks resulting from Early Proterozoic arc magmatism and related sedimentation. These rocks were accreted to the southern edge of the Archean Wyoming craton over an interval of 130 m.y., beginning at about 1,790 Ma (Reed and others, 1987). Basement rocks include complexly folded and interleaved quartzo-feldspathic gneiss, amphibolite, biotite schist, gneiss, and migmatite, commonly metamorphosed to high T, low P upper-amphibolite facies assemblages. The

---

*Present address: Chevron, 1013 Cheyenne Drive, Evanston, Wyoming 82930, United States.

Erslev, E. A., Kellogg, K. S., Bryant, B., Ehrlich, T. K., Holdaway, S. M., and Naeser, C. W., 1999, Laramide to Holocene structural development of the northern Colorado Front Range, *in* Lageson, D. R., Lester, A. P., and Trudgill, B. D., eds., Colorado and Adjacent Areas: Boulder, Colorado, Geological Society of America Field Guide 1.

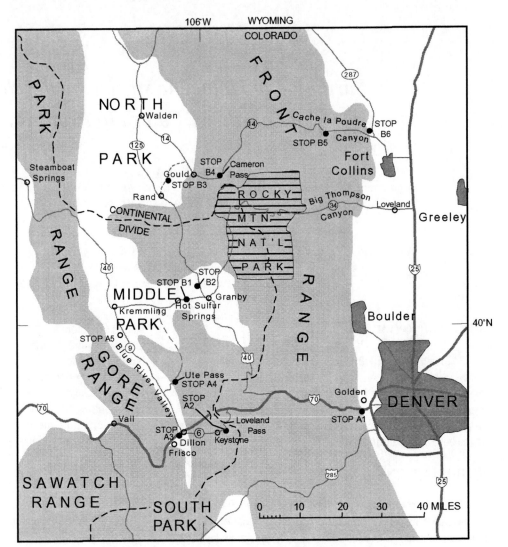

Figure 1. Road map for field trip showing major topographic features and stop locations.

biotite-rich rocks locally contain layers and lenses of marble, quartzite, and conglomerate, indicating sedimentary protoliths. Compositions and textures from less deformed and metamorphosed amphibolites and some quartzo-feldspathic gneisses in a few areas suggest metavolcanic protoliths. In the Big Thompson and nearby Cache La Poudre River drainages on the east side of the Front Range (Fig. 1), initial metamorphic pressures of 8-10 kb were interpreted to be due to subduction-related metamorphism between 1758 Ma and 1726 Ma (Selverstone and others, 1997).

The supracrustal rocks are cut by calc-alkaline plutons, chiefly granodiorite and monzogranite, but ranging from peridotite to granite. These plutons are part of the Routt Plutonic Suite and range in age from 1,780 to 1,650 Ma (Tweto, 1987; Reed and others, 1987; Reed and others, 1993). They are commonly foliated and concordant, suggesting syntectonic, synmetamorphic intrusion. Unfoliated, discordant plutons in some areas are older than foliated, concordant plutons in other areas, showing that deformation and metamorphism were not synchronous throughout the region. The correspondence of intrusive and metamorphic ages, the high temperature-low pressure metamorphism,

and the structural patterns around the major plutons led Reed and others (1987) to suggest that much of the deformation and metamorphism was related to emplacement of the plutons rather than subsequent tectonic and metamorphic events.

Widespread mid-crustal reheating at 1.4 Ga (Shaw and others, 1999) was accompanied by re-equilibration at pressures similar to those of the main ~1.7 Ga mineral assemblage (Selverstone and others, 1997) as well as intrusions of dikes, stocks, and discordant plutons of weakly foliated to non-foliated two-mica granite, hornblende dacite, granodiorite and gabbro. This Berthoud Plutonic Suite (Tweto, 1987) is part of a continent-scale igneous event at about 1.4 Ga (Anderson, 1983). In the southern Front Range, these rocks were cut by the 1,092-1,074 Ma (Unruh and others, 1995) Pikes Peak batholith which intruded at depths as shallow as 5 km (Barker and others, 1976).

During the Proterozoic, the region was cut by an anastomosing swarm of northeast-trending zones of recurrent ductile deformation and cataclasis. These shears were zones of weakness that locally controlled intrusive activity and mineralization during the Late Cretaceous and Tertiary. For example, a swarm of shear zones through the central Front Range is parallel to the concentration of mostly Late Cretaceous to Eocene intrusions along the Colorado Mineral Belt (Tweto and Sims, 1963). Northwest- and north-northeast-trending fault zones of late Proterozoic age cut the ductile shear zones and also were reactivated during Phanerozoic deformation (Tweto and Sims, 1963).

During the early Paleozoic, thin platformal deposits of quartz-rich sands and carbonate rocks covered the region. In the late Paleozoic, northwest-trending mountain ranges and basins formed during the basement-involved Ancestral Rocky Mountain orogeny. In the area traversed by the field trip, erosion during Ancestral Rocky Mountain uplift removed the earlier Paleozoic

sedimentary cover, but these strata are preserved in adjacent basins in which as much as 5 km of Pennsylvanian synorogenic strata accumulated (Fig. 3). The later Laramide arches partly coincide with the Paleozoic uplifts, but the Laramide arches have a more northerly trend. Some of the faults active in the late Paleozoic were reactivated during Laramide and Neogene deformation.

CENOZOIC TECTONIC PROBLEMS AND HYPOTHESES

1) What were the styles of basement deformation during the Laramide?

The multitude of Laramide structural geometries has resulted

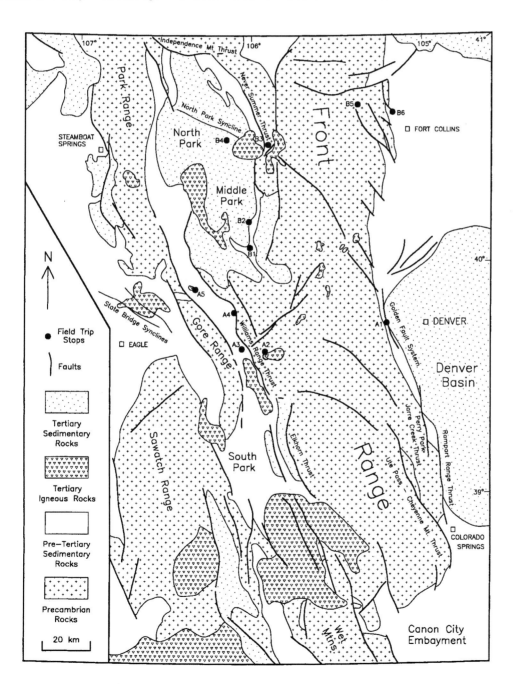

Figure 2. Simplified geologic map of the northern Front Range after Tweto (1979), with stop localities shown as filled dots.

in a mirroring multitude of kinematic hypotheses. In the 1970s and 1980s, geoscientists were polarized into opposing horizontal-compression and vertical-tectonics schools. The vertical-tectonics school was dominant in the 1970s, represented by upthrust (Prucha and others, 1965) and block uplift models (Stearns, 1971; 1978; Matthews and Work, 1978). In the 1980s, incontrovertible evidence for thrusting of Precambrian basement over Phanerozoic sediments (e.g., Smithson and others, 1979; Gries, 1983; Lowell, 1983; Stone, 1985) swung opinion back towards models invoking horizontal shortening and compression. Seismic profiles (Smithson and others, 1979; Gries and Dyer, 1985), subthrust petroleum

drilling at Laramide basin margins (Gries, 1983), and balancing constraints (Stone, 1984; Erslev, 1986; Erslev and Rogers, 1993; Brown, 1988; Spang and others, 1985) have demonstrated that the Laramide was the result of lateral shortening due to horizontal compression. High-angle, dip-slip faults do occur in numerous locations in the Laramide foreland, however, indicating distinct differences between the basement-involved Laramide and synchronous thin-skinned Sevier (Schmidt and Perry, 1988) orogens of the Rocky Mountain foreland.

The diversity of Laramide structural trends, with faults, folds, and arches trending in nearly every direction, has been attributed

Northeastern flank of Front Range near Morrison

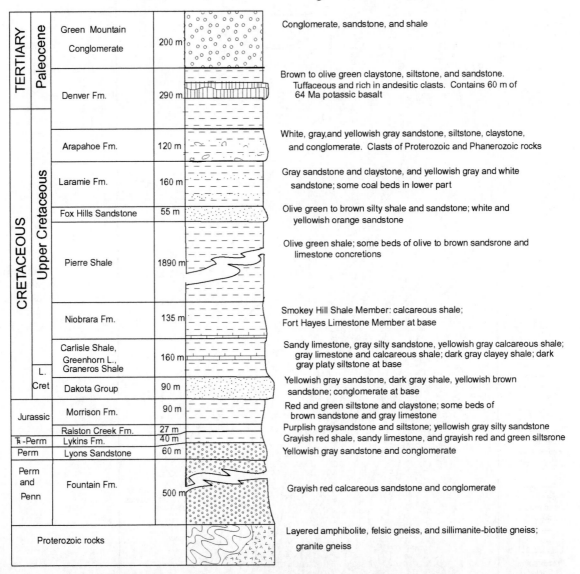

Figure 3 (this and opposite page). Stratigraphic columns for (a) northeastern flank of the Front Range at stop A1, after Scott (1972), and (b) Middle Park and North Park (composite), after Izett (1968), Izett and Barclay (1973), and Tweto (1976).

to multiple stages of differently oriented shortening and compression (Gries, 1983; Chapin and Cather, 1981; Bergh and Snoke, 1992), reactivation of pre-existing weaknesses in the basement (Hansen, 1986; Blackstone, 1991; Chase and others, 1993; Stone, 1986, 1995), transpressive motions (Wise, 1963; Sales, 1968), rotation and indentation by the Colorado Plateau (Hamilton, 1988), and detachment of the crust (Lowell, 1983; Brown, 1988; Kulik and Schmidt, 1988; Oldow and others, 1990; Erslev, 1993).

These hypotheses may all be valid for individual areas within the foreland but their regional significance is not clear.

2) Why are the Laramide structures of the eastern, northeastern, and western margins of the Front Range so dissimilar?

Laramide structures of the Front Range basement arch are extremely diverse. This field trip begins and ends on the eastern flank of the range, where the ENE-directed Golden thrust system

North Park and Middle Park

North Park Formation (North Park) - Tuffaceous sandstone, conglomeratic sandstone, limestone, claystone, and tuff

Troublesome Formation (Middle Park) - Siltstone, tuff, sandstone, and conglomerate

White River Formation (North Park) - Light-gray to white tuffaceous siltstone and claystone

Coalmont Formation (North Park) - Sandstone, conglomerate, carbonaceous shale, and coal

Middle Park Formation (Middle Park)
Upper member (Paleocene) - Arkosic grit, sandstone, and conglomerate; abundant andesitic debris in lower part

Windy Gap Volcanic Member - Andesitic volcanic breccia and conglomerate

Dark-gray marine shale and a few beds of fine-grained sandstone

Gray, platy-weathering calcareous shale and shaly limestone; gray micritic limestone at base

Black shale; limestone and fine-grained sandstone at top

Upper quartzite member, middle shale member, and lower quartzite member

Light greenish-gray and maroon, locally calcareous claystone sandstone and limestone beds in lower part

Light-tan cross-bedded calcareous quartz sandstone.

Red and locally green claystone, shale, and platy sandstone

Coarse-grained arkosic maroon sandstone and conglomerate

Mostly amphibolite-facies gneiss and granitic to gabbroic intrusive rocks west and southwest of Denver (Berg, 1962) changes northward into an array of WSW-directed, higher-angle reverse faults (Erslev, 1993). While earlier interpretations of vertical uplift and gravity sliding provided a temptingly integrated hypothesis (Fig. 4a; Boos and Boos, 1957; Matthews and Work, 1978; Tweto, 1975,1979), the faults are not vertical and horizontal shortening is clearly indicated (Berg, 1962; Erslev, 1993; Erslev and Selvig, 1997; Weimer and Ray, 1997; Holdaway, 1998). Symmetric, concave-downward upthrusts on the arch margins were proposed by Jacobs (1983; Fig. 4b), but major thrust overhangs have never been documented on the northeastern side of the Front Range. Hypotheses of strike-slip deformation have also been suggested (Fig. 4b), but recent analyses of minor faults generally indicate shortening perpendicular to the strike of the major faults (Selvig, 1994; Holdaway, 1998).

On the western side of the range, however, low-angle faults displace Precambrian basement considerable distances over sedimentary strata. Large basement overhangs are documented in several areas where post-Laramide stocks bow up hanging-wall basement overhangs and expose underlying sedimentary strata. The vertical tectonic school explained these fault windows as landslides, which are common in areas where basement rocks overlie Cretaceous shale. But the magnitude of the overhangs, their consistency along strike, and associated shortening structures indicate that these faults are thrusts rooted in basement.

Asymmetrical "chip models" (Fig. 4c,d, Kluth and Nelson, 1988; Raynolds, 1997; Erslev, 1993) can explain the differences between the major fault geometries on either side of the Front Range arch. To date, however, the details of the structural transition between east-directed thrusting on the southeastern margin of the arch and west-directed thrusting on the northeastern margin of the arch have not been fully addressed. In addition, the effects of changes in Laramide shortening directions through time are not clear.

3) What was the tectonic regime during mid-Tertiary igneous activity?

The deformation that occurred between Eocene Laramide shortening and Neogene Rio Grande rift extension remains enigmatic. Tectonic activity during this interval is indicated by distinct late Oligocene - early Miocene and middle Miocene - Holocene pulses of sedimentation (Chapin and Cather, 1994; Baldridge and others, 1995) and igneous activity (Steven, 1975; Christiansen and Yeats, 1992). In New Mexico, broad Oligocene sedimentary basins with thick clastic and volcanic sequences parallel the current Rio Grande rift and have been attributed to an early stage Rio Grande extension (Brister and Gries, 1994).

Recent analysis of faults cutting Oligocene volcanic and intrusive rocks (T. Wawrzyniec, personal commun., 1996; Erslev, unpub. data, 1998) indicate a phase of north-south strike-slip faulting that post-dates the Eocene close of the Laramide orogeny. Mid-Tertiary rocks as young as 24 Ma were cut by north-south strike-slip faults paralleling the Rio Grande rift in north-central New Mexico (Erslev, unpub. data, 1998), and in the west side of the Front Range at Cripple Creek (T. Wawrzyniec, personal commun., 1996) and eastern North Park (Erslev, unpub.

data, 1998). Mid-Tertiary strike-slip deformation in the southern Rocky Mountains may have paralleled the current Rio Grande rift, with transtensional tectonics forming mid-Tertiary pull-apart basins and facilitating magmatic intrusion into the crust.

4) What is the relationship between Laramide and Tertiary deformation and the regional uplift of the southern Rockies and adjoining high plains?

Central Colorado is part of a 1,000-kilometer-long topographic high named the Alvarado Ridge by Eaton (1986). It parallels the Rio Grande rift, which is closely related structurally and temporally to regional extension in the Basin and Range Province to the west. The high elevations can be attributed to numerous mechanisms, including thickening of the crust during Laramide shortening and/or thinning of the mantle lithosphere during low-

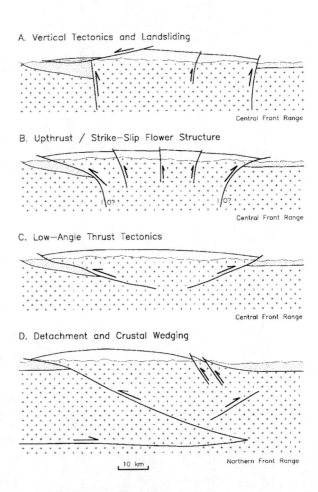

Figure 4. Simplified, unexaggerated east-west cross sections illustrating tectonic models for the Front Range: (A) vertical uplift with gravity sliding on western flank, (B) symmetric, concave-downward fault geometry used for thrust (Jacobs, 1983) and strike-slip (Chapin, 1983) interpretations, (C) low-angle thrust interpretations (Kluth and Nelson, 1988; Raynolds, 1997), and (D) back-thrust basement chip (Erslev, 1993; Holdaway, 1998).

angle Laramide subduction, mid-Tertiary igneous activity, and/or Neogene Rio Grande extension. Eaton (1986) attributed uplift to Neogene rifting coeval with extension and lithospheric thinning along the crest of the ridge. The age of this topographic high has been questioned (Gregory and Chase, 1992; Gregory, 1994) based on the leaf morphology of late Eocene and early Oligocene floras from Florissant in the southern Front Range. Gregory and Chase interpret these flora to have lived at essentially the same altitude as they are found today. Radical river incision and reorganization throughout the Rocky Mountain region, however, strongly suggest Neogene regional uplift (Steven and others, 1997), although these features have also been attributed to climatic changes.

FIELD TRIP ROAD LOG AND STOP DESCRIPTIONS

STOP A1. I-70 roadcut north of Morrison (Fig. 5)

Driving instructions: Take I-70 west out of Denver to exit 259 for Morrison. After exiting, turn south on Rte. 26 (under the interstate) and take an immediate left into the parking lot. We will climb the ridge east of the parking lot. After an overview at the top of the ridge, participants can walk northeast down to the paved path which traverses the Jurassic and Cretaceous units exposed in the ridge.

This locality exposes Jurassic Morrison Formation and Cretaceous Dakota Formation (Fig. 3a) folded above the Golden fault (Fig. 2) near a critical transition in structural styles along the eastern boundary of the Front Range. Permo-Pennsylvanian Fountain Formation and Precambrian rocks underlie the mountains to the west, whereas younger Cretaceous and Tertiary rocks underlie the Denver Basin to the east. Rapid exposure of the basement at the start of the Laramide orogeny is indicated by the minimal time between the youngest marine ammonite zone in the Pierre Shale (69 Ma) and the basement clasts in the latest Cretaceous-Paleocene Arapahoe Formation, which is overlain by the Denver Formation that includes the 64 Ma basalt flows of Table Mountain. The widespread late Eocene erosional surfaces and minimal elevation differences between exposures of the Late Eocene Wall Mountain Tuff (Leonard and Langford, 1994) allow for minimal post-Laramide faulting along the southeastern margin of the Front Range.

Structural styles of Laramide faulting and folding change radically along the flank of the range. Near the I-70 roadcut at stop A1, an emergent splay of the Golden fault system lies roughly at the base of the Dakota hogback. This thrust fault cuts out much of the Benton Formation, all of the Niobrara Formation, and much of the Pierre Shale. The Golden fault system probably connects with the Perry Park-Jarre Creek thrust to the south, which puts Proterozoic basement against Paleocene rocks and has a stratigraphic separation of over 2.8 km (Scott, 1963). The Golden fault system has been interpreted as a series of parallel, WSW-dipping reverse faults (Berg, 1962; Weimer and Ray, 1997) which thrust basement eastward over the western edge of the southern Denver Basin. The highly asymmetrical shape of the southern and central Denver Basin can be attributed to thrust loading by this thrust system.

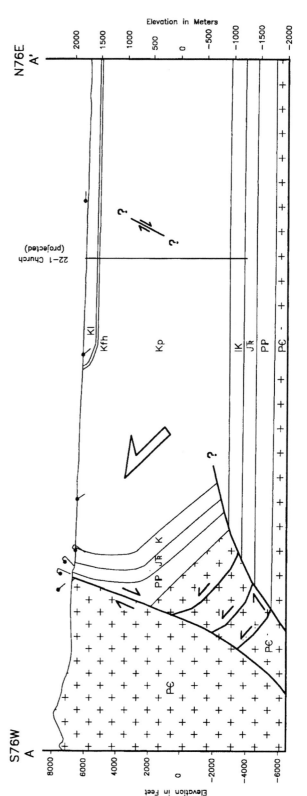

Figure 5. Seismically constrained (DOE, 1993) structure section through the eastern flank of the Front Range at the Rocky Flats Plant 20 km north of stop A1 (from Selvig, 1994).

About 40 km north of stop A1, the eastern margin of the Front Range steps eastward, giving it a more northerly average orientation. These steps are created by northwest-striking, northeast-dipping reverse faults that expose basement rocks in the core of fault-propagation folds (Erslev and Rogers, 1993). Relative to the southern Denver Basin, the northern Denver Basin is both shallower and more symmetric, with the basin axis farther from the range.

STOP A2. *Contact metamorphosed rocks in the Williams Range thrust zone, which places Precambrian gneiss over Pierre shale (Fig. 6)*

Driving instructions: Take I-70 43 miles west to the Loveland Pass exit (Exit 216), continue 13.5 miles southwest on Rte. 6 over Loveland Pass. Turn left on Gondola Road (Fig. 6), cross Montezuma Road, jogging slightly left onto North Fork Drive; park on east side of the circle at the end of drive. Note: this parking area is on private land and access permission must be granted by the owners. Follow a faint trail through the woods toward the cliffs on the ridge to the east.

The Williams Range thrust forms the west-central structural boundary of the Front Range (Fig. 2) and extends from Middle Park just north of Kremmling to South Park, where it is probably continuous with the Elkhorn thrust to the south (Bryant and others, 1981). North of Middle Park, the low-angle, en echelon Never Summer thrust steps east from the Williams Range thrust and defines the east side of North Park.

The age of thrusting on the west side of the Front Range is not precisely known, although the onset of Laramide deformation in this area is generally regarded as the age of the 70 Ma Pando Porphyry near Minturn, about 30 km to the west (Tweto and Lovering, 1977). Synorogenic, late Paleocene units in South Park are overridden by the Elkhorn thrust, indicating that if the Elkhorn and Williams Range thrusts are synchronous, movement along the Williams Range thrust probably continued into early Eocene. The Never Summer thrust cuts rocks of the Paleocene and Eocene Coalmont Formation (Fig. 3b, O'Neill, 1981), but the age of the part of the formation overridden by the fault is not precisely known.

Unlike the higher angle faults of the eastern margin of the Front Range, the Williams Range thrust is low angle to nearly horizontal in most places. At this locality, the Montezuma stock, a monzogranite porphyry with an age of 38-39 Ma (Marvin and others, 1989; McDowell, 1971) domed Williams Range thrust along normal faults, forming a thrust window (Fig. 6; Ulrich, 1963). A basement overhang of at least 9 kilometers is indicated by the distance from the thrust window to the frontal exposure of the thrust to the west combined with the angle of the thrust plane and the thickness of the sedimentary section. Contact metamorphism of both the cataclastic basement rocks in the hanging wall and the Pierre Shale in the footwall allows what may the best exposure of a Laramide thrust in the entire Front Range!

The nearly horizontal Williams Range thrust is exposed at the base of the cliff, where Pierre Shale hornfels is overlain by 2-3 m of basement hornfels grading upward from cataclasite to biotite-quartz-feldspar gneiss. Pervasive chlorite-epidote alteration is almost certainly related to contact metamorphism by the Montezuma stock. Numerous normal faults cut the Williams Range thrust zone and may be related to late-stage stock emplacement or Neogene extension.

STOP A3. *Faulting in the Dakota Sandstone*

Driving instructions: Rejoin Rte. 6, drive 7.2 miles west to Dillon Dam Road, just after the town of Dillon. Turn left, proceed 0.4 miles and park in small turnout at west end of the first segment of the dam.

The Blue River valley follows part of the belt of sedimentary rocks that extends along the west flank of the Front Range. To the northeast, on the west side of the Williams Fork Mountains, the Williams Range thrust defines the structural margin of the Front Range. The Williams Range thrust is buried beneath an extensive landslide complex, except where the thrust is exposed along Interstate 70 about 2.5 km to the northeast of stop A3 (Kellogg, 1997a). There, the thrust dips about 35° east and contains a 5-meter-thick zone of brecciated Precambrian gneiss overlying Pierre Shale. The buried trace of the thrust climbs along the west side of the range to the north and tops the range at Ute Pass, which we will visit at stop A4.

The landslide deposits may be as thick as several hundred meters and contain blocks of Proterozoic rocks tens of meters long. They are deeply incised and no longer retain hummocky topography, suggesting a late Tertiary or early Pleistocene age (Kellogg, 1997a,b). Initial brecciation of these rocks may have occurred during fault-bend folding as the Williams Range thrust flattened from relatively steep dips at depth, where Proterozoic rocks occupy both the hanging wall and the footwall, to gentle dips where the footwall is mostly Cretaceous shale (Kellogg, 1997b). Evidence for pervasive fracturing of the basement hanging-wall rocks includes the smooth and rounded crest of the Williams Fork Mountains to the northeast. This mountain crest is in stark contrast to the rugged crest of the Gore Range, which is only slightly higher than the Williams Fork Mountains and is underlain by similar but less fractured Proterozoic rocks. Starting in late Neogene time, incision of the Blue River undercut the shattered Proterozoic rocks, which Kellogg (1997b) inferred caused much of the west side of the Williams Fork Mountains to slide.

The Blue River valley is a 5-to-9-km-wide half graben, bounded on the west by the east-dipping Blue River normal fault of Neogene age. The fault defines the abrupt east margin of the Gore Range. A complex zone of faulted and fractured Proterozoic rock 0.5-1.0 km wide lies in its footwall (Tweto and others, 1970; West, 1978). The Blue River fault has a minimum displacement of 1.2 km based on the relief and the absence of Phanerozoic sedimentary rocks in the Gore Range west of the fault. The latest movement along the fault was probably no younger than Pliocene or early Pleistocene (West, 1978). An extensive apron of glacial deposits now covers most of the fault trace and the valley floor west of the Blue River.

The half graben is cut by numerous north-striking normal

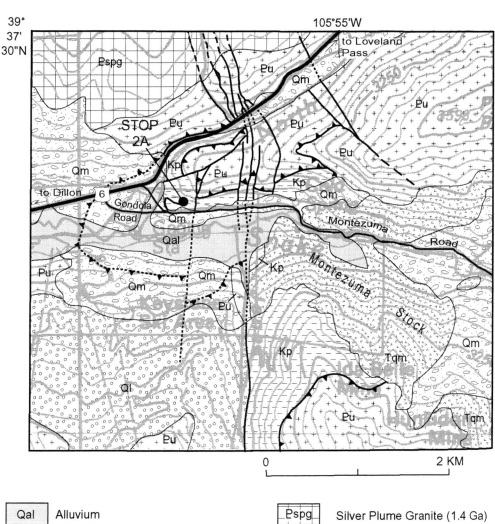

Figure 6. Geologic map of the Keystone area around stop A2, showing a portion of the Williams Range thrust window. Adapted from Ulrich (1963).

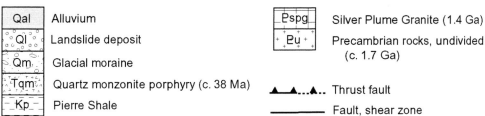

Qal	Alluvium
Ql	Landslide deposit
Qm	Glacial moraine
Tqm	Quartz monzonite porphyry (c. 38 Ma)
Kp	Pierre Shale

Pspg	Silver Plume Granite (1.4 Ga)
Pu	Precambrian rocks, undivided (c. 1.7 Ga)
▲▲...▲..	Thrust fault
————	Fault, shear zone

faults that are almost entirely east dipping, suggesting that the Blue River graben is a west-tilted structure above a listric fault at depth. Kellogg (1999) suggested that the west-directed Gore fault, a reverse fault along the west side of the Gore Range with significant movement in both late Paleozoic and Laramide times, is listric and provided the surface along which Neogene reactivation occurred (Fig. 7).

Faults in the outcrop of folded Dakota Formation include normal, strike-slip and thrust faults. Multiple slickenlines on the thrust planes suggest multidirectional thrusting. Data collected from several localities in the area indicate two distinct directions (NNE and NE) of horizontal shortening in the Tertiary.

STOP A4. Williams Range thrust at Ute Pass (Fig. 8)

Driving instructions: Return to Rte 6, which turns into Rte 9, and continue 19 miles north from I-70 underpass. Turn right on Ute Pass Road and climb to the pass, stopping at the first large exposure of fractured Precambrian rock beyond the pass.

This locality is just east of the Williams Range thrust (Fig. 8), whose trace is covered in the gully just to the southwest. The low-angle attitude of the thrust was demonstrated by a small outcrop of Pierre Shale in the valley to the north, which was exposed when the road over Ute Pass was improved to serve the mill for the Henderson molybdenum mine (Tweto and Reed, 1973). As discussed at the last stop, extensive fracturing of the Precambrian gneiss is possibly related to fault-bend folding of the hanging-wall rocks when they moved through a bend in the thrust (Kellogg, 1997b).

A short distance to the west, a low-angle footwall imbricate of the Williams Range thrust places calcareous shale of the Niobrara Formation above Pierre Shale. Farther west, a spectacular vista of the Gore Range and Blue River half graben is opposite a prominent roadcut in Pierre Shale containing numerous minor thrust faults. Just above the road to the southeast, the Gunsight Pass Member of the Pierre Shale (W.A. Cobban, personal commun., 1998) is prominently exposed in an overturned footwall syncline.

West of Ute Pass, Proterozoic granite and migmatite form the high peaks of the Gore Range, which is bounded on the west by the Gore fault (Fig. 7). This fault also marks the eastern margin of the Central Colorado trough, which formed during the late Paleozoic Ancestral Rocky Mountain orogeny. Post-Cretaceous slip on the Gore fault is at least 300 m. The Blue River normal fault marks the east face of the Gore Range facing Ute Pass. The impression from the topography is that the Gore Range is a block that had a more or less flat surface prior to being uniformly elevated and dissected. North of the high peaks, faults cut across the range, which becomes is a smooth-topped ridge capped by about 200 m of Dakota Sandstone and underlying strata (Fig. 9). Here, total throw on the Blue River fault is only 100 to 200 m, a substantial decrease from the minimum 1,200 m throw to the south.

Apatite fission-track geochronology

Apatite fission-track (AFT) dates from both sides of the Blue River graben (Fig. 9) show that uplift and cooling of the Proterozoic basement rocks occurred in late Paleogene and Neogene

time. In addition, young AFT dates adjacent to the graben generally become older away from the graben. The underlying assumptions and results are outlined in the following discussion.

At the beginning of Laramide deformation (about 70 Ma), the Proterozoic rocks of the Gore range were overlain by about 3.1 km of Phanerozoic sedimentary rocks under an epicontinental sea in which the Pierre Shale was being deposited. Assuming a thermal gradient of 25°/km and a sea-floor temperature of 4°C, a temperature of 110°C, the annealing temperature for fission tracks in apatite, would be attained at 4.2 km below sea level. Most of the apatite fission-tracks at depths greater than 4.2 km below this sea floor would have been annealed in Late Cretaceous time, and the upper 1.1 km of basement rock would have been in the Laramide partial annealing zone and have had AFT dates greater than 70 Ma.

The presence of Cretaceous sedimentary rocks on the low, northern part of the Gore Range demonstrates that less than 3.1 km of rock has been stripped away from this area since the close of the Laramide, yet AFT dates from basement rocks in this area are 20-30 Ma. AFT dates of apatite near Ute Pass are similar to those in the low part of the Gore Range, except for one older date of 35 Ma from a down-faulted block forming the bottom of the Williams Fork Valley (Tweto and Reed, 1973).

AFT dates from a 1.4 km vertical interval in the high, central part of the Gore Range, where only Proterozoic basement rocks are exposed, are from 6 to 32 Ma (Fig. 9), with younger dates from the base of the Blue River fault scarp and older dates from high altitudes on the west side of the range (Naeser and others, 1999). Isothermochrons (surfaces of constant age) dip west, away from the east flank of the Gore Range, and are discordant to the even-topped spurs that extend to the Blue River fault from the crest of the range. If the ridge crests at the same altitude represent the remains of an erosional surface, this surface must be younger than 16 Ma (late Neogene), which is an AFT date from the highest sample along the east face of the range.

Uplift and erosion alone cannot account for the young AFT

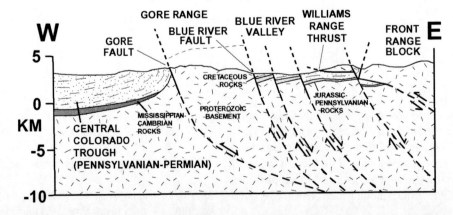

Figure 7. Schematic, east-west cross section across the Blue River graben showing the prevalent east-facing normal faults and the suggestion that they sole into the Gore fault, a high-angle contractional structure with significant displacement during the late Paleozoic. Adapted from Kellogg (1999).

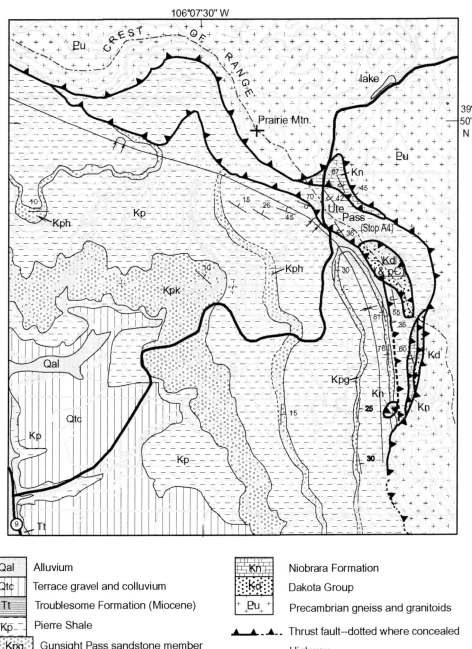

Figure 8. Geologic map of the Ute Pass area around stop A4. Adapted from Holt (1961).

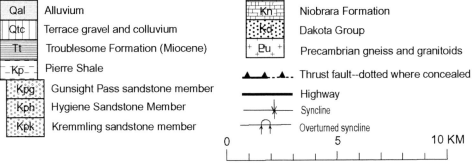

Qal	Alluvium
Qtc	Terrace gravel and colluvium
Tt	Troublesome Formation (Miocene)
Kp	Pierre Shale
Kpg	Gunsight Pass sandstone member
Kph	Hygiene Sandstone Member
Kpk	Kremmling sandstone member

Kn	Niobrara Formation
Kd	Dakota Group
Pu	Precambrian gneiss and granitoids

▲▲ ..▲.. Thrust fault--dotted where concealed
━━━━━ Highway
Syncline
Overturned syncline

0   5   10 KM

dates; the rocks along the axis of the Rio Grande rift must have been heated in late Tertiary time. Magma at high levels in the crust probably caused the elevated heat flow, which wiped out the Laramide annealing or partial annealing zones. Quantitative information on Laramide uplift from the fission-track data in the Blue River region, therefore, cannot be obtained. The pattern of young AFT dates adjacent to the Blue River valley is similar to those from the Sawatch Range (Bryant and Naeser, 1980; Shannon, 1987; Kelly and Chapin, 1995) and Sangre de Cristo Mountains (Lindsay and others, 1986; Kelly and Chapin, 1995), which also border the Rio Grande rift. The considerable Neogene heating, uplift, and erosion along the northern Rio Grande rift also may be coeval with the Miocene tilting of the Front Range postulated by Steven and others (1997).

STOP A5. Green Mountain intrusive center in Pierre Shale

Driving instructions: Return to Rte 9 and head north 12 miles to the northern end of Green Mountain Reservoir. Turn left on Heney turnoff and drive to good exposures east of the dam.

A complex of laccoliths and sills surround a small stock on the west side of Green Mountain. Dikes related to this center intrude the Pierre Shale 3 to 10 km east and north of Green Mountain. Some dikes in the northern part of the high Gore Range have been correlated with this center (Tweto and Lovering, 1977). Basaltic or andesitic lava flows to the east and south may have had their source here at Green Mountain. The southernmost flows are in a tilted fault block that dips 25° south into a fault. Other flows rest unconformably on gently east-dipping Cretaceous rocks.

This porphyritic latite or trachyte contains sanidine, andesine, hornblende and augite phenocrysts. A fission-track analysis on zircon gave an age of 29.9 +/- 2.4 Ma (Naeser and others, 1973). Faults of small displacement cut the intrusive rocks and adjacent rocks are altered to kaolinite. Pyrite and, rarely, sphalerite and galena occur along the faults in the intrusive rock and sedimentary country rock (Taggert, 1962).

Driving instructions to evening accommodations at Hot Sulphur Springs Resort: Return to Rte 9 and head north to Kremmling. Turn east on Rte. 40 at Kremmling, turning left on the first road in Hot Sulphur Springs. The springs emanate from a fault separating the Jurassic Morrison Formation from the Cretaceous-Paleocene Middle Park Formation. Enjoy!

DAY 2

STOP B1. Overview of Middle Park, Mt. Bross fault, and the Breccia Spoon syncline

Driving instructions: Rejoin Rte 40 going east toward Granby. Turn right on Grand County Road 55 at the east side of Hot Sulphur Springs. Turn around at the cemetery and park at crest of hill overlooking Hot Sulphur Springs.

This stop gives a panoramic view of structural relationships in Middle Park. Middle Park is a section of the axial foreland basin that probably stretched along the entire western margin of the Front Range during the Laramide orogeny. To the south, synorogenic strata are exposed in South Park but have been stripped by erosion in the intervening Blue River half graben. Here and to the north, the Middle Park basin is continuous with the North Park basin, with the different names resulting from the topographic high separating the two topographic basins. This area shows the complexity of the Tertiary structural evolution in the west flank of the Front Range. Igneous and volcanoclastic units of Laramide,

mid-Tertiary and Neogene age are faulted and either folded or tilted during multiple deformation events spanning the Tertiary.

The Mt. Bross reverse fault is exposed east of Mount Bross and dips east, bringing Pierre Shale over Middle Park Formation (Izett, 1968). Further east, the Pierre Shale is overlain by the Middle Park Formation in a slight angular unconformity marked by the truncation of a sandy unit within the Pierre Shale. The cliffy exposures above the unconformity expose the volcanoclastic Windy Gap Volcanic Member of the Middle Park Formation and define the western limb of the Breccia Spoon syncline. This north-trending syncline appears to be folded by the WNW-trending basement uplift west of Hot Sulfur Springs, suggesting changes in shortening directions through time. The Middle Park Formation is overlain by the tuffaceous Rabbit Ears volcanic rocks which are Oligocene to earliest Miocene in age (Izett, 1968). These beds are overlain by the Miocene Troublesome Formation, which were folded and subsequently overlain by tilted Pliocene(?) basalt flows (Izett, 1968).

The aptly named Troublesome Formation is well dated by mammal fossils and volcanic ash with fission track ages between 20 and 13 Ma (Izett and Barclay, 1973). To the west near Kremmling, it blankets the Precambrian rocks in the hanging wall of the Williams Range thrust. This suggests that the Troublesome Formation may have been deposited in localized basins caused by backsliding on this thrust fault. Fold axes in the Troublesome Formation parallel the Williams Range thrust and are defined by rotated strata with slight to locally steep (80°) dips (Izett and Barclay, 1973). The origin of this folding is unclear. Later tilting of Pliocene basalt flows is consistent with Neogene normal faulting seen to the south in the Blue River half graben.

The cliffs across the river are made up of the Windy Gap Volcanic Member of the Middle Park Formation. The Middle Park Formation and the correlative Coalmont Formation in North Park are synorogenic with the Laramide orogeny. Up to 2 kilometers of Middle Park Formation are exposed in this area. The lowest part of the Middle Park Formation is the Cretaceous-Paleocene Windy Gap Volcanic Member. The diversity of rock types within the Windy Gap Volcanic Member shows the complexity of early Laramide events. It contains lenticular zones of fragmental andesitic rocks that thicken to the south that were probably extruded from volcanoes to the east and southeast along the Colorado Mineral Belt (Izett, 1968). Clasts of Precambrian basement rocks in arkosic sedimentary rocks, both within and underneath the andesitic rocks, indicate unroofing of the Front Range in the early Laramide. Arkosic sedimentary rocks become dominant at the top of the Windy Gap Volcanic Member and then fine upward into siltstones near the top of the Middle Park Formation (Izett, 1968).

STOP B2. Minor faulting in the Middle Park Formation

Driving instructions: Return to Rte 40 and travel east, turning north on Rte 125, stopping at the large road cut at the bends in the road one kilometer north of its intersection with Grand County Road 408.

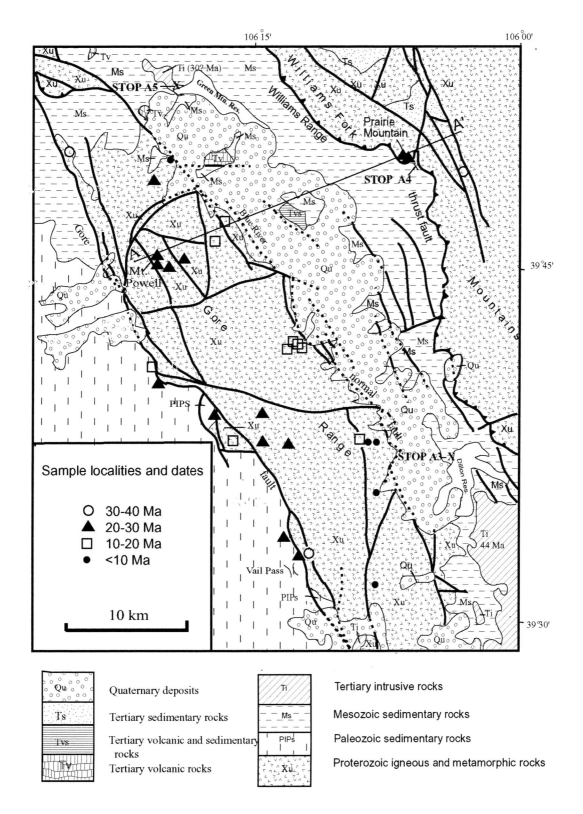

Figure 9. Map showing locations of fission-track samples and ranges of data from the Gore Range and the Ute Pass vicinity. Line of section shown on Fig. 10. Geology simplified from Tweto and others (1970).

34

Both the Windy Gap Volcanic Member, which forms the cliffs one kilometer to the south, and the overlying arkosic sands of the Middle Park Formation contain a multitude of slickensided faults. Initial observations indicate multiple distinct populations of thrust and strike-slip faults. At several localities in the Windy Gap Volcanic Member, thrust slickenlines trend from N-S to E-W, with more east-west oriented slickenlines cut by more northeast-southwest slickenlines. At this locality, spectacular thrust faults indicating E-W shortening appear to be cut by northeast- and north-striking strike-slip faults.

STOP B3. North Park syncline

Driving instructions: Continue north on Rte 125 over Willow Creek Pass. At Rand, turn northeast on Jackson County Road 27. Park on overlook (Owl Ridge) into North Park.

This locality is on the southwest margin of the North Park syncline, a large, northwest-striking structure that is discordant to the major Laramide thrusts and folds on the eastern border of North Park. The North Park syncline parallels the Independence Mountain thrust system that defines the north end of North Park, where the thrust system truncates NNW-trending Laramide folds in the Coalmont Formation (Blackstone, 1977). The core of the North Park syncline contains Miocene North Park Formation, which is roughly equivalent to the Troublesome Formation to the south and Browns Park Formation to the west.

The origin of the North Park syncline is problematic. Early workers linked it to normal faulting on the northeast side of the structure. Later subsurface work indicates stratal shortening across the fault zone, suggesting a northeast-dipping thrust fault or a transpressive structure (D. Stone, personal commun., 1999). The age of the fold is also problematic since it includes Miocene strata deposited long after the end of Laramide shortening.

Recent seismic profiles and well information indicate that the southwest side of the anticline is detached and suggest periods of both shortening and extension (Fig. 11. Preliminary studies of mid-Tertiary strike-slip faulting discussed earlier do suggest that north-south shortening and compression roughly normal to the fold axis could have occurred during the mid-Tertiary. Another possibility is that gravitational sliding away from a

volcanic center may have caused localized thrust faulting and folding, trapping the synclinal keel of the Miocene North Park Formation. Concentric thrust faults caused by gravity spreading of the Marysvale Volcanic Field in southern Utah (Merle and others, 1993) could be analogous structures.

STOP B4. Laramide thrusting and mid-Tertiary igneous activity at Cameron Pass

Driving instructions: Continue northeast on Jackson County Road 27 and turn right (east) at T intersection with Rte 14 toward Gould. Stop at exposures of red sandstone 1 mile west of Cameron Pass.

The Cameron Pass area exposes the east-dipping Never Summer thrust system of Laramide age overprinted by mid-Tertiary intrusions of the Mt. Richthofen intrusive complex and later high-angle faults. The roadcuts expose overturned Mesozoic strata beneath Precambrian rocks brought up by a splay of the Never Summer thrust system. South of the road, the rugged Nokhu Crags expose contact-metamorphosed Pierre Shale adjacent to the 29 Ma (K-Ar; Corbett, 1964) Mt. Richthofen batholith (O'Neill, 1981). Upward bowing by the batholith has formed a window in Never Summer thrust system in which Pierre shale is exposed 10 kilometers from the nearest exposure of the thrust front, where Precambrian rocks are thrust on top of the Coalmont Formation. The overhang may have been expanded by the intrusion of the Mt. Richthofen batholithic complex, but the Laramide overhang still must have been quite substantial. The similarities between the Never Summer and the Williams Range (STOP A2) thrust windows are striking!

The Mt. Richthofen batholith is cut by high-angle, north-striking faults with early strike-slip slickenlines overprinted by extensional dip-slip slickenlines. Chapin (1983) proposed that several major north-striking fault zones along the western margin of the Front Range formed during late Laramide right-lateral strike-slip faulting. The fact that strike-slip slickenlines cut the 29 Ma Mt. Richthofen batholith suggests either unusually prolonged Laramide shortening or a separate phase of post-Laramide strike-slip faulting. The existence of similar strike-slip faults in southern Colorado, where they cut the mid-Tertiary Cripple Creek intrusions (T. Wawrzyniec, personal commun., 1996), and in northern New Mexico, where they cut late Oligocene igneous rocks (Erslev, unpub. data, 1998), suggests that strike-slip faulting may have been regionally important in mid-Tertiary time.

Strike-slip faulting on the western margin of the Front Range could provide an explanation for the Independence Mountain thrust system that forms the northern boundary of North Park. Right-lateral motion on the eastern margin of North Park could have moved the basin northward, providing the impetus for stuffing the northern end of the basin underneath the Independence Mountain thrust system. Accommodation of right-lateral shear along thrusts would explain why evidence for right-lateral shear has not been documented north of the Colorado-Wyoming border.

The causal relationships between mid-Tertiary sedimentation, igneous activity and faulting are unclear. Laramide and/or

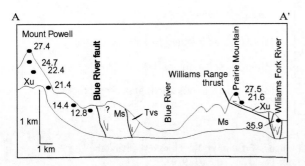

Figure 10. Cross section from the Williams Fork to Mt. Powell showing locations of apatite fission-track dates. 5X vertical exaggeration.

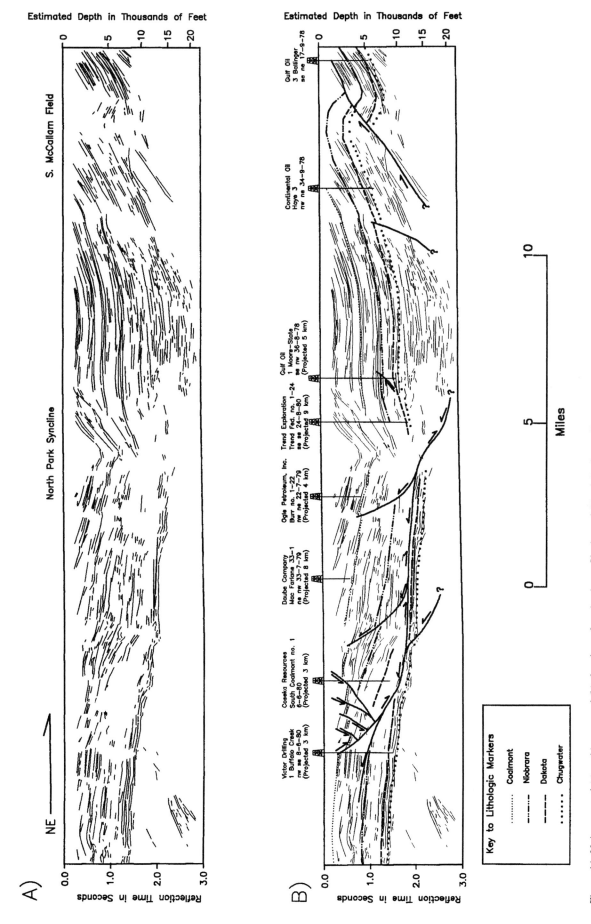

Figure 11. Uninterpreted (a) and interpreted (b) sketches of a seismic profile through North Park syncline near stop B3. Profile is oriented NE-SW from approximately 10 km SSE of Coalmont, Colorado, to approximately 15 km ENE of Walden, Colorado. Well logs indicate that the Mesozoic section is thicker in the center than on the flanks, suggesting thrust faulting repeated and thickened section. Later extension may have reactivated the low-angle thrusts, resulting in normal faulting in the hanging wall and tightening of the North Park syncline. Seismic data provided by Seismic Exchange, Inc.; interpretation is that of the authors.

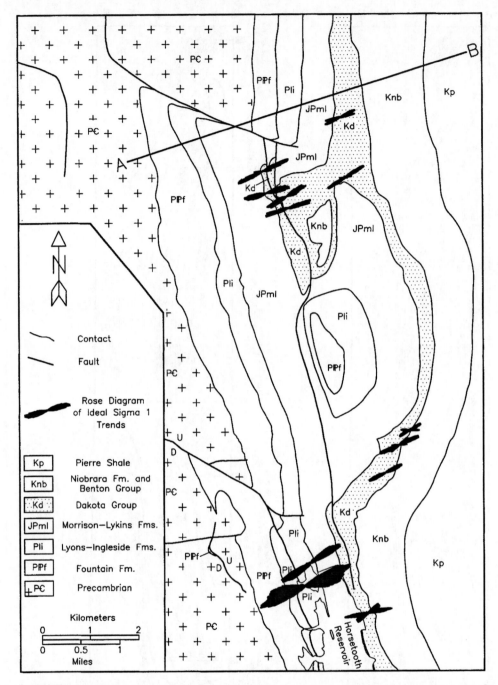

Figure 12. Simplified geologic map (after Braddock and others, 1988a, 1988b) of the eastern margin of the Front Range northeast of Fort Collins, Colorado, with rose diagrams showing the trends of maximum compression directions indicated by minor faults (Holdaway, 1998)

earlier deformation may have produced a fundamental zone of crustal weakness along the current Rio Grande rift that was exploited by intrusion and strike-slip faulting in the mid-Tertiary. The transtensional to transpressive environment indicated by mid-Tertiary faulting may have also facilitated magmatic intrusion in the crust. Alternatively, thermal softening due to mid-Tertiary igneous activity may have focused strike-slip faulting driven by the gravitational spreading of the extending

Cordilleran orogeny to the southwest (Christiansen and Yeats, 1992). In either case, the resultant zone of structural and magmatic weakening may be responsible for subsequent Rio Grande rifting synchronous with Basin and Range extension. Thus, the coherence of the Colorado Plateau during Neogene extension may not be due to the abnormal strength of its underlying lithosphere so much as the unusual weakness of the lithosphere underlying the current Rio Grande rift.

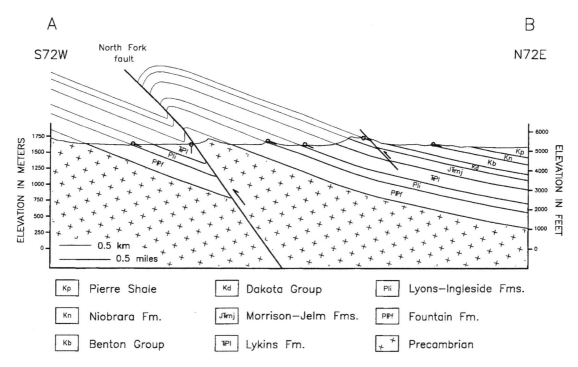

Figure 13. Cross section through the North Fork fault and fold (Holdaway, 1998). Line of section is shown in Fig. 12.

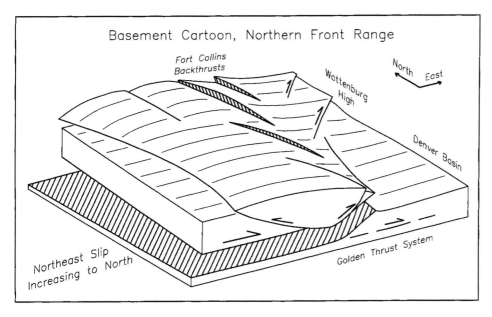

Figure 14. Diagrammatic block diagram of the Laramide basement structure of the central and northern Front Range.

STOP B5. Precambrian mylonitized pegmatite in the Skin Gulch Shear Zone

Driving instructions: Continue east on Rte 14. 0.6 miles after the small tunnel through finer-grained, mylonitic basement rocks, park on north side of road beyond the 35 MPH sign.

    This exposure exhibits excellent shear textures that are readily apparent in outcrop due to the coarse grain size of the mylonite's pegmatite protolith. C-S surfaces and sigma porphy-

roclasts indicate high-angle reverse faulting at this outcrop, but studies in progress at University of New Mexico suggest multiple stages and ages of Precambrian ductile shearing on shear zones throughout the Front Range.

STOP B6: North Fork Fault and associated fault-propagation fold

Driving instructions: Continue east on Rte 14. At the T intersection with Rte 287, turn left (north). After 1.2 miles, take a sharp

*left turn on to private property. Permission is needed for access,
but the view from the nearby roadcut of Morrison Formation is
worthy of a stop.*

The northwest-striking North Fork Fault causes a two kilometer right-lateral separation of the basement-Fountain Formation unconformity (Fig. 12, Braddock and others, 1988a). This
structure is typical of the major, southeast-plunging fault-propagation folds in the northeastern Front Range (Fig. 13). Trenching
of this fault along the irrigation ditch shows that the fault dips
approximately 45° northeast, suggesting thrust/reverse motion.
Overlying strata are truncated by the fault at the level of the Permian Ingleside Formation, whose bleaching adjacent to the fault
in the hanging-wall anticline suggests iron reduction during
hydrocarbon migration through these rocks. At the level of the
Cretaceous Dakota Group, the strata form a nearly continuous
fold, indicating fault-propagation folding where basement faulting dies upward into cover folding. This fold is highly conical,
dying out to the southeast.

Most minor faults in the northeastern margin of Front Range
indicate ENE shortening and compression, with almost no suggestion of late extensional faulting common on the west side of
the Front range (Holdaway, 1998). At this locality, however, the
Ingleside Formation in the hanging wall anticline is elongated by
normal faults. Bedding elongation during fault-propagation folding probably generated these faults.

The thrust and reverse faults in the northeastern Front Range
are not directly responsible for the uplift of the range because
their northeasterly dip brings the basin up relative to the range.
Erslev and Selvig (1997) explained this geometry as the resulting
from backthrusting off a blind, east-northeast-directed thrust in
the basement. More recent work by Holdaway (1998) has shown
that an additional component of deep crustal wedging is necessary to uplift the range (Figs. 4d, 14). In this case, thrust on the
east side of the Front Range may simply be surface tightening
due to wedging at depth, a situation analogous to structures on
the northeast side of the Wind River Mountains (Erslev, 1986).

*Driving instructions back to Denver: Follow Rte 14 south and
east through Fort Collins to I-25 south.*

ACKNOWLEDGMENTS

The compilation of this field trip guide was a greatly assisted
by the meticulous road log of Dudley Bolyard (1997). Reviews
by Dave Lageson, Jack Reed, Bob Scott, and Bruce Trudgill
improved this manuscript.

REFERENCES CITED

Anderson, J.L., 1983, Proterozoic anorogenic plutonism in North America, in
    Medaris, L.G., and others, eds., Proterozoic geology: Geological Society
    of America Memoir 161, p. 133-154.
Baldridge, W.S., Keller, G.R., Haak, V., Wendlandt, E., Jiracek, G.R., and Olsen,
    K.H., 1995, The Rio Grande Rift, in Olsen, K.H., (ed.), Continental rifts:
    evolution, structure, tectonics: Elsevier, p. 233-275.

Barker, Fred, Hedge, C.E., Millard, H.T., and O'Neil, J.R., 1976, Pikes Peak
    batholith: geochemistry of some minor elements and isotopes, and impli-
    cations for magma genesis, in Epis, R.S., and Weimer, R.J., eds., Studies
    in Colorado field geology: Colorado School of Mines Professional Con-
    tributions 8, p. 44-56.
Berg, R.R., 1962, Mountain flank thrusting in Rocky Mountain foreland,
    Wyoming and Colorado: American Association of Petroleum Geologists
    Bulletin, v. 46, p. 2019-2032.
Bergh, S., and Snoke, A., 1992, Polyphase Laramide deformation in the Shirley
    Mountains, south central Wyoming foreland: Mountain Geologist, v. 29,
    p. 85-100.
Bird, P., 1988, Formation of the Rocky Mountains, a continuum computer model:
    Science, v. 239, p. 1501-1507.
Bird, P., 1998, Kinematic history of the Laramide orogeny in latitudes 35o-49o,
    western United States: Tectonics, v. 17, p. 780-801.
Blackstone, D.L., Jr, 1977, Independence Mountain thrust fault, North Park basin,
    Colorado: Contributions to Geology, v. 16, p. 1-15.
Blackstone, D.L., 1991, Tectonic relationships of the southeastern Wind River
    Range, southwestern Sweetwater Uplift, and Rawlins Uplift, Wyoming:
    Report of Investigation No. 47, Geological Survey of Wyoming, 24 p.
Bolyard, D.W., 1997, Colorado Front Range Field Trip Road Logs, in Bolyard,
    D.W., and Sonnenberg, S.A., eds., Geologic history of the Colorado Front
    Range: Rocky Mountain Association of Geologists Field Trip Guidebook,
    189 p.
Boos, C.M., and Boos, M.F., 1957, Tectonics of eastern flank and foothills of
    Front Range: American Association of Petroleum Geologists Bulletin, v.
    41, p. 2603-2676.
Braddock, W.A., Connor, J.J., Swann, G.A., and Wolhford, D.D., 1988a, Geo-
    logic map of the Laporte quadrangle, Larimer County, Colorado: U.S.
    Geological Map GQ-1621.
Braddock, W.A., Wohlford, D.D., and O'Connor, J.T., 1988b, Geologic map of
    the Horsetooth reservoir quadrangle, Larimer County, Colorado: U.S.
    Geological Survey Map GQ-1625.
Brewer, J.A., Allmendinger, R.W., Brown, L.D., Oliver, J.E., and Kaufman, S.,
    1982, COCORP profiling across the Rocky Mountain front in southern
    Wyoming, Part 1, Laramide structure: Geological Society of America
    Bulletin, v. 93, p. 1242-1252.
Brister, B.S., and Gries, R.R., 1994, Tertiary stratigraphy and tectonic develop-
    ment of the Alamosa basin (northern San Luis Basin), Rio Grande rift,
    south-central Colorado, in Keller, G.R., and Cather, S.M., eds., Basins of
    the Rio Grande rift—structure, stratigraphy, and tectonic setting: Boulder,
    Colorado, Geological Society of America Special Paper 291, p. 39-58.
Brown, W.G., 1988, Deformation style of Laramide uplifts in the Wyoming fore-
    land, in Schmidt, C.J., and Perry, W.J., Jr., (eds.), Interaction of the Rocky
    Mountain foreland and the Cordilleran thrust belt: Geological Society of
    America Memoir 171, p. 53-64.
Bryant, B., and Naeser, C.W., 1980, The significance of fission-track ages of
    apatite in relation to the tectonic history of the Front and Sawatch Ranges:
    Geological Society of America Bulletin, v. 91, pt. 1, p. 156-164.
Bryant, B., Marvin, R.F, Naeser, C.W., and Mehnert, H.H., 1981, Ages of igneous
    rocks in the South Park-Breckenridge region, Colorado, and their relation
    to the tectonic history of the Front Range uplift, in Shorter Contributions
    to Isotope Research in the western United States, 1980: U.S. Geological
    Survey Professional Paper 1199-C, p. 15-35.
Chapin, C.E., 1983, An overview of Laramide wrench faulting in the southern
    Rocky Mountains with emphasis on petroleum exploration, in Lowell,
    J.D., ed., Rocky Mountain foreland basins and uplifts: Denver, Rocky
    Mountain Association of Geologists, p. 169-179.
Chapin, C.E., and Cather, S.M., 1981, Eocene tectonism and sedimentation in the
    Colorado Plateau-Rocky Mountain area: Arizona Geological Digest, v.
    14, p. 175-198.
Chapin and Cather, 1994, Tectonic setting of the axial basins of the northern and
    central Rio Grande rift, in Keller, G.R., and Cather, S.M. , eds., Basins of
    the Rio Grande rift—structure, stratigraphy, and tectonic setting: Boulder,
    Colorado, Geological Society of America Special Paper 291, p. 5-25.

Chase, R., Genovese, P., and Schmidt, C., 1993, The influence of Precambrian rock compositions and fabrics on the development of Rocky Mountain Foreland folds, in Schmidt, C.J., Chase, R., and Erslev, E.A., (eds), Basement-cover kinematics of Laramide foreland uplifts: Geological Society of America Special Paper 280, p. 45-72.

Christiansen, R.L., and Yeats, R.S., 1992, Post-Laramide geology of the U.S. Cordilleran region, in Burchfiel, B.C., Lipman, P.W., and Zoback, M.L., eds., The Cordilleran Orogen: Conterminous U.S.: Boulder, Colorado, Geological Society of America, The Geology of North America, v. G-3, p. 261-406.

Corbett, M.K., 1964, Tertiary igneous petrology of the Mt. Richthofen-Iron Mountain area, north-central Colorado: University of Colorado Ph.D. thesis, 115 p.

DOE, 1993, Phase II geologic characterization data acquisition high resolution deep seismic; revised final report: U.S. Department of Energy, Rocky Flats Plant, 155 p.

Eaton, G.P., 1986, A tectonic redefinition of the southern Rocky Mountains: Tectonophysics, v. 132, p. 163-193.

Erslev, E. A., 1986, Basement balancing of Rocky Mountain foreland uplifts, Geology, v. 14, p. 259-262.

Erslev, E.A., 1993, Thrusts, back-thrusts, and detachment of Laramide foreland arches, in Schmidt, C.J., Chase, R., and Erslev, E.A., (eds.), Laramide basement deformation in the Rocky Mountain foreland of the western United States: G.S.A. Special Paper 280, p. 339-358.

Erslev, E.A., and Rogers, J.L., 1993, Basement-cover kinematics of Laramide fault-propagation folds: in Schmidt, C.J., Chase, R., and Erslev, E.A., eds., Basement-cover kinematics of Laramide foreland uplifts: Geological Society of America Special Paper, p. 125-146.

Erslev, E.A., and Selvig, B., 1997, Thrusts, backthrusts and triangle zones: Laramide deformation in the northeastern margin of the Colorado Front Range, in Bolyard, D.W., and Sonnenberg, S.A., Geologic history of the Colorado Front Range, Rocky Mountain Association of Geologists, Denver, Colorado, p. 65-76.

Gregory, K.M., 1994, Paleoclimate and paleoelevation of the Florissant flora, in,

Evanoff, E., ed., Late Paleocene geology and paleoenvironments of central Colorado with emphasis on the geology and paleontology of Florissant Fossil Beds National Monument: Guidebook for field trip Rocky Mountain Section of the Geological Society of America at Durango Colorado, p. 57-66.

Gregory, K.M., and Chase, C.G., 1992, Tectonic significance of paleobotanically estimated climate of the late Eocene erosion surface, Colorado: Geology, v. 20, p. 581-585.

Gries, R.R., 1983, North-south compression of the Rocky Mountain foreland structures, in Lowell, J.D., and Gries, R.R., eds., Rocky Mountain foreland basins and uplifts: Rocky Mountain Association of Geologist, Denver, Colorado, p. 9-32.

Gries, R., and Dyer, R.C., 1985, Seismic exploration of the Rocky Mountain region: Denver, Colorado, Rocky Mountain Association of Geologists and Denver Geophysical Society, p. 1139-1142.

Hamilton, W., 1988, Laramide crustal shortening, in Perry, W.J., and Schmidt, C.J., eds., Interaction of the Rocky Mountain foreland and the Cordilleran thrust belt: Geological Society of America Memoir 171, p. 27-39.

Hansen, W.R., 1986, History of faulting in the eastern Uinta Mountains, Colorado and Wyoming, in Stone, D.S. (ed.), New interpretations of northwest Colorado Geology: Rocky Mountain Association of Geologists, p. 19-36.

Holdaway, S.M., 1998, Laramide deformation of the northeastern Front Range, Colorado: Evidence for deep crustal wedging during horizontal compression: M.S. thesis, Colorado State University, 146 p.

Izett, G.A., 1968, Geology of the Hot Sulphur Springs quadrangle, Grand County, Colorado: U.S. Geological survey Professional Paper 586, 79 p.

Izett, G.A., and Barclay, C.S.V., 1973, Geologic map of the Kremmling quadrangle, Grand County, Colorado: U.S. Geological Survey Geologic Quadrangle Map GQ 1115, scale 1:24,000.

Jacobs, A.F., 1983, Mountain front thrust, southeastern Front Range and northwestern Wet Mountains, in Lowell, J.D., ed., Rocky Mountain foreland basins and uplifts: Rocky Mountain Association of Geologists, p. 229-244.

Kelley, S.A., and Chapin, C.E., 1997, Internal structure of the southern Front Range, Colorado, from an apatite fission-track thermochronology perspective, in Bolyard, D.W., and Sonnenberg, eds., Geologic History of the Colorado Front Range, 1997 RMS-AAPG Field trip #7: Rocky Mountain Association of Geologists, Denver, p. 19-30.

Kellogg, K.S., 1997a, Geologic map of the Dillon quadrangle, Summit and Grande Counties, Colorado U.S. Geological Survey Open-File Report 97-738, scale 1:24,000.

Kellogg, K.S., 1997b, The Williams Range thrust near Dillon, Colorado - hanging wall fracturing during Laramide thrusting preparing bedrock for major Neogene and Pleistocene landsliding: Geological Society of America Abstracts with Programs, v. 29, no. 6, p. A163.

Kellogg, K.S., 1999, Neogene basins of the northern Rio Grande rift—partitioning and asymmetry inherited from Laramide and older uplifts: Tectonophysics, v. 305, p. 141-152.

Kluth, C.F., and Nelson, S.N., 1988, Age of the Dawson Arkose, southwestern Air Force Academy, Colorado, and implications for the uplift history of the Front Range: The Mountain Geologist, v. 25, no. 1, p. 29-35.

Kulik, D.M., and Schmidt, C.J., 1988, Regions of overlap and styles of interaction of Cordilleran thrust belt and Rocky Mountain foreland, in Schmidt, C.J., and Perry, W.J., Jr., eds., Interaction of the Rocky Mountain foreland and the Cordilleran thrust belt: Geological Society of America Memoir 171, p. 75-98.

Leonard, E.M., and Langford, R.P., 1994, Post-Laramide deformation along the eastern margin of the Colorado Front Range - a case against significant deformation: The Mountain Geologist, v. 31, p. 45-52.

Lindsey, D.A., Andriessen, P.A.M., and Wardlaw, B.R., 1986, Heating, cooling, and uplift during Tertiary time, northern Sangre de Cristo Range, Colorado: Geological Society of America Bulletin, v. 97, p. 1133-1143.

Lowell, J.D., 1983, Foreland deformation, in Lowell, J.D., ed., Rocky Mountain foreland basins and uplifts: Rocky Mountain Association of Geologists, Denver, Colorado, 392 p.

Marvin, R.F., Mehnert, H.H., Naeser, C.W., and Zartman, R.E., 1989, U.S. Geological survey radiometric ages-Compilation "C", Part 5. Colorado, Montana, Utah, and Wyoming: Isochron/West, no. 11, p. 1-42.

Matthews, V., III, and Work, D.F., 1978, Laramide folding associated with basement block faulting along the northeastern flank of the Front Range, Colorado, in Matthews, V., III (ed.), Laramide folding associated with basement block faulting: Geological Society of America Memoir 151, p. 101-124.

McDowell, F.W., 1971, K-Ar ages of igneous rocks from the western United States: Isochron/West no. 2, p. 1-16.

Merle, O.R., Davis, G.H., Nickelsen, R.P., and Gourlay, P.A., 1993, Relation of thin-skinned thrusting of Colorado Plateau strata in southwestern Utah to Cenozoic magmatism: Geological Society of America Bulletin, v. 105, p. 387-398.

Naeser, C.W., Izett, G.A., and White, W.H., 1973, Zircon fission-track ages from some middle Tertiary igneous rocks in northwestern Colorado: Geological society of America Abstracts with Programs, v. 5, no. 6, p. 498.

Naeser, C.W., Bryant, Bruce, Kellogg, Karl, and Perry, W.J., Jr., 1999, Middle to late Tertiary cooling of the Gore and western Front Ranges, Central Colorado, from apatite fission-track data: Geological Society of America Abstracts with Programs, v. 31, National Meeting, Denver.

Oldow, J.S., Bally, A.W., and Lallemant, H. G., 1990, Transpression, orogenic float, and lithospheric balance: Geology, v. 18, p. 991-994.

O'Neill, J.M., 1981, Geologic map of the Mount Richthofen quadrangle and the western part of the Fall River Pass quadrangle, Grand and Jackson Counties, Colorado: U.S. Geological Survey Miscellaneous Investigations Series Map I-1291.

Prucha, J.J., Graham, J.A., and Nickelson, R.P., 1965, Basement-controlled deformation in Wyoming Province of Rocky Mountain foreland: American Association of Petroleum Geologist Bulletin, v. 49, p. 966-992.

Raynolds, R.G. , 1997, Synorogenic and post-orogenic strata in the central Front Range, Colorado, in Bolyard, D.W., and Sonnenberg, S.A., eds, Geologic history of the Colorado Front Range, 1997 RMS-AAPG Filed Trip #7:

Denver, Colorado, Rocky Mountain Association of Geologists, p. 43-48.

Reed, J.C., Jr., Bickford, M.E., Premo, W.R., Aleinikoff, J.N., and Pallister, J.S., 1987, Evolution of the Early Proterozoic Colorado province: Geology, v. 15, no, 9, p. 861-865.

Reed, J.C., Jr., Bickford, M.E., and Tweto, O, 1993, Proterozoic accretionary terranes of Colorado and southern Wyoming, in Van Schmus, W.R. and Bickford, M.E., eds., Transcontinental Proterozoic provinces, in Reed, J.C., Bickford, M.E., Houston, R.S., Link, P. K., Rankin, D.W., Sims, P.K., and Schmus, W.R., eds, Precambrian: Conterminous U.S.: Geological Society of America Decade of North American Geology, v. C-2, p. 211-228.

Sales, J.K., 1968, Cordilleran foreland deformation: American Association of Petroleum Geologists Bulletin, v. 52, p. 2000-2015.

Schmidt, C.J., and Perry, W.J., Jr., 1988, Interaction of the Rocky Mountain foreland and the Cordilleran thrust belt: Geological Society of America Memoir 171, 582 p.

Scott, G.R., 1963, Bedrock geology of the Kassler quadrangle, Colorado: U. S. Geological Survey Professional Paper 421-B, p. 71-125, scale 1:24,000.

Scott, G.R., 1972, Geologic map of the Morrison quadrangle, Jefferson County, Colorado: U.S. Geological Survey Miscellaneous Geologic Investigations Map I-790-A, scale 1:24,000.

Selvig, B.W., 1994, Kinematics and structural models of faulting adjacent to the Rocky Flats Plant, central Colorado: M.S. thesis, Colorado State University, 133 p.

Selverstone, J., Hodgins, M., Shaw, C., Aleinikoff, J.N., and Fanning, C.M., 1997, Proterozoic tectonics of the northern Colorado Front Range, in Bolyard, D.W. and Sonnenberg, S.A., eds, Geologic history of the Colorado Front Range, 1997 RMS-AAPG Field Trip #7: Rocky Mountain Association of Geologists, Denver, p. 9-18.

Shaw, C.A., Snee, L.W., Selverstone, Jane, and Reed, J.C., Jr., 1999, 40Ar/39Ar thermochronology of mesoproterozoic metmorphism in the Colorado Front Range: Journal of Geology, v. 107, p. 49-67.

Smith, J.H., 1964, Geology of the sedimentary rocks of the Morrison quadrangle: U.S. Geological Survey Miscellaneous Geologic Investigations Map I-428, scale 1:24,000.

Smithson, S.B., Brewer, J.A., Kaufman, S., Oliver, J.E., and Hurich, C.A., 1979, Structure of the Laramide Wind River uplift, Wyoming, from COCORP deep reflection data and from gravity data: Journal of Geophysical Research, v. 84, p. 5955-5972.

Spang, J.H., Evans, J.P., and Berg, R.R., 1985, Balanced cross sections of small foldthrust structures: Mountain Geologist, v. 22, p. 3746.

Steven, T. A., 1975, Middle Tertiary volcanic field in the southern Rocky Mountains, in Curtis, B.F., ed., Cenozoic history of the Southern Rocky Mountains: Geological Society of America Memoir 144, p. 75-94.

Steven, T.A., Evanoff, E., Yuhas, R.H., 1997, Middle and late Cenozoic tectonic and geomorphic development of the Front Range of Colorado, in Bolyard, D.W., and Sonnenberg, S.A., eds., Geologic history of the Colorado Front Range, 1997 RMS-AAPG Filed Trip #7: Rocky Mountain Association of Geologists, Denver, p. 115-124.

Stone, D.S., 1984, The Rattlesnake Mountain, Wyoming, debate: a review and critique of models: The Mountain Geologist, v. 21, p. 37-46.

Stone, D.S., 1986, Seismic and borehole evidence for important pre-Laramide faulting along the axial arch in northwest Colorado, in Stone, D.S. (ed.), New interpretations of northwest Colorado Geology: Rocky Mountain Association of Geologists, p. 19-36.

Stone, D.S., 1995, Structure and kinematic genesis of the Calla wrench duplex: Transpressional reactivation of the Precambrian Cheyenne belt in the Laramie basin, Wyoming: American Association of Petroleum Geologists Bulletin, v. 79, p. 1349-1376.

Stearns, D.W., 1971, Mechanisms of drape folding in the Wyoming province: Wyoming Geological Association, 23rd Annual Field Conference, Wyoming Tectonics Symposium Guidebook, p. 82-106.

Stearns, D.W., 1978, Faulting and forced folding in the Rocky Mountain foreland: in Matthews, V., III, ed., Laramide folding associated with basement block faulting in the western United States: Geological Society of America Memoir 151, p. 1-37.

Taggert, J.N., 1962, Geology of the Mount Powell quadrangle, Colorado: Harvard University unpublished Ph.D. thesis, 239 p.

Tweto, O., 1975, Laramide (Late Cretaceous-early Tertiary) orogeny in the Southern Rocky Mountains, in Curtis, B.F., ed., Cenozoic history of the southern Rocky Mountains: Geologic Society of America Memoir 144, p. 1-44.

Tweto, O., 1976, Geologic map of the Craig 1ox2o quadrangle, northwestern Colorado: U.S. Geological Survey Miscellaneous Investigations Series map I-972.

Tweto, O., 1978, Northern rift zone guide 1, Denver-Alamosa, Colorado, in Hawley, J.W., compiler, Guidebook to the Rio Grande rift in New Mexico and Colorado: New Mexico Bureau of Mines and Mineral Resources Circular 163, p. 13-32.

Tweto, O., 1979, Geologic map of Colorado: U.S. Geological Survey, scale 1:500,000.

Tweto, O., 1987, Rock units in the Precambrian basement in Colorado: U.S. Geological Survey Professional Paper 1321, 54 p.

Tweto, O., Bryant, B., and Williams, F.E., 1970, Mineral resources of the Gore Range-Eagles Nest Primitive Area and vicinity, Summit and Eagle Counties, Colorado: U.S. Geological Survey Bulletin 1319-C, p. C1-C127.

Tweto, O., and Lovering, T.S., 1977, Geology of the Minturn 15-minute quadrangle, Eagle and Summit Counties, Colorado: U.S. Geological Survey Professional Paper 956, 96 p.

Tweto, O., and Reed, J.C., Jr., 1973, Reconnaissance of the Ute Peak 15-minute quadrangle, Grand and Summit Counties, Colorado: U.S. Geological Survey Open-File Map, scale: 1:62,500.

Tweto, O., and Sims, P.K., 1963, Precambrian ancestry of the Colorado mineral belt: Geological Society of America Bulletin, v. 74, p. 991-1014.

Unruh, D.M., Snee, L.W., Foord, E.E., and Simmons, W.B., 1995, Age and cooling history of the Pikes Peak batholith and associated pegmatites [abs]: Geological Society of America Abstracts with Programs, v. 27, no. 6, p. 468.

Ulrich, G.E., 1963, Petrology and structure of the Porcupine Mountain area, Summit County, Colorado: University of Colorado Ph.D. thesis, 205 p.

Weimer, R.J., and Ray, R.R., 1997, Laramide mountain flank deformation and the Golden fault zone, in Bolyard, D.W. and Sonnenberg, S.A., Geologic history of the Colorado Front Range, 1997 RMS-AAPG Field Trip #7: Rocky Mountain Association of Geologists, Denver, p. 49-64.

West, M.W., 1978, Quaternary geology and reported surface faulting along the east flank of the Gore Range, Summit County, Colorado: Colorado School of Mines Quarterly, v. 73, no. 2, 66 p.

Wise, D.U., 1963, Keystone faulting and gravity sliding driven by basement uplift of the Owl Creek Moutnains, Wyoming: Americal Association of Petroleum Geologists, v. 80, p. 1397-1432.

Printed in U.S.A.

Geological Society of America
Field Guide 1
1999

Laramide faulting and tectonics of the northeastern Front Range of Colorado

Eric A. Erslev and Steven M. Holdaway*

Department of Earth Resources, Colorado State University, Fort Collins, Colorado 80523, United States

TRIP OVERVIEW

The northeastern Front Range provides an excellent test of Laramide tectonic hypotheses. This trip will feature moderate hikes to excellent exposures of basement-involved structures. We will integrate new fault and balancing data into a model for backlimb deformation in basement-cored foreland arches.

INTRODUCTION

The excellent exposures of the plunging structures in the northeastern Front Range of Colorado have made the area a classic locality in the Rocky Mountain foreland (Fig. 1). The southeastern plunge of the structures allows the observation of different structural levels within single structures. Well-preserved slickensided fractures are common in the quartz arenites of the Ingleside, Lyons and Dakota sandstones. These fractures can be used to test kinematic and tectonic models for the Laramide foreland.

This field trip will examine the structural geometries and fracture patterns in the northeastern flank of the Front Range arch. We will illustrate an overall model for deformation in the backlimb of asymmetric basement-involved foreland arches by using the plunge of the structures to examine the structural patterns at different structural levels and at varying distances from the Precambrian core of the Front Range.

LARAMIDE STRUCTURAL MODELS FOR THE NORTHEASTERN FRONT RANGE

The accessibility of the northeastern Front Range and its proximity to academic institutions and the Denver petroleum industry have made the area an active contributor to debates concerning Laramide structural style. Early interpretations

based on field geometries were initially discounted by advocates of vertical tectonics but have been verified as the primary importance of horizontal shortening and compression during the Laramide orogeny have been reaffirmed (Fig. 2).

Ziegler (1917; Fig. 2a) proposed motion on a planar, 50°-dipping reverse fault underlying the Milner Mountain anticline. Boos and Boos (1957; Fig. 2b) also showed a planar reverse fault in an analogous section, but steepened the fault dips to around 80°. Prucha and others (1965) introduced the concept of concave-downward upthrusts to the area. By increasing fault dips downward, from 30°-dipping thrusts in the sedimentary strata to vertical faults cutting basement, low-angle thrusts in the cover were explained within a vertical tectonic framework. This "upthrust" geometry was adopted by Braddock and others (1970) and Le Masurier (1970; Fig. 2c) for the Milner Mountain anticline, which they interpreted as being underlain by a 60° dipping reverse fault at the surface that becomes vertical at depth. The most extreme interpretation of fault dips was contributed by Matthews and Sherman (1976; Fig. 2d) and Matthews and Work (1978) who suggested that a planar, 80°-dipping normal fault formed the western margin of the Milner Mountain anticline.

These interpretations of fault angle are in clear conflict with the evidence for horizontal shortening and compression in the Rocky Mountain foreland. Recent work at C.S.U. (Erslev and Rogers, 1993; Erslev, 1993; Holdaway, 1998) has shown that fault dips in the Fort Collins area range from 20° to 70° to the northeast, consistent with the first cross sections of the area by Ziegler (1917). These angles of faulting are compatible with the lateral shortening indicated by seismic information to the north in the Laramie Range (Brewer and others, 1982) and to the south along the Golden Fault (Jacob, 1983; Erslev and Selvig, 1997; Weimer and Ray, 1997), which indicate large west-dipping thrust faults overlapping the western margin of the Denver basin. But the fault dip directions in these areas are opposite to those in the northeastern Front Range, where exposed thrust and reverse faults dip to the northeast.

*Present address: Chevron, 1013 Cheyenne Drive, Evanston, Wyoming 82930, United States.

Erslev, E. A., and Holdaway, S. M., 1999, Laramide faulting and tectonics of the northeastern Front Range of Colorado, *in* Lageson, D. R., Lester, A. P., and Trudgill, B. D., eds., Colorado and Adjacent Areas: Boulder, Colorado, Geological Society of America Field Guide 1.

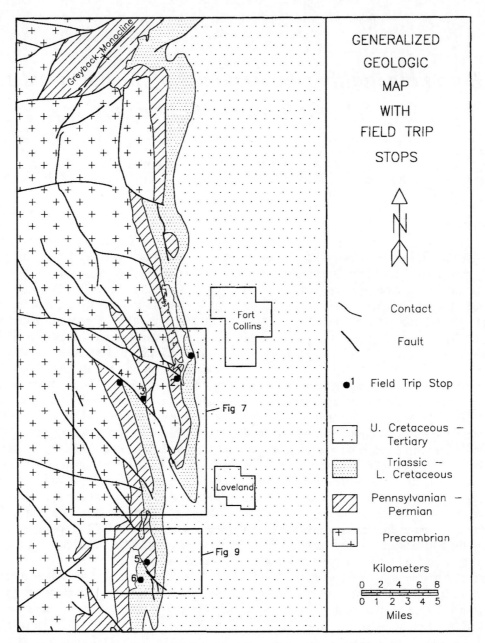

Figure 1. Simplified geologic map of the northern Front Range showing field trip stops and the locations of Figures 7 and 9.

This switch in fault dip directions along the eastern margin of the Front Range reveals a fundamental structural problem: what is the relationship of the faults and folds in the foothills of the northeastern Front Range to the overall structure of the Front Range? How did they contribute to the uplift of the Front Range basement-cored arch? These faults and folds exhibit the opposite shear sense of the regional structure, bringing the basin side up and the mountain side down, in direct conflict with the overall uplift of the range. Restoration of faulting exposed at the surface using simple vertical uplift (Fig. 3a), strike-slip (Fig. 3b), or upthrust (Fig. 3c) models cannot explain the uplift of the

northeastern Front Range because these faults uplift the basin, not the range.

Erslev and Selvig (1997) addressed the origin of this along-strike variability of fault dip direction by suggesting that the west-dipping thrusts of the Golden fault system are the primary faults and that northeast-dipping faults are backthrusts above a blind section of the Golden fault system that never emerges from the basement (Fig. 3d). In this model, basin-ward tilt of the northeastern margin of the Front Range occurred by domino-style rotation of backthrust-bounded blocks in a wide zone of shear. Holdaway (1998) tested this model with regional

a) Milner Mtn. Anticline (Ziegler, 1917)

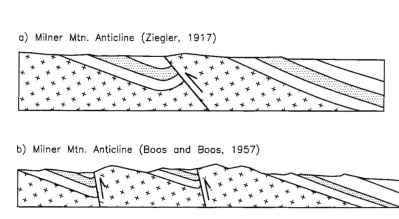

b) Milner Mtn. Anticline (Boos and Boos, 1957)

c) Milner Mtn. Anticline (LeMasurier, 1970)

Figure 2. Evolution of structural interpretations of the northeastern Front Range.

d) Greyback Monocline (Matthews and Sherman, 1976)

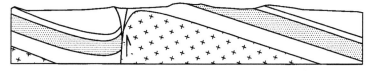

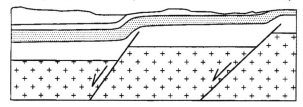

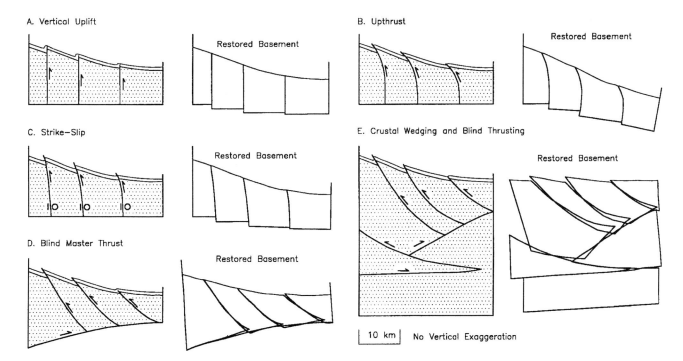

A. Vertical Uplift

Restored Basement

B. Upthrust

Restored Basement

C. Strike-Slip

Restored Basement

E. Crustal Wedging and Blind Thrusting

Restored Basement

D. Blind Master Thrust

Restored Basement

10 km   No Vertical Exaggeration

Figure 3. Structural models and basement block restorations for the northeastern Front Range.

balancing and found that whereas simple backthrusting can explain a portion of the uplift in this part of the range, it cannot by itself explain all of the uplift of the arch margin because the amount of exposed faulting is not sufficient to account for all of the observed uplift (Fig. 3d). He suggested that uplift of the northeastern margin of the Front Range arch was due to the insertion of a lower crustal wedge beneath the arch (Fig. 3e). This hypothesis is consistent with the larger slip amounts on thrust faults (Elkhorn thrust, Williams Range thrust, Never Summer thrust) bounding the western margin of the Front Range arch. While the geometry of the deep crustal wedge is unknown, this model can explain all of the uplift of the northeastern Front Range.

FIELD TRIP ROAD LOG AND STOP DESCRIPTIONS

This field trip follows paved roads in Larimer County, Colorado, from Horsetooth Reservoir west of Fort Collins to Carter Lake Reservoir southwest of Loveland. The location of stops are shown on the geologic sketch map in Fig. 1. The first four stops are in the Horsetooth Reservoir Quadrangle (Braddock and others, 1988c), the middle stops are in the Masonville Quadrangle (Braddock and others, 1970) and the final two stops are in the Carter Lake Quadrangle (Braddock and others, 1988b).

STOP 1. Northern dam embankment of Spring Creek dam on Horsetooth Reservoir

Driving instructions: Drive north on I-25 to the Harmony Road exit in south Fort Collins. Exit and head west on Harmony Road, which will turn into County Road 38E, through Fort Collins to the top of Spring Creek dam. Turn north on County Road 23 and park on north side of dam. Cross the road and walk west to the cliff overlooking Horsetooth Reservoir.

From the lip of Duncan's ridge, a popular Fort Collins climbing cliff in the Lytle Sandstone west of the parking lot, the rugged and irregular topography underlain by Proterozoic crystalline rocks lies to the west and is overlain by a late Paleozoic to Mesozoic clastic sequence (Fig. 4). This sequence contains three resistant sandstone units, the Permian Ingleside Formation, Permian Lyons Formation and Cretaceous Dakota Group, which form well-defined hogbacks generally dipping 15° to 30° east toward the northern Denver Basin. Strike-valleys are formed by the less resistant Fountain, Owl Canyon and Triassic-Jurassic formations. The Dakota Group forms a sandwich of units, with the Lytle and Plainview Sandstones below the Skull Creek Shale, which is overlain by the Muddy Sandstone which consists of the Fort Collins and Horsetooth Sandstones.

ERA	PERIOD	FORMATION	THICKNESS (m)
Mesozoic	Cretaceous	Pierre (not to scale)	2000
Mesozoic	Cretaceous	Niobrara	96–107
Mesozoic	Cretaceous	Benton Group	125–152
Mesozoic	Cretaceous	Dakota Group	88–108
Mesozoic	Jurassic	Morrison	70–98
Mesozoic	Jurassic	Sundance	6–18
Mesozoic	Triassic	Jelm	23–43
Mesozoic	Triassic	Lykins	137–213
Paleozoic	Permian	Lyons	9–18
Paleozoic	Permian	Owl Canyon	30–75
Paleozoic	Permian	Ingleside	21–61
Paleozoic	Pennsylvanian	Fountain	245–345
Precambrian			

Figure 4. Stratigraphic column of Phanerozoic units in the northeastern Front Range (from Graham and Ethridge, 1995).

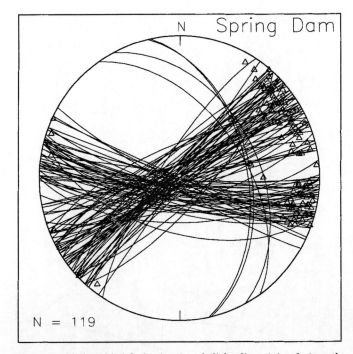

Figure 5. Slickensided faults (arcs) and slickenlines (triangles) northeast of Spring Dam indicating ENE-WSW shortening and compression.

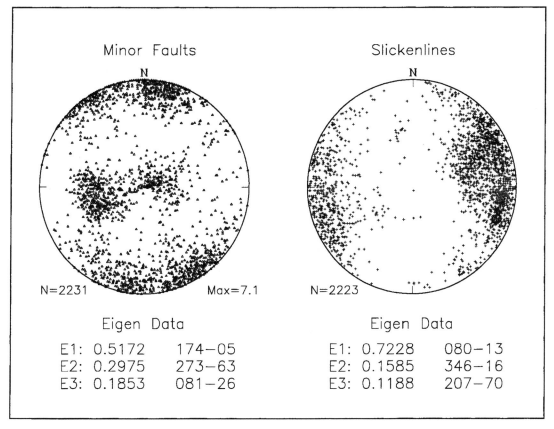

Figure 6. Stereonets of minor fault poles and slickenlines for the northeastern Front Range with average eigenvector orientations and eigenvalues. The 4 fault pole maxima correspond to conjugate thrust and strike-slip faults. The average slip direction (080-13) is given by the first eigenvector of the slickenline orientations.

To the south, the section of road built over the Skull Creek Shale collapsed in the Spring of 1995 when unusually wet conditions saturated the road bed and accelerated slumping of the road. Repair of the roadbed included the installation of a culvert to drain the Skull Creek strike valley, replacement of the earth fill and regrading of the slope below the road. An intense rainstorm in July 1997 did not damage the road, although it did trigger numerous landslides and caused major flooding in Fort Collins.

After the overview at Duncan's ridge, the trip will walk east across the road to the Horsetooth sandstone which has excellent exposures of strike-slip fractures indicating near-horizontal shortening and compression trending N79E (Fig. 5). The silica-cemented quartz arenites of the Ingleside, Lyons and Dakota formations preserve slickensided fractures throughout the region. Shear fractures in these units rarely follow bedding planes, suggesting that they form on ideal, "Andersonian" planes symmetric to the stress axes. Fault surfaces show excellent slickenlines in the fine-grained, low-porosity cataclasite which coat the surfaces. The multitude of shear fractures suggest that these planes are strain hardened, annealing after slip to be stronger than the original sandstone. This means that early phases of slip can be preserved along with later phases of slip. As a result, if multiple

slip direction existed during Laramide deformation, they should be recorded by the fractures. Shear sense on the fractures is usually clearly indicated by Riedel fractures (Petit, 1987) which intersect the major fault planes at a low angle, giving a roughness to the plane in the direction of slip.

Fault studies at CSU have focused on Laramide faults in strata deposited before the Laramide orogeny and after the Ancestral Rocky Mountain orogeny. Once the faults are divided into subsets, strain and stress orientations can be determined using average slickenline orientations, conjugate relationships, P-T dihedra (Allmendinger and others, 1989), and reduced stress tensor methods (Angelier, 1990). The possibility of multiple compression directions is evaluated by using the Compton (1966) method which calculates the ideal σ_1 orientation for each fault based on the average angle between conjugate pairs. This method can discriminate different compression directions when conventional conjugate, Angelier (1990) and P-T dihedral methods just average the compression directions.

Holdaway (1998) compiled 2232 minor fault and slickenline orientations to determine the kinematic development of the northeastern Front Range (Fig. 6). When the ideal σ_1 trends are plotted in rose diagrams for the area around Horsetooth Reservoir and Milner Mountain anticline (Fig. 7), they show a uni-

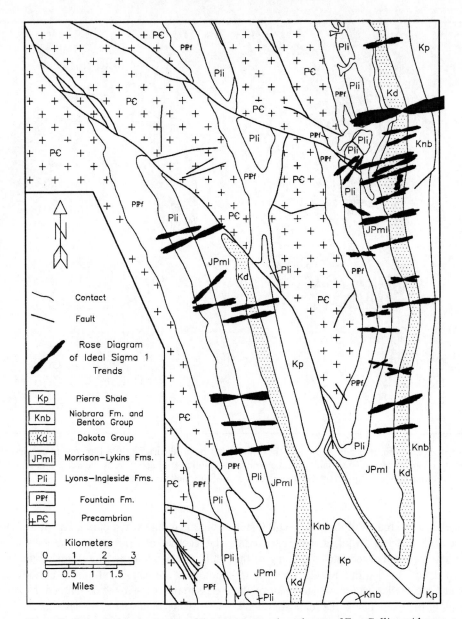

Figure 7. Expanded geologic map of the area west and southwest of Fort Collins with rose diagrams showing ideal σ_1 trends calculated by the Compton (1966) method assuming an angle of 50° between conjugate minor faults.

form ENE compression direction. The general sparsity of multi-modal ideal σ_1 trends suggests one direction of regional compression and shortening (Holdaway, 1998).

STOP 2. The Perch

Driving instructions: Drive south back across the dam to County Road 38E and turn west. Park at the parking lot at the Perch, which is on the first ridge crest after you drive around the southern arm of Horsetooth Reservoir.

This stop shows the form of many of the basement-cored fault-propagation folds in the northeastern Front Range. To the east, the Dakota Group sandstones form a smoothly-arcuate

outcrop pattern defining a southeast-plunging fold axis. The Lyons Formation, which forms the ridge under the Perch, is both folded and faulted, with 5 fault splays defining tilt domains. Farther west, the Precambrian basement-Fountain Formation unconformity was broken by two major faults but was not folded. This indicates that basement block movements at deeper structural levels transitions upward into arcuate fault-propagation folding.

All workers in this area have proposed basement block motion, with detailed analyses (LeMasurier, 1970; Erslev and Rogers, 1993) showing only minor (20°) basement rotations in the hanging wall tip of the largest structure in the area, the Milner Mountain anticline. This can be seen in the excellent geologic maps of the region by Braddock and others (1970, 1988a, 1988b, 1998c, 1989), which show planar basement contacts offset by faults with minimal folding of the contact in the vicinity of the faults. These faults progressively lose displacement upward in the folded sedimentary strata, forming fault-propagation folds (Erslev, 1991; Erslev and Rogers, 1993). Of the six major faults that cut the Precambrian-Pennsylvania contact near Fort Collins, only the Milner Mountain fault cuts the Dakota Group.

The strata form doubly-plunging anticlines, like Bellvue dome to the north, or, more commonly, southeasterly-plunging anticlines cored by northwest-striking faults. Trenching of the major faults near Fort Collins show that the faults dip 20° to 70° northeast, consistent with the earliest interpretations of Ziegler (1917). Absolute slip directions, which have been interpreted as everything from strike slip to dip slip, are hard to determine from the major faults because their strain-softened gouge zones only preserve the last motion on the fault. In general, however, the largest faults strike perpendicular to compression directions indicated by minor faults (Fig. 7), suggesting dip-slip on thrust and reverse faults.

STOP 3. Basement hanging wall immediately adjacent to the Fletcher Hill fault

Driving instructions: Continue west on County Road 38E and turn west. Cross the saddle between Milner Mountain and Horsetooth Mountain and park across from first intersection with a paved road coming in on the left.

This locality exposes the hanging wall tip of the Fletcher Hill fault, which places basement to the northeast over the Cretaceous Dakota Group. The Precambrian-Fountain unconformity is cut by reverse faults that dip close to 60°NE, with the largest splay dipping 52°NE. The exposed stair-step offsets on the east-tilted Precambrian-Fountain unconformity indicate that there has been very little folding of the unconformity in the hanging-wall tip of the Fletcher Hill structure. This lack of folding indicates that folding did not precede faulting, as predicted by the popular fold-thrust model of structural development (Berg, 1992). To the northwest, along the aqueduct, the footwall cut-off of the Dakota group shows local bedding detachment and buckle folding (see Fig. 30a in Prucha and others, 1965, for a photograph of this exposure).

STOP 4. Syncline below the Fletcher Hill fault north of Masonville

Driving instructions: Continue west on County Road 38E, which turns into County Road 27 at the town of Masonville. North of Masonville at a sharp left turn, park opposite of a large syncline in the Lyons Formation. Note: This is private land for which access permission is required.

The syncline in the Lyons Formation is angular at this locality, with local out-of-the-hinge thrusting. To the south, the

Dakota Group sandstones show more rounded fold hinges, probably due to distributed deformation in the shales and evaporites of the Lykins Formation. The Fletcher Hill fault is well-exposed in a valley incised into the resistant schists to the northeast. The Precambrian-Fountain gouge contact within the fault zone has a 60°NE dip, suggesting reverse faulting on the Fletcher Hill fault. At this locality, the lack of steeply-inclined sedimentary beds at the fault suggests a localized sub-thrust basement wedge, a style evident elsewhere in the Rocky Mountain foreland. Because the formation of a wedge steepens the fault plane, the 60°NE dip should be interpreted as a maximum dip for the fault. Less than a kilometer to the north, the sedimentary beds are overturned, indicating the lack of an underlying basement wedge.

The possibility of basement folding was investigated at this locality and at Milner Mountain to the southeast (Erslev and Rogers, 1993). In both places, schistose rocks strike parallel to the fault. This orientation should provide an optimal configuration for flexural slip on foliation planes in the basement. Dip measurements, measured from the west, were taken at regular intervals from the fault. These orientations show a 50 meter

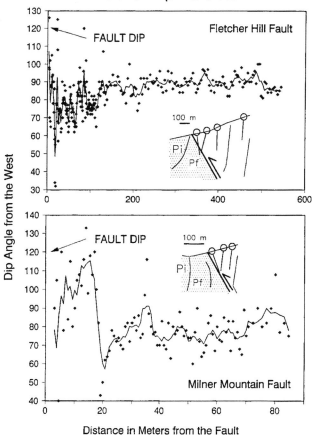

Figure 8. Dip profiles of basement schistosity adjacent to the Fletcher Hill (at Stop 4) and Milner Mountain (1 km east of Stop 3) faults.

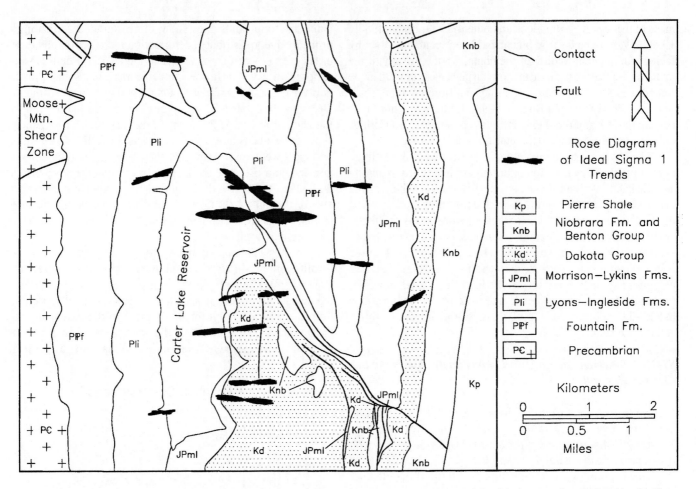

Figure 9. Expanded geologic map of the area around Carter Lake Reservoir with rose diagrams showing ideal σ_1 trends calculated by the Compton (1966) method assuming an angle of 50° between conjugate minor faults.

zone of possible basement rotation associated with the fault zones (Fig. 8). Thus, extensive folding in these Laramide fault-propagation folds is largely restricted to the sedimentary strata.

STOP 5. Overview of Carter Lake anticline

Driving instructions: Return south on County Road 27 and turn right at Masonville. Turn right on Rte 34 and then take the first left on County Road 29. Turn right on County Road 18 and then left on County Road 31, which will take you to Carter Lake Reservoir. Park at the northern abutment of the second dam and climb to the ridge for an overview. Note: The ridge directly above this spot is on public land, but the extension of the ridge to the south is on private land for which permission is required.

The area surrounding Carter Lake Reservoir exposes a zone of interaction between distinctly different structural trends. The top of this ridge is in the forelimb of the asymmetric Carter Lake anticline from which we can view the intersection of north- and northwest-trending structures. The Pennsylvanian Fountain Formation is exposed in the core of the anticline to the east and the resistant Lyons and Ingleside Sand-

stones form the ridge on which we are standing as well as the ridge to the east across the fold axis. The Carter Lake anticline is slightly asymmetric and trends north along most of its length until it abruptly becomes highly asymmetric and takes a north-westerly trend south of where we are standing.

Multiple fold trends are found throughout the northeastern Front Range, but the Carter Lake Reservoir area best shows the interaction of distinct north- and northwest-trending structures. The variety of structural trends may be caused by multiple stages of differently oriented shortening and compression, an edge effect that refracted maximum shortening and compression axes to become nearly perpendicular to the arch margin, or reactivation of pre-existing weaknesses. Rose diagrams of ideal σ_1 axis trends for individual localities in the Carter Lake Reservoir area (Fig. 9) show dominantly uni-modal, east-west compression perpendicular to north-trending structures.

Immediately north of the Carter Lake Reservoir area, Precambrian dikes parallel several northwest-striking faults (Braddock and others, 1970). This suggests that northwest-trending structures in this area may be localized on pre-existing weaknesses, which can explain their obliquity to east-west Laramide shortening and compression. Several other northwest- and west-northwest- striking faults are both oblique to Laramide contraction directions and parallel to Precambrian foliations or mylonite zones in the northeastern Front Range (Braddock and others 1988a, 1988b). This suggests that reactivation of pre-existing basement weaknesses can explain at least some of the structural variability in this area.

STOP 6. *South side of south dam, Carter Lake*

Driving instructions: Continue south on County Road 31, parking at southeast corner of the third dam.

This locality exposes two sets of thrusts with radically different trends. This can be explained by localized effects, but the presence of multistage, multidirectional shortening elsewhere in the Laramide orogen (Gries, 1983; Chapin and Cather, 1983; Bergh and Snoke, 1992; Erslev, in review) suggests that these faults may represent distinct periods of shortening.

Driving instructions back to Denver: Continue south on County Road 31 and turn left on County Road 8E at the first intersection. Go east until the road ends, turn right on County Road 23 and then left on Rte 56, which will take you through Berthoud to I-25.

ACKNOWLEDGMENTS

This manuscript was improved by reviews by Dave Lageson and Bruce Trudgill.

REFERENCES CITED

Allmendinger, R.W., Aydin, A., Engelder, T., and Pollard, D.D., 1989, Quantitative interpretation of joints and faults: G.S.A. Short Course.

Angelier, J., 1990, Inversion of field data in fault tectonics to obtain the regional stress - III. A new rapid direct inversion method by analytical means: Geophysical Journal International, v. 103, p. 363-379.

Berg, R.R., 1962, Mountain flank thrusting in the Rocky Mountain foreland, Wyoming and Colorado: AAPG Bull. v. 46, p. 2010-2032.

Bergh, S., and Snoke, A., 1992, Polyphase Laramide deformation in the Shirley Mountains, south central Wyoming foreland: Mountain Geologist, v. 29, p. 85-100.

Boos, C.M., and Boos, M.F., 1957, Tectonics of eastern flank and foothills of Front Range: A.A.P.G. Bull., v. 41, p. 2603-2676.

Braddock, W.A., Connor, J.J., Swann, G.A., and Wolhford, D.D., 1988a, Geologic map of the Laporte quadrangle, Larimer County, Colorado: U.S. Geological Map GQ-1621.

Braddock, W.A., Calvert, R.H., Gawarecki, S.J., and Nutalaya, P., 1970. Geologic map of the Masonville quadrangle: U.S. Geol. Survey Geol. Quad. Map GQ-832.

Braddock, W.A., Calvert, R.H., O'Connor, J.T., and Swann, G.T., 1989, Geologic map of the Livermore quadrangle, Larimer County, Colorado: U.S. Geological Survey Map GQ-1618.

Braddock, W.A., Nutalaya, P., and Colton, R.B., 1988c, Geologic map of the Carter Lake quadrangle: U.S. Geol. Survey Geol. Quad. Map GQ-1628.

Braddock, W.A., Wohlford, D.D., and O'Connor, J.T., 1988b, Geologic map of the Horsetooth reservoir quadrangle, Larimer County, Colorado: U.S. Geological Survey Map GQ-1625.

Brewer, J.A., Allmendinger, R.W., Brown, L.D., Oliver, J.E., and Kaufman, S., 1982, COCORP profiling across the Rocky Mountain front in southern Wyoming, Part 1, Laramide structure: Geological Society of America Bulletin, v. 93, p. 1242-1252.

Chapin, C.E., and Cather, S.M., 1983, Eocene tectonics and sedimentation in the Colorado Plateau; Rocky Mountain area, in Lowell, J.D., ed., Rocky Mountain foreland basins and uplifts: Rocky Mountain Association of Geologists, p. 33-56.

Compton, R.R., 1966, Analyses of Pliocene-Pleistocene deformation and stresses in northern Santa Lucia Range, California: Geological Society of America Bulletin, v. 77, p. 1361-1380.

Erslev, E.A., 1991, Trishear fault-propagation faulting: Geology, v. 19, p. 617-620.

Erslev, E.A., 1993, Thrusts, backthrusts and detachment of Laramide foreland arches, in Schmidt, C.J., Chase, R., and Erslev, E.A., eds., Laramide basement deformation in the Rocky Mountain foreland of the western United States: Geological Society of America Special Paper 280, p. 125-146.

Erslev, E.A., in review, Multi-stage, multi-directional Tertiary shortening and compression in north-central New Mexico: submitted to GSA Bulletin.

Erslev, E.A., and Rogers, J.L., 1993, Basement-cover kinematics of Laramide fault-propagation folds: in Schmidt, C.J., Chase, R., and Erslev, E.A., eds., Laramide basement deformation in the Rocky Mountain foreland of the western United States: Geological Society of America Special Paper 280, p. 339-358.

Erslev, E.A., and Selvig, B., 1997, Thrusts, backthrusts and triangle zones: Laramide deformation in the northeastern margin of the Colorado Front Range, in Bolyard, D.W., and Sonnenberg, S.A., Geologic history of the Colorado Front Range, Rocky Mountain Association of Geologists, Denver, Colorado, p. 65-76.

Flores, R.M., Roberts, S.B., and Perry, W.J., 1995, Paleocene paleogeography of the Wind River, Bighorn, and Powder River Basins, Wyoming, Wyoming Geological Association Guidebook.

Graham, J., and Ethridge, F.G., 1995, Sequence stratigraphic implications of gutter casts in the Skull Creek Shale, lower Cretaceous, northern Colorado: Mountain Geologist, v. 32, p. 81-94.

Gries, R.R., 1983, North-south compression of Rocky Mountain foreland structures, in Lowell, J.D., ed., Rocky Mountain foreland basins and uplifts: Rocky Mountain Association of Geologists, p. 9-32.

Holdaway, S.M., 1998, Laramide deformation of the northeastern Front Range, Colorado: Evidence for deep crustal wedging during horizontal compression: M.S. thesis, Colorado State University, 146 p.

Jacobs, A.F., 1983, Mountain front thrust, southeastern Front Range and northwestern Wet Mountains, in Lowell, J.D., ed., Rocky Mountain foreland basins and uplifts: Rocky Mountain Association of Geologists, p. 229-244.

LeMasurier, W.E., 1970, Structural study of a Laramide fold involving shallow seated basement rock, Front Range Colorado: Geological Society of America Bulletin, v. 81, p. 435-450.

Matthews, V., III, and Sherman, G.D., Origin of monoclinal folding near Livermore, Colorado: Mountain Geologist, v. 13, p. 61-66.

Matthews, V., III, and Work, D.F., 1978, Laramide folding associated with basement block faulting along the northeastern flank of the Front Range, Colorado, in Matthews, V., III (ed.), Laramide folding associated with basement block faulting: Geological Society of America Memoir 151, p. 101-124.

Petit, J.P., 1987, Criteria for the sense of movement on fault surfaces in brittle rocks: Journal of Structural Geology, v. 9, p. 597-608.

Prucha, J.J., Graham, J.A., and Nickelson, R.P., 1965, Basement-controlled deformation in Wyoming Province of Rocky Mountain foreland: American Association of Petroleum Geologist Bulletin, v. 49, p. 966-992.

Weimer, R.J., and Ray, R.R., 1997, Laramide mountain flank deformation and the Golden fault zone, in Bolyard, D.W. and Sonnenberg, S.A., Geologic history of the Colorado Front Range, 1997 RMS-AAPG Field Trip #7: Rocky Mountain Association of Geologists, Denver, p. 49-64.

Ziegler, V., 1917, Foothills structure in northern Colorado: Journal of Geology, v. 25, p. 715-740.

Geological Society of America
Field Guide 1
1999

Hydrogeology and wetlands of the mountains and foothills near Denver, Colorado

K. E. Kolm and J. C. Emerick
Division of Environmental Science and Engineering, Colorado School of Mines, Golden, Colorado 80401, United States

ABSTRACT

This trip integrates the geomorphic, botanic, and hydrogeologic aspects of diverse ground-water and surface-water systems of the Geneva Creek sub-basin near Guanella Pass, Colorado. Fracture-flow crystalline bedrock systems, and alluvial, colluvial, and glacial systems are integrated for slope and riverine wetland structure and function. Current research regarding the development of the Wetlands Integrated Hydrologic Analysis method for delineating wetlands structure and function in the southern Rocky Mountains is discussed with respect to various sites observed along the field trip route.

INTRODUCTION

This field trip integrates the geomorphic, botanic, and hydrogeologic aspects of the diverse ground-water and surface-water systems, and wetlands of the Geneva Creek subbasin, and part of the North Fork of the South Platte River System from Guanella Pass to Bailey, Colorado. Fractured-flow bedrock ground-water systems and flow through unconsolidated deposits and soils occur in this landscape. The geomorphic systems and materials affecting the hydrogeology of the area include weathering and pedogenic processes and deposits, Holocene mass movements and deposits, and Pleistocene glacial and Holocene fluvial erosional and depositional features. The route passes through three life zones: the alpine, subalpine, and montane. Along the way, many of the plant communities typical of the southern Rocky Mountains are seen. The various plant communities mentioned in the following road log are discussed in greater detail in Mutel and Emerick (1992). The wetlands observed along the field trip result from the combined effects of surface water and ground-water systems, as well as local geology and soils.

The road log starts at Guanella Pass (see Figure 1), which can be reached from Georgetown, Colorado, located along Interstate 70 west to Denver, Colorado. In Georgetown, follow the roadsigns to the Guanella Pass road. Alternatively, Guanella pass may be reached by driving west from Denver along US 285. At

Grant, turn right, following the all-weather road up the valley, past the Geneva Basin Ski Area, to the top of Guanella Pass. Either route requires approximately 1.5 to 2 hours driving time.

GENERAL BACKGROUND

Wetlands are important features in the landscape, particularly in mountain valleys where they occur in association with streams, rivers, and ground-water discharge at seeps and springs. Wetland definitions are numerous and often complicated, however, one proposed by Tarnocai (1979) works well for our purposes here: "Wetland is defined as land having the water table at, near, or above the land surface or which is saturated for a long enough period to promote wetland or aquatic processes as indicated by hydric soils, hydrophytic vegetation, and various kinds of biological activity which are adapted to the wet environment." During the last decade or so, there has been increased recognition of the ecological and hydrological importance of wetlands (see for example, the excellent text of Mitsch and Gosselink, 1993). For the past three years, we have been involved in studies to characterize the hydrologic functions of wetlands, which has resulted in the Wetland Integrated Hydrologic Analysis (WIHA) method. This field trip provides an opportunity to demonstrate the use of the WIHA method.

Based on studies by Kolm (1996); Kolm, et al.(1996); Kolm, et al. (1998); and Harper-Arabie and Kolm (1998), the

Figure 1. View of Guanella Pass from Squaretop Mountain looking east.

Wetland Integrated Hydrologic Analysis (WIHA) method was designed to identify the hydrologic functions that are present in the wetland being evaluated. In addition to providing a standard methodology for wetland hydrologic analysis, the advantages of incorporating the conceptualization and characterization approach of Kolm (1996) and Kolm, et al (1996) are that: 1) the method is non-invasive; 2) in using the approach, the subsurface framework and ground-water flow system can be estimated with minimal time spent on-site; and 3) WIHA provides a clear basis for model design when used for mathematical modeling of the wetland system.

METHOD DEVELOPMENT

The WIHA method was developed in an iterative manner that began with a theoretical concept of the common wetland systems in Colorado. This was followed by data collection, as needed, which led to a better understanding of the wetland systems. This process resulted in the development of several models of wetland structure and hydrology that were tested with field studies (Kolm, et al., 1998). The field sites included five wetlands that were identified as being typical of wetland types in the southern Rocky Mountains. These included two that were initially classified according to Brinson (1993) as hydrogeomorphic (HGM) slope wetlands and three that were initially classified as HGM riverine wetlands.

Prior to the field studies, existing data, including both the natural and anthropogenic features of the five wetland sites, were collected and analyzed. Field reconnaissance was conducted, as necessary, to relate the preliminary analysis of the information collected to field study site conditions. In areas where field data were sparse, basic photointerpretation and terrain analysis techniques were applied to remote sensing data, aerial photography, and topographic maps to acquire informa-

tion, and to quantify and distribute wetland system parameters. Based upon the information that was collected and measured, the appropriate wetland structure, physical and chemical characteristics, distribution, and continuity of stratigraphic and lithologic units (soil and rock) were determined for each field site. The type, properties, and distribution of geomorphic materials, landforms, slope, and other geomorphic processes and characteristics were interpreted with respect to the surface water and ground-water.

Each type of wetland system was then characterized and classified using surface and subsurface characterization, and hydrogeologic and ground-water system characterization and quantification, as appropriate (Kolm et al, 1998). The importance of hydrogeology and ground-water systems was determined for each wetland class.

RESULTS AND DISCUSSION

The basic steps involved in the Wetland Integrated Hydrologic Analysis method include: 1) problem definition; 2) data base development; 3) preliminary conceptualization; 4) surface and subsurface characterization; 5) ground-water system characterization; and 6) wetland system characterization and classification based on function, including water supply (Figure 2). Based upon the WIHA assessment, the hydrodynamics and predominant water source of a wetland can be characterized, and the wetland hydrologic functions determined (Table 1). The WIHA method will be demonstrated during the field trip.

The three-dimensional hydrogeologic structure of the wetland system has a dominant control over hydrologic functions (Kolm et al. (1998); Harper-Arabie and Kolm, 1998). Therefore, the wetland classes in the southern Rocky Mountains should be based upon the wetland hydrogeologic structure, with the standard HGM classes of Brinson (1993) regarded as subclasses:

slope wetlands and riverine wetlands. Slope wetlands occur at ground water discharge sites, and typically are observed at the base of slopes along valley bottoms. Riverine wetlands occur adjacent to streams and rivers, which are the main source of surface and soil water in this wetland type. Wetlands in valley bottoms are often combinations of these two types. The structural classes predominantly observed, based on this hydrologic assessment, are three (plus)-layer, two-layer, and one-layer (Figure 3). However, it is common in the southern Rocky Mountains to have a single wetland composed of a combination of structural systems adjacent to, and grading into, each other.

ROAD LOG

Mileage

0.0   **Stop 1: Top of Guanella Pass, elevation 3557 meters (11,669 ft.).** Fracture-flow ground-water systems in the alpine region can be observed here. The climate at high elevations above treeline is cold and windy (see Figure 4).

There is no weather station at Guanella pass. However, one exists on Berthoud Pass at 3449 m elevation (11,314 ft.) approximately 22 kilometers to the north, where similar climate occurs. Averages from a nineyear weather record at that station indicate a mean annual temperature of –2°C (28.5°F), and an average precipitation of 91 cm (36 inches) per year, with 57 percent of the precipitation falling during the period December through May. The Berthoud Pass station received an average of 950 cm (374 inches) of snow per year during the period of record. Winds in the alpine are severe and exceptionally turbulent. Winds as high as 58.1 m/s (130 mph) have been recorded at Berthoud Pass, and during the period from January through May, 1974, that site had an average wind speed of 7.6 m/sec (17 mph). These windy conditions redistribute the snowfall such that there are often deep snow banks on the lee side of ridges, while the windward side and the tops of ridges are blown free of snow. Westerly winds are most common, although local topography channels winds in a variety of directions close to the ground, around ridges, and down valleys.

Geomorphology: Glacial cirques are observed on Squaretop and Bierstadt peaks to the west and southeast, respectively. Other glacial features include overlysteepened Ushaped valleys, and scattered till in Geneva and Duck Creeks. Creep and debris flow or earth flow processes occur on over-steepened valley sides to the east, west, and north. Freezethaw/wetdry weathering processes result in periglacial features such as rock polygons, stone stripes, and stonebanked terraces in soils on the uplands. Modern fluvial processes

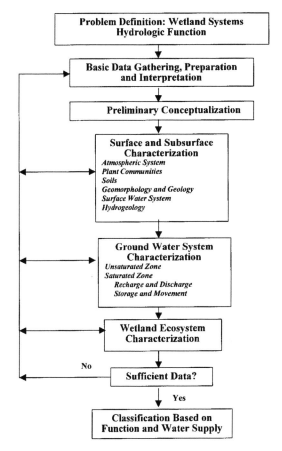

Figure 2. Steps involved in the Wetland Integrated Hydrologic Analysis (WIHA) method.

occur in the valleys; stream baseflow is probably maintained by groundwater discharge.

Vegetation: Land above upper treeline is in the alpine life zone. Wind, snow accumulation, and soil moisture directly influence patterns of plant community distribution in this life zone. Shrubby willow carrs cover much of the ground near Guanella Pass and are indicative of winter snow accumulation depths of 1 to 2 m and usually high soil moisture, often in groundwater discharge areas. Predominant willow species include *Salix planifolia, S. glauca,* and *S. brachycarpa.* Dry, windblown sites are dominated by stands of elk sedge (*Kobresia myosuroides*), and more mesic sites by tufted hairgrass (*Deschampsia caespitosa*). Deep snow bank sites that may be snowfree only for a few weeks of the year are almost devoid of vascular plant species, although sibbaldia (*Sibbaldia procumbens*), a short cloverlike plant, grows on less severe sites of this type. Where there is standing water during much of the summer, wet meadow communities are typically dominated by aquatic sedge (*Carex aquatilis*).

TABLE 1. DEFINITIONS OF HYDROLOGIC FUNCTIONS IDENTIFIED IN SOUTHERN ROCKY MOUNTAIN WETLANDS (after Kolm et al., 1998)

Function Abbreviation	Function Name	Function definition
ET	Evapotranspiration	The removal of water from the wetland to the atmosphere.
GWinter	GW interception	Water entering wetland from a ground-water source outside of the wetland boundaries.
GWoutss	GWoutsprings/seeps	The removal of water from the groundwater system to the surface water system via springs and seeps.
GWoutr	GWoutriver	The removal of water from the groundwater system to the surface water system via rivers and lakes (results in a gaining stretch of river or lake).
GWstorelt	GW long term storage	Storage of ground-water in subsurface of wetland during long periods of time.
GWstoredyn	GW dynamic storage	Short term storage of water in the subsurface of wetland during extreme events, such as flooding.
GWmove	GW movement	Moderation of the direction and velocity of groundwater within a wetland.
SWstoredyn	SW dynamic storage	Short term storage of surface water. i.e. during flooding events
SWstorelt	SW storage long term	Storage of surface water for long periods of time. Water may be stored because recharge to a ground-water system is retarded by a low hydraulic conductivity layer, or is retained by microtopography or dam like structures.
SWin	SW in	Surface water entering wetland from a surface water system.
SWout	SW out	The removal of water from the wetland to a surface water system.
ATMin	Atmosphere in	Water entering the wetland from the atmosphere.
ED	Energy Dissipation	Degree to which wetland can reduce the surface water energy.

Hydrogeology: Groundwater recharge occurs from precipitation distributed throughout these high elevations. Recharge can be particularly high along the ridgetops during the summer with the melting of large snow accumulations; note the snow cornices on the east, southeast- and northeast-facing slopes. Ground-water flow paths include flow through fractured bedrock from ridge tops to valley bottoms or other discharge areas, as well as through soil or weathering materials, usually as infiltration to the water table. These shallow and fractured-flow systems are unconfined. Ground-water discharge appears as springs or seeps in gullies along fracture zones, at the base of over-steepened valley walls (such as in the Mt. Spalding area to the east), and in stream channels where it is an important component of the surface water baseflow. It is hypothesized that the directions of preferred ground-water flow are oriented parallel with the 1) northwest-southeast and 2) northeast-southwest alignment of observed stream channels.

Wetlands: The wetlands observed at the crest of Guanella Pass are primarily of the 3-layer and 2-layer hydrogeologic structure class, and include fractured crystalline bedrock systems supporting the overlying saturated glacial and mass wasting materi-

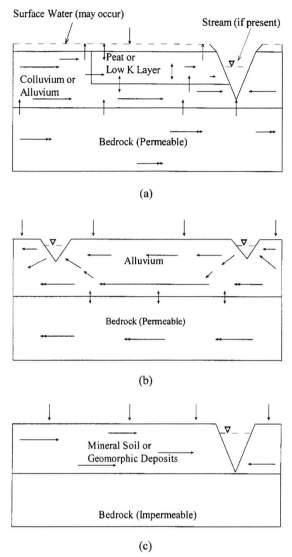

Figure 3. Conceptual models of: (a) a typical three (plus) hydrologic layer slope wetland, peat or low K-dominated; (b) a typical two hydrologic layer riverine wetland, unidirectional flow; and (c) a typical one hydrologic layer slope wetland. Arrow indicates direction of local ground-water flow. Double arrow indicates direction of regional ground-water flow (after Harper-Arabie and Kolm, 1998).

als. Peat is present in the 3-layer wetlands. These wetlands would be classified as slope wetlands.
(0.2 mile)

0.2 To the northwest, there is a good view of a slumping hill slope with a willow community (discharge area).
(0.1 mile)

0.3 Upper treeline, where "flag trees" (*krummholz*) are observed, is usually at an elevation of approximately 3475 m (11,400 ft) in this region of the Rocky Mountains (although this can vary considerably according to local wind and soil conditions). An approximate "rule of thumb" is that upper tree line generally is located at the elevation where the mean July temperature is 10°C. The tree species here is Engelmann spruce (*Picea engelmanii*). The dominant wind direction is from the west, and snowpack accumulation is on the east side of these trees. Treeline marks the transition from the alpine life zone to the forested subalpine life zone.
(0.6 mile)

0.9 A fracture zone-controlled drainage and groundwater discharge zone is observed to the west, with flow moving from the Squaretop cirque. To the south, a moraine-dammed lake (Duck Lake) is seen. The lake is impounded artificially by a dam located at the breach in the recessional glacial moraine (probably Pinedaleaged). Note the Geneva Basin Ski Area and the east and northeast-facing ski runs that take advantage of snowpack accumulation.
(0.6 mile)

1.5 **Stop 2: Overlook of Duck Lake and Pinedale moraine, at an approximate elevation of 3383 m (11,100 ft).**

Geomorphology: Morainal features are observed including kettles, hummocky topography, and a moraine-dammed lake. This stop offers a good view of the over-steepened, U-shaped valley containing till "plastered" on the valley side. Deeper winter snowpack on the west side of the valley results in greater mass movement.

Vegetation: Moraines are well drained. The resulting dry soils support grass-dominated herbaceous plant communities. Spruce or willows grow in lower mounds of till where soils are more stable or moist.

Hydrogeology: A decreased river gradient above the lake has promoted increased sedimentation. The steeper river gradient below the dam has resulted in channel downcutting and increased erosion of formerly deposited materials. The east-facing valley walls are a major groundwater discharge zone; groundwater also discharges into the lake and seeps through the moraines. Increased sedimentation above the lake, as well as higher soil moisture and a shallow water table, are contributing factors for the willow communities seen there.

Wetlands: The slope wetlands observed at this location are primarily 2-layer hydrogeologic systems consisting of fractured crystalline bedrock overlain by saturated glacial or alluvial materials.
(0.2 mile)

1.7 Note the dam constructed across the breached moraine.
(0.1 mile)

1.8 Note the narrow drainage channel; groundwater discharge is observed at the base of the moraine.
(0.2 mile)

2.0   The road cut exposes a cross section of glacial till; the materials are well graded and poorly sorted. This location is the first observance of bristlecone pine (*Pinus aristida*), indicating windy conditions and dry and rocky soils. Ski runs are seen to the west. Note the avalanche/debris avalanche chute to the north of the last ski run to the right. The willows indicate that the soils in the chute are wet and the water table is near the surface.
(0.3 mile)

2.5   This is the entrance to Geneva Basin Ski Area; the Tarryall Mountains are seen to the south (downvalley).
(0.2 mile)

2.7   Note the willow-covered debris flow/avalanche materials in the valley that have flowed to the south. To the west, note rock fall (weathering) talus where the absence of soil prevents plant growth.
(0.5 mile)

3.2.1   Springs (discharge) are seen above the road from here to stop 3.
(0.4 mile)

3.6   **Stop 3: Overlook of Geneva Creek, elevation approximately 3200 m (10,500 ft).**
Geomorphology: Note the glacial moraine at treeline up-valley toward stop 2. Below the moraine lie debris avalanche/debris flow deposits derived from the over-steepened valley walls. Below the overlook is an outwash plain with a younger landslide/debris avalanche fan stratigraphically and topographically on top. Directly beyond the road switchback is the sear (release area) of the debris avalanche/landslide. It appears that the release area is a ground-water discharge zone, which, in turn, may have triggered the debris avalanche. Ground-water seeps through the till and colluvium located on the U-shaped, over-steepened valley sides. The debris fan has pushed the river to the west side of the valley. Falling colluvium on the road indicates active creep.
Vegetation: The following plant community relationships are observed here: a) spruce is located on the sites where soils are relatively shallow and more stable; b) aspen (*Populus tremuloides*) occurs on unstable or disturbed sites, often near ground-water discharge areas; c) willows grow on ground-water discharge areas and on wet disturbed sites. Willows are also growing on the outwash plain and the debris flow fan (both deposits are wet areas).
Hydrogeology: Groundwater is seeping from fractured bedrock and along the colluvium/bedrock contact. The roadcut has created a springline discharge area. Groundwater probably has triggered the debris avalanches; ground-water is currently draining at these mass movement sites.
(0.4-0.6 mile)

4.0-4.2   **Stop 4: Outwash plain and top of debris-flow fan at an approximate elevation of 3139 m (10,300 ft.).**
Geomorphology: This stop is on a glacial terrace/outwash plain; a younger debris avalanche/landslide fan covers the top of the outwash plain.
Vegetation: Note the extensive willow communities indicating nearsurface ground water. Beavers prefer valley bottoms that are wider and relatively flat such as seen here. Aspen and willow are favorite items for food and building materials. Beaver occupation in these sites is cyclic. When enough aspen and willow are present, beavers invade the area and construct dams, raising the water table. This often kills conifers as a result of over-saturation of the root zone. Over time, the beavers remove all of the large aspen and willow, and eventually leave the area. The abandoned dams disintegrate and the water table is lowered, allowing reinvasion of conifers. Once the aspen and willow recover (perhaps in 10 to 20 years), the beaver return. There is evidence that beaver populations have recently returned to this area.
Hydrogeology: Groundwater tables are near the surface in the outwash materials; the river channel and stage probably control the groundwater level. Groundwater discharge in the fan deposits is evident by the presence of willows.
Wetlands: The wetlands observed at this stop are a mixture of 3-layer and 2-layer hydrogeologic systems. Slope wetlands occur near the valley sides, grading into riverine wetlands adjacent to the stream.
(0.9-1.3 miles)

5.3   A lodgepole pine (*Pinus contorta*) forest appears on colluvium/till pockets; this indicates drier conditions and past disturbance from a forest fire.
(0.3 mile)

5.6   Note the medial moraine on the other side of the drainage; this deposit supports a sparse lodgepole pine-Engelmann spruce forest. Outwash materials supporting a willow community indicate a shallow water table or a discharge zone.
(0.4 mile)

6.0   Note the burned area along the road that is covered with a young lodgepole forest. In the Rocky Mountains, lodgepole pine and aspen are socalled "successional" tree species, invading forested areas that have been disturbed by fire, logging, or mass movements. As time passes, and barring further disturbance, the lodgepole and aspen are gradually replaced by Engelmann spruce, subalpine fir (*Abies lasiocarpa*), and Douglasfir (*Pseudotsuga menziesii*), which form climax forest stands.
(0.2 mile)

6.2   The main road intersects Bruno Gulch Road. Note two levels of outwash plains. The willows (on wet

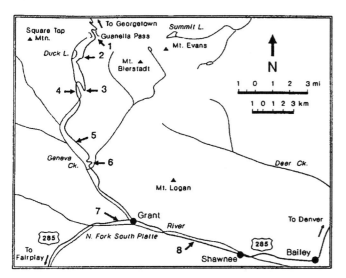

Figure 4. Map of the road log.

soils) and shrubby cinquefoil (*Pentaphylloides floribunda*) (on drier soils) are located in the moist channels. The tops of the outwash plains are well-drained, have dry soils, and are covered by grasses.
(0.9 mile)

7.1.1 **Stop 5: Geneva Park at an approximate elevation of 2957 m (9,700 ft). Note the remnants of a former peat mine.**

Geomorphology: The flat, parklike topography indicates a valley filled with glacial, glacial-lacustrian, and fluvial sediments.

Vegetation: The presence of peat deposits and soils high in iron sulfide indicates that bog conditions once existed. Cold, waterlogged soils allowed an accumulation of peat up to five feet thick at this site over a period of several thousand years. During that time, the vegetation growing on the peat probably consisted mainly of aquatic sedge and other herbaceous vegetation similar to the wet meadow site on Guanella Pass. Currently, willows predominate, indicating that the water table is still near the surface (but probably not as wet as it was in previous times). Ground-water levels are controlled by nearby stream channels. Note the bristlecone pines on dry, rocky soils above the road.

Hydrogeology: The entire park is probably a regional discharge zone; high water-table conditions predominate. Flow is toward the middle of the park to the modern stream channel. Ground-water recharge to the valley bottom is by precipitation and infiltration on surrounding uplands. Ground-water flow is through colluvium and fractured bedrock, discharging into the channel in the center of the park. The presence of peat indicates that the fractured bedrock system maintains the long-term discharge,

and hence sustains the Geneva Park wetlands system. Peat accumulation occurs at the rate of perhaps 1-3 cm per century and requires saturated soil conditions. Surface water is derived from the drainage basin of Geneva Creek, and is related to direct precipitation and runoff, to the release of snowpack, and to ground-water discharge (see Figure 5).

Wetlands: The wetlands observed at this stop are a mixture of 3-layer and 2-layer hydrogeologic systems. The 3-layer systems are characterized by the low hydraulic conductivity peat layer, which was mined. Slope wetlands are observed near the valley sides, and riverine wetlands are seen along the stream channel.
(0.6 mile)

7.7 Burning Bear campground.
(0.1-0.2 mile)

7.8-7.9 The top of the terminal moraine. Note the view of the entire glacial/lacustrian area. Mt. Evans is up-valley to the northeast.
(0.3-0.6 mile)

8.2-8.4 Note the V-shaped fluvial valley below.
(0.6 mile)

8.4 **Stop 6: Cross section of the terminal moraine at an approximate elevation of 2865 m (9400 ft).**

Geomorphology: Note the well-graded, poorly sorted terminal moraine; the materials and features are well-drained, hummocky, and have good soil development. A fluvially derived V-shaped valley containing a downcutting, high gradient river is below the moraine.

Vegetation: Sagebrush (*Seriphidium vaseyanum*) grows on dry, well-drained sites. Several bristlecone pines can be seen growing on the moraine. Bristlecone are among the longest-lived tree species. In California, specimens have been found that are over 4,000 years old, although in Colorado the maximum recorded age of this species is only about 1,200 years.

Hydrogeology: Ground-water seeps through the moraines. Geneva Creek is a gaining stream due to groundwater input as baseflow.

Wetlands: The wetlands observed at this stop are a mixture of 2-layer and 1-layer hydrogeologic systems. Most of these wetlands are slope wetlands, although some riverine wetlands are nearby.
(0.3-0.5 mile)

8.7 The road crosses an outwash plain/terrace here.
(0.6-0.7 mile)

9.3-9.4 Tumbling River Ranch.
(0.4-0.5 mile)

9.8 Note the Terrace. Colorado blue spruce (*Picea pungens*) grow along the road in the valley bottom.
(0.8 mile)

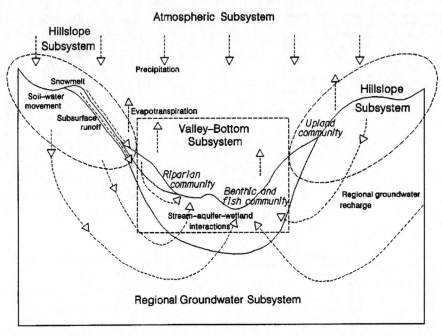

Figure 5. Schematic cross-section of the Geneva Valley showing the interaction of the surface water, ground-water, and wetlands systems.

10.6   Whiteside Camp Ground is located here. Note the ponderosa pine (*Pinus ponderosa*) and Douglas fir on drier sites. The appearance of these conifer species and that of the blue spruce indicates that we are leaving the subalpine life zone and entering the montane life zone. Phreatophytes include river birch (*Betula fontinalis*), aspen, thin-leafed alder (*Alnus tenuifolia*), and willow species.
      (0.7 mile)

11.3   Geneva Creek Picnic Ground.
      (0.7 mile)

12.0   Terrace and colluvial materials are seen along the road. Soils are welldrained and thick, and support mountain mahogany (*Cercocarpus montanus*) shrub communities.
      (1.0 mile)

13.0   Junction with U.S. 285 and the North Fork of the South Platte River at Grant, Colorado. Turn right (toward the southwest).
      (0.3-0.8 mile)

13.3-13.8   Excavated material from the Roberts Tunnel is on the south side of road.
      (0.0 miles)

13.8   **Stop 7: Harold D. Roberts Tunnel at an approximate elevation of 2652 m (8700 ft).** The Harold D. Roberts Tunnel was constructed in 1962 by the Army Corps of Engineers as part of a trans-mountain water diversion project. The tunnel delivers water from Lake Dillon, located on the western slope of the Continental Divide, to the South Platte watershed on the eastern slope.
      Geomorphology: Note the river channel (mod-

ern) with waste material from the tunnel excavation on the opposite bank.
      Vegetation: The wider valley of the North Fork of the South Platte River supports narrowleaved cottonwood (*Populus angustifolia*). Intensive land use, including farming and grazing, has eliminated much of the natural riparian shrub communities that once proliferated on both sides of the river channel.
      Hydrogeology: Because of the water diversion, a more constant stream flow is maintained. The South Platte River is still gaining by groundwater baseflow and is the ultimate groundwater discharge area from fractureflow in the surrounding crystalline rocks. Near-surface water tables in the South Platte alluvium are indicated by cottonwoods. Turn around and proceed down-valley (northeast) along US 285 toward Denver, Colorado.
      (0.7 mile)

14.5   Grant, Colorado (elevation 2642 m 8667 ft.). The climate at Grant is considerably different than that at Guanella Pass. Measurements from a weather station located here indicate an average annual temperature of 3.7°C (38.6°F), and an average precipitation of 40 cm (15.6 inches) per year, 65 percent of which falls during the months from June through November.
      (0.6 mile)

15.1   Terrace gravels can be observed on the south side of the road.
      (0.5 mile)

15.6   U.S. 285 crosses a terrace.
      (1.1 miles)

16.7   Narrow-leaved cottonwoods, aspen, and willows indicate a fluvial terrace with a near-surface ground-water table.
(1.3 miles)

18.0   **Stop 8: The former Silvertip Lodge located along US 285 at an approximate elevation of 2500 m (8200 ft).**

Geomorphology: Terraces and a modern floodplain are present here. Today, the stream is meandering; terrace gravels indicate an earlier braided stream history.

Vegetation: Sparse vegetation on sunny and dry southfacing slopes consists of Rocky Mountain juniper (*Juniperus scopulorum*), ponderosa pine, mountain mahogany, and several grass species. Douglasfir grow on moist, northfacing slopes. Riparian species here include lanceleaf cottonwoods (*Populus X acuminata*), narrowleaved cottonwoods, balsam poplar (*Populus balsamifera*) and sandbar willow (*Salix interior*). This location is near the southern edge of balsam poplar distribution. The balsam poplar, more widely distributed in the northern latitudes, occurs only in scattered riparian locations in Colorado.

Hydrogeology: Recharge of groundwater is by precipitation and infiltration on upland slopes; recharge is greatest on northfacing sites. Groundwater flow is through colluvium and in bedrock fractures; groundwater discharge is into drainages such as the South Platte River. Scattered seeps on valley sides are observed. Groundwater is stored in the alluvium and terraces near the river. This region is part of an "isohyetal bullseye" (38-41 cm of precipitation per year; 1516 inches per year) and is drier than most of the surrounding region. Most wells are located in fractured areas (drainage areas or valleys usually) or in Platte River alluvium.

Wetlands: The wetlands observed at this site are a mix of 2-layer and 1-layer hydrogeologic systems. Most of these are slope wetlands.
(0.4-1.0 miles)

18.6-19.0   Channelization of the river can be seen. Note that old channels or oxbow lakes support willows and other phreatophytes.
(1.8-2.2 miles)

20.8   Shawnee Store.
(0.5 mile)

21.3   Note the fan on the north side of the river; cottonwoods are growing in this channel; intermittent channel flow occurs here.
(1.0 mile)

22.3   Platte Canyon High School has solar panels (part of the isohyetal bullseye phenomena: less clouds, therefore, higher insolation).

(3.2 miles)

25.5   Bailey, Colorado. US 285 leaves the North Fork valley at this point. End of road log.

SUMMARY

The Wetland Integrated Hydrologic Analysis (WIHA) is a new method to assess wetland hydrology based upon an integrated approach. This approach identifies the system attributes that should be characterized with field visits and readily available data. These attributes can be interpreted to determine wetland hydrology and confirmed with modeling or intensive field data, if needed.

One of the benefits of WIHA is that the three-dimensional hydrogeologic structure of the wetland is assessed, which can provide insight into the long-term sustainability of the wetland. Long-term sustainability is becoming a significant wetland management issue. The methods used to develop the WIHA for the southern Rocky Mountains of Colorado can be applied to other geographic regions and can be demonstrated in the Guanella Pass region. WIHA can be altered to represent geologic, geomorphic, and hydrologic processes, and other characteristics that are specific to different regions.

REFERENCES CITED

Brinson, M.M. 1993. A Hydrogeomorphic Classification for Wetlands. U.S. Army Corps of Engineers, Washington, D.C., USA. Wetlands Research Program Technical Report, WRP-DE-4.

Harper-Arabie, R. and K.E. Kolm. 1998. An integrated approach to wetlands characterization in determining hydrologic functions: Proceedings of the First International Symposium on Integrated Technical Approaches to Site Characterization; Argonne National Laboratory, Chicago, Illinois, pp. 103 - 118.

Kolm, K.E. 1996. Conceptualization and Characterization of Ground-Water Systems Using Geographic Information Systems. Engineering Geology 42:111-118.

Kolm, K.E., P.K.M. van der Heijde, J. S. Downey and E. D. Gutentag. 1996. Conceptualization and Characterization of Ground-water Flow Systems: Subsurface Fluid-flow (Ground-water and VadoseZone) Modeling. American Society for Testing and Materials, Philadelphia, PA, USA. ASTM STP 1288.

Kolm, K.E., R.M. Harper-Arabie, and J.C. Emerick. 1998. A Stepwise, Integrated Hydrogeomorphic Approach for the Classification of Wetlands and Assessment of Wetland Hydrological and Geochemical Function in the Southern Rocky Mountains of Colorado: Colorado Geologic Survey, Colorado Department of Natural Resources Technical Report, Denver, CO. 241p.

Mitsch, W.J., and J.G. Gosselink. 1993. Wetlands, 2nd edition. Van Nostrand Reinhold, New York. 722 p.

Mutel, C.F., and J.C. Emerick. 1992. From Grassland to Glacier: the Natural History of Colorado and the Surrounding Region. Johnson Books, Boulder, CO. 290p.

Tarnocai, C. 1979. Canadian wetland registry, in Proceedings of a Workshop on Canadian Wetlands Environment, C.D.A. Rubec and F.C. Pollett, eds., Canada Land Directorate, Ecological Land Classification Series, No.12, pp. 9-38.

Printed in U.S.A.

Geological Society of America
Field Guide 1
1999

Field trip to Manitou Springs, Colorado, with specific emphasis on the sediments of Cave of the Winds and their relationship to nearby alluvial deposits and spring sediments

Fred G. Luiszer

Department of Geological Sciences, University of Colorado, Boulder, Colorado 80309, United States

ABSTRACT

Allogenic and authigenic sediments at Cave of the Winds can be correlated to nearby sediments. The allogenic sediments, which consist of clay, silt and sand, are correlated to the Nussbaum Alluvium. Magnetostratigraphy of the cave sediments combined with the aminostratigraphy of alluvial terraces in the Manitou Springs area indicate that the Nussbaum Alluvium is ~1.9 Ma and that the major episode of sedimentation in Cave of the Winds started ~4.5 Ma and stopped ~1.5 Ma.

The authigenic sediments, which consist of iron and manganese oxides, are correlated to similar modern spring sediments. The spring sediments along with the chemistry of the water issuing from the springs of Manitou can be used to show that a mixing zone under Manitou Springs is presently dissolving ~71 tonnes of limestone per year. The cave itself along with the iron- and manganese-rich sediments at the cave indicate that the process of dissolution taking place at the springs is the same process responsible for the dissolution of the cave. Magnetostratigraphic analysis indicates that most of the cave was dissolved from the limestone starting ~4.5 Ma and ceased ~1.5 Ma.

INTRODUCTION

Manitou Springs, which is located 10 km west of Colorado Springs (Fig. 1), is well known for the mineral springs that issue from several locations within the city (Fig. 2). Cave of the Winds, which is 1.5 km north of Manitou Springs, was first discovered by Arthur B. Love in the early 1870's. In 1880 and 1881 large portions of the cave became known from efforts of George W. Snider, who dug open many of the sediment-filled passages; soon thereafter the cave was commercialized. Strieby (1893) was the first to suggest that Cave of the Winds and the springs of Manitou Springs were genetically related. He correlated red clay found in limestone cavities near the springs to the red clay in Cave of the Winds. He also presented hypotheses regarding the origin of the CO_2 and mineral water that were remarkably sound considering the paucity of available data. In his famous paper, "Features of Limestone Caverns," Bretz (1942) states that Cave of the Winds is phreatic with no evi-

dence of vadose activity. Morgan (1950) suggested that the origin of Cave of the Winds was associated with circulation of CO_2-rich groundwater through fractures and joints formed during the Laramide Orogeny. Bianchi (1967) indicated that the Cave of the Winds was exposed during an erosional event related to oscillating climatic conditions during the Pleistocene.

STOP 1: CAVE OF THE WINDS

At Cave of the Winds there are several geological features that will be looked at and discussed: the general geology of the region, the Nussbaum Alluvium, and sediments in Cave of the Winds. For those using this guide as step-by-step guide should note that the sequence of the stops is not important because all of the stops are within a few kilometers of each other. The only thing that should be remembered is that the process that is taking place under Manitou Springs is the same process that formed Cave of the Winds millions of years ago.

Luiszer, F. G., 1999, Field trip to Manitou Springs, Colorado, with specific emphasis on the sediments of Cave of the Winds and their relationship to nearby alluvial deposits and spring sediments, *in* Lageson, D. R., Lester, A. P., and Trudgill, B. D., eds., Colorado and Adjacent Areas: Boulder, Colorado, Geological Society of America Field Guide 1.

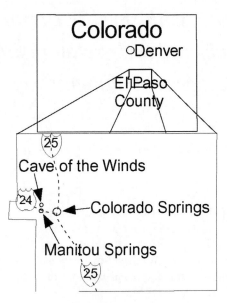

Figure 1. Location of study area.

The Cave of the Winds

The Cave of the Winds is developed for the most part in the Ordovician Manitou Formation, composed primarily of dolomitic marine limestone. In a few places in Cave of the Winds and in other nearby caves, cave passages can be found that extend into the overlying Mississippian Williams Canyon Formation and in even fewer places one to two meters into the Mississippian Leadville Formation. Most of the cave passages are developed along joints and fractures that were produced during the early Laramide Orogeny. Some of the cave passages are developed along Pennsylvanian paleokarst features, which are evident along the commercial trail. The cave passages tend to have larger dimensions in upper and lower portions of the Manitou Formation where the bedrock has a higher calcite content (Table 1).

Geology of Manitou Springs region

The springs of Manitou issue from, and Cave of the Winds is hosted by, limestones of a block of Paleozoic sediments (Fig. 3). Historically, this block has been called the Manitou Embayment. The Ute Pass Fault, a reverse fault that is the southwest boundary of the embayment, juxtaposes Precambrian crystalline rocks and Paleozoic sediments. The northwest boundary

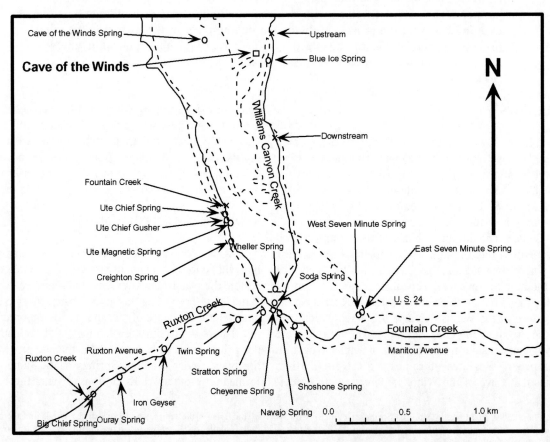

Figure 2. Location map of springs, wells, streams, and other sampling sites.

Table 1.   Compositions of predominant rock types that react with the spring water at Manitou Springs.

	Pike Peak Granite*	Pike Peak Granite biotite*	Manitou Formation #1 (Base)	Manitou Formation #2	Manitou Formation #3	Manitou Formation #4	Manitou Formation #5 (Top)	Williams Canyon Formation	Leadville Formation (Base)
SiO_2	73.00**	34.70	5.09	13.59	19.20	7.10	1.87	8.74	11.49
Al_2O_3	13.37	18.34	0.63	2.01	0.95	0.42	0.20	0.82	1.98
Fe_2O_3	0.55	3.27	0.29***	0.56	1.16	0.56	0.09	0.39	1.67
FeO	1.71	25.68	NM	NM	NM	NM	NM	NM	NM
MgO	0.09	0.97	4.13	5.21	17.54	7.36	0.53	0.72	19.74
CaO	1.05	0.46	48.63	39.62	24.58	44.06	54.86	51.55	24.89
Na_2O	3.41	0.68	0.13	0.04	0.09	0.01	0.20	0.15	0.06
K_2O	5.58	9.26	0.89	0.86	1.58	0.55	0.26	0.38	0.58
TiO_2	0.22	2.15	0.01	0.04	0.07	0.02	0.01	0.04	0.06
P_2O_5	0.02	0.03	0.29	0.38	0.21	0.29	0.36	0.27	0.20
MnO	0.05	0.55	0.02	0.06	0.09	0.10	0.02	0.02	0.05
F	0.46	2.75	NM	NM	NM	NM	NM	NM	NM
CO_2****	NL	NL	42.67	36.77	38.44	42.62	43.63	41.24	41.09
Cl (ppm)	400	1700	147	198	454	339	130	45	332
S (ppm)	NL	NL	211	42	29	51	128	1016	588
Li (ppm)	48	2200	NM	NM	NM	NM	NM	NM	NM
Pb (ppm)	49	30	NM	NM	NM	NM	NM	NM	NM
Sr (ppm)	108	10	211	98	65	82	123	92	81
Total	99.53	99.06	102.76	99.12	103.91	103.09	102.02	104.32	101.81

*Averge composition of the Pikes Peak Granite wholerock and biotite from Hawley and Wobus (1977).

Values in Wt% except were noted.   * Total iron in the limestones is reported as Fe_2O_3

NM = Not Measured   NL = Not Listed   **** Calculated by assuming that all the MgO and CaO are in dolomite and calcite.

between the Precambrian crystalline rocks and Paleozoic sediments underlies an erosional surface. The eastern boundary is the Rampart Range Fault. The Paleozoic sediments in the block generally dip gently to the southeast. The 1.05 Ga Pikes Peak Granite and 1.7 Ga metamorphics crop out to the south, west, and north of and underlie the embayment. The sediments in the embayment consist of limestone, siltstone, sandstone, conglomerate, and shale (Fig. 3). Almost all of the subsurface water flow in the embayment occurs in the very permeable Manitou and Williams Canyon Formations and to a lesser extent the Leadville Formation.

The origin of Cave of the Winds is very dependent upon the geometry of the Paleozoic sediments, the faults and Fountain Creek. The Rampart Range Fault, which juxtaposes permeable limestone beds and impermeable shale, creates a very effective barrier to the east. This barrier prevents any water that enters the limestone beds of the embayment from draining out to the east. The northwestern exposed up-dip side of the embayment acts as a collection area that channels meteoric water into the limestone beds. The limestone beds along the Ute Pass Fault act as a collection gallery that channels ascending CO_2-rich mineral water into the embayment. The water that enters the embayment drains out at the lowest point possible, where Fountain Creek intersects the limestone. Thus, we find all of the modern springs located along Fountain Creek in Manitou Springs.

Nussbaum Alluvium

The presence of the Nussbaum Alluvium, which caps the ridge east of Cave of the Winds, indicates that in the past Fountain Creek flowed to the east over Cave of the Winds. At that time the springs that are now located in Manitou would have been located along the paleo-Fountain Creek above Cave of the Winds. Subsequent discussion will show that there is a corrosive mixing zone below the springs of Manitou. Apparently, in the past, below the paleo-Fountain Creek, there was a similar mixing zone, which was dissolving Cave of the Winds. Aminostratigraphy of snail shells present in the Nussbaum and other radiometrically dated alluvial terraces in the Manitou Springs area was used to date the Nussbaum Alluvium at ~1.9 Ma (Luiszer, 1997). The age of the Nussbaum Alluvium was then used to calibrate the magnetostratigraphy of the sediments in Cave of the Winds.

Sediments at Cave of the Winds

There are two categories of sediments at Cave of the Winds: chemical and detrital. The chemical sediments include flowstone, stalactites, stalagmites, and helictites. While most of these speleothems consist of calcite, the helictites and frostwork consist mostly of aragonite. Gypsum is also found in the cave in the form of selenite needles, flowers, crusts, and starbursts. The above speleothems are commonly found in caves and are not indicative of early speleogenesis. The manganese- and iron-rich sediments present throughout Cave of the Winds, which are evidence of the mixing zone that existed at the cave, are much more important for understanding the genesis of the cave. These sediments are especially prominent in Thieves Canyon (Fig. 4). Though not as obvious, these same sediments can be seen along the tourist trail in the Guides Rest area (Fig. 5) and outside of the cave on an outcrop in the northwest corner of the upper parking lot. Evidence of the modern counterpart can be seen in Manitou Springs, where Fe- and Mn-rich sediments are being

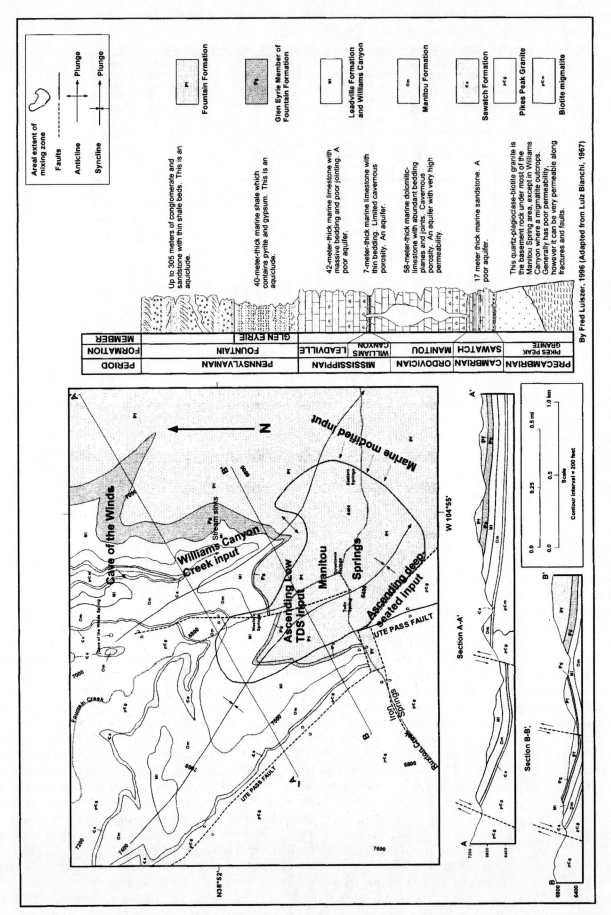

Figure 3. Geologic map of the Manitou Springs area with suggested extent of mixing zone.

deposited as a result of the mixing of oxygen-rich near-surface meteoric and metal-rich deep-seated waters.

Another chemical sediment that is apparently related to the mixing zone is calcite spar that is found with the Fe- and Mn-rich sediments (Fig. 4). The co-precipitation of these minerals was probably a function of the depth of the mixing zone. Apparently the CO_2-rich and the metal-rich waters mixed at depth where the hydrostatic pressure was high enough to keep the CO_2 in solution. As the water ascended to the paleo-springs, the decreasing pressure allowed the CO_2 to outgas. The resultant pH increase caused the co-precipitation of the calcite and metal oxides. Carbon dioxide bubble trails in Thieves Canyon and near the Guides Rest are evidence of CO_2 outgassing (Fig. 5).

To the chagrin of cave management and explorers alike is the detrital sediments that fill many of the passages to the ceiling throughout the cave. The oldest of these sediments is debris from the solution of the limestone. This bed is normally only a few centimeters thick, however, a ~10-cm thick bed of the debris does occur in Thieves Canyon (Fig. 4). Fe- and Mn-rich sediments, which represent an early stage of the mixing zone, overlie the solution debris. Overlying the Fe- and Mn-rich sediments is a thick sequence of finely bedded clay (Fig. 4). The shear volume of the clay found in the cave along with the clay mineralogy indicates that the cave clay was derived from nearby soils (Fig. 6), which were eroded and transported into the cave by surface streams. Overlying the clay is a coarsening

upward sequence of silt, sand and gravel, which in parts of the cave can be up to a few meters thick. Magnetostratigraphy of the sediments in the Grand Concert Hall (Luiszer, 1997) indicates that major dissolution at Cave of the Winds started ~4.5 Ma ago and that major detrital sedimentation ceased ~ 1.5 Ma ago (Fig. 7). In the last 1.5 Ma detrital sedimentation at the cave has been limited to the introduction of sediments near the cave entrances from rainstorms and snowmelt. In this same interval the chemical sedimentation has been dominated by the formation of the generic speleothems consisting of calcite, aragonite and gypsum.

THE MANITOU SPRINGS STOPS

The stops in Manitou Springs will allow participants to look at many of the springs with discussions that will emphasize how the mixing zone located below Mantiou Springs controls the content of the different springs. The calcite, Mn-rich and Fe-rich sediments that precipitate from different springs and their relation to the sediments present at Cave of the Winds will also be looked at and discussed.

STOP 2: THE IRON SPRINGS

Below Manitou Springs two major types of water are mixing, one is a deep-seated, CO_2-rich mineral water, the other a

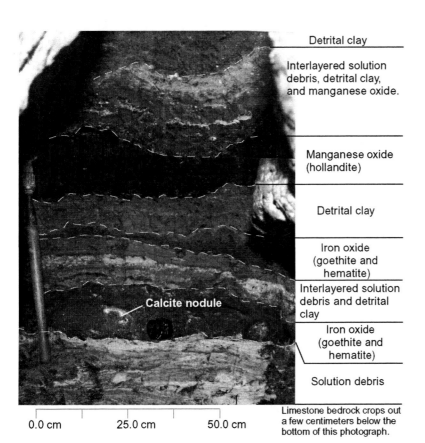

Detrital clay

Interlayered solution debris, detrital clay, and manganese oxide.

Manganese oxide (hollandite)

Detrital clay

Iron oxide (goethite and hematite)

Interlayered solution debris and detrital clay

Iron oxide (goethite and hematite)

Solution debris

Calcite nodule

0.0 cm   25.0 cm   50.0 cm

Limestone bedrock crops out a few centimeters below the bottom of this photograph.

Figure 4. Photograph of outcrop at Electric Coolaid Acid Test Alcove, which is located at south end of Thieves Canyon in Cave of the Winds. Dashed lines are drawn on contacts of sediment units.

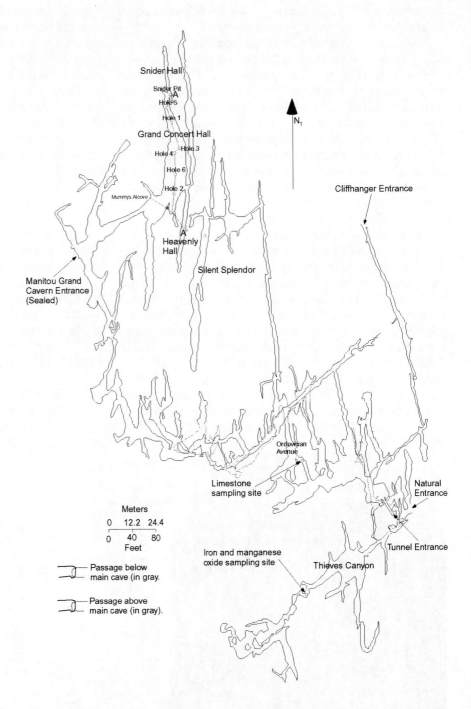

Figure 5. Map of Cave of the Winds, Manitou Springs, Colorado, showing locations of samplings sites. Modified from Paul Burger, 1996.

near-surface meteoric water. Because of the nonlinear nature of the solubility of calcite in water with varying CO_2 content, the mixing of waters that contain different amounts of and that are saturated with dissolved calcite results in a solution that is undersaturated with calcite. The undersaturated water below Manitou Springs is dissolving ~70 tonnes of limestone every year. Study of the springs at Manitou has shown that these two major waters type can be further broken down into four distinct sources: Williams Canyon Creek and subsurface waters that have chemistry similar to the Iron Geyser, the Cave of the Winds Spring and the Seven Minute Spring (Luiszer, 1997).

There are several lines of evidence that indicate that the Iron Geyser represents water that has been modified solely by contact with the Pikes Peak Granite. The elements that are

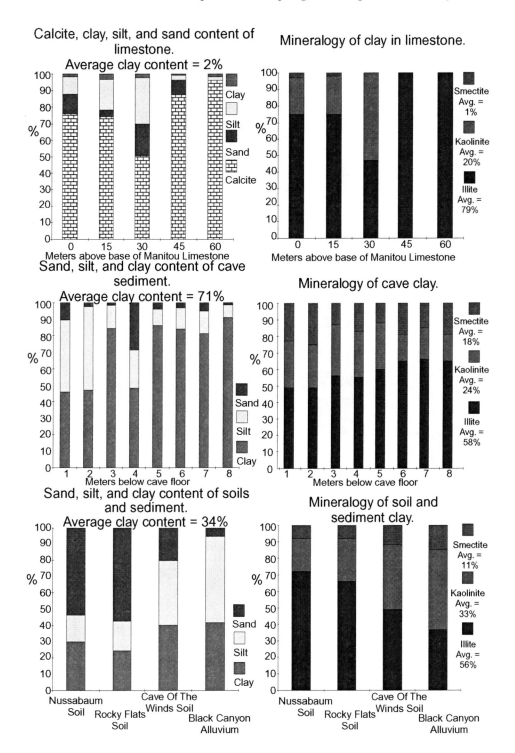

Figure 6. Grain size distribution and clay mineralogy.

relatively abundant in the granite are sodium, potassium, lithium, iron, and silicon (Table 1). Accordingly, the water of the Iron Geyser contains elevated amounts of all of these elements (Table 2). The high concentrations are probably related to the elevated temperature at which the water-rock interaction took place and the extended period of time that the water was in contact with the rock. The quartz-with-no-steam-loss geothermometer indicates that the water issuing from the Iron Geyser obtained a temperature of 126 °C. Assuming an average geothermal gradient of 30 °C/km, the calculated temperature is equivalent to a depth of ~4 km. Radio carbon dating indicates that some of the water issuing from the downtown springs can be up to ~30 ka. The Iron Geyser may be much older, because the downtown springs are actually a mixture

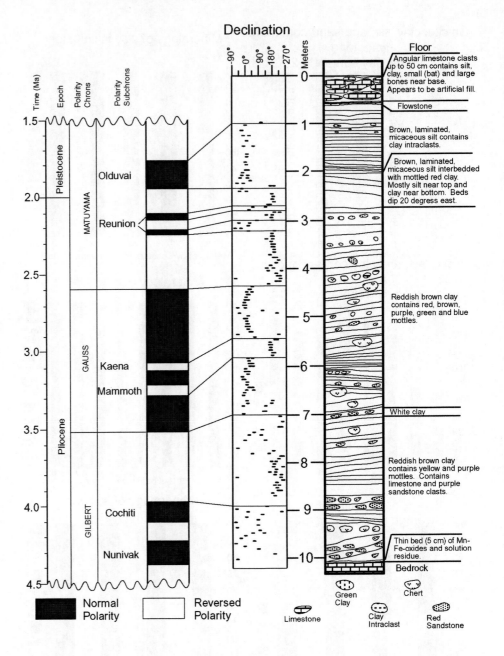

Figure 7. Paleomagnetic correlation and stratigraphy of Grand Concert Hall Hole 5. Paleomagnetic time scale adapted from Harland and others (1982).

of near-surface modern water and water similar to the Iron Geyser.

All of the iron springs are actively precipitating iron oxide sediments. This is a result of the oxidation as the nearly oxygen-free spring water contacts the atmosphere and absorbs oxygen. The iron-rich sediment contains a fairly high amount of lead and arsenic, which is similar to the sediments at Cave of the Winds (Table 3). Apparently, water with a composition similar to the Iron Geyser is entering the mixing zone at depth below Manitou Springs. When this water mixes with the oxygenated near-surface meteoric water iron oxide is precipitated in the dissolving cave system. This is why most of the other springs in Manitou have a low iron content. The high arsenic

and lead content of the iron is related to the ability of iron oxide to absorb these and other metals as it precipitates.

STOP 3: THE WESTERN SPRINGS

The western springs include the Ute Chief Magnetic, Gusher, Ute Chief, and the Creighton Springs. These springs have the lowest total dissolved solids (TDS) of any of the springs in Manitou because they are composed of a large amount of near-surface waters. The high nitrate content of the western springs, especially the Gusher Spring (Table 2), indicates that Williams Canyon Creek is one of the near-surface waters. Dye tracing has shown that it takes ~85 days for dye to

Table 2. Water Data

	Williams Canyon Creek	Cave of the Winds Spring	Iron Geyser	7 Minute Spring	Gusher Spring	Creighton
Bicarbonate (HCO₃)	232	225	1726	2538	1281	1004
Fluoride (F)	1.5	3.6	6.6	4.8	4.0	3.9
Chloride (Cl)	8.9	31	195.0	547.0	87	69
Bromide (Br)	0.05	0.18	0.96	2.10	0.39	0.27
Nitrate (NO₃)	15.80	0.41	0.14	1.14	6.17	1.45
Sulfate (SO₄)	56	52	227.0	366.0	84	106
Nitrite (NO₂)	N.D.	N.D.	N.D.	N.D.	N.D.	N.D.
Phosphate (PO₄)	0.04	N.D.	N.D.	N.D.	N.D.	N.D.
Boron (B)	0.2	N.D.	1.0	2.1	1.0	N.M.
Iron (Fe total)	0.01	0.09	11.40	0.08	0.05	0.02
Manganese (Mn)	0.03	0.07	1.20	1.60	0.13	0.08
Lithium (Li)	0.02	0.06	0.98	0.77	0.24	0.16
Sodium (Na)	14	39	617.7	530.9	136	118
Potassium (K)	2.0	1.1	94.1	41.67	19	15
Magnesium (Mg)	24	9.1	34.5	134.8	52	47
Calcium (Ca)	67	48	354.7	558.9	296	187
Lead (Pb)	N.M.	N.M.	0.006	0.006	<.001	<.001
Arsenic (As)	N.M.	N.M.	0.122	0.022	0.023	N.M.
TDS	422	410	3272	4729	1967	1550
Cond. (mS/cm)	0.54	0.61	3.61	5.42	2.08	1.68
Temp. °C	5.6	11.4	7.8	11.6	12.4	11.1
pH	8.25	7.64	6.27	6.44	6.13	5.97
Eh (volts)	0.495	0.501	0.317	0.520	0.559	0.571
DOX	9.0	0.1	0.30	3.40	1.20	2.00
DCO₂-gr/l	N.M.	N.M.	2.82	2.68	2.56	1.30
SiO₂	19	28	82	18	21	18
Total alk.	191	185	1415	2080	1050	823
Sample Date	5/4/91	4/27/91	4/7/91	4/3/91	4/23/91	9/18/91
Flow l/min	40	15	0.3	0.3	40	270
Type	Creek	Spring	Well	Well	Well	Well

N.D. = Not Detectable (<0.01 ppm)   N.M. = Not Measured

Concentrations in ppm except where noted.

Table 3. Analyses of cave and spring sediments.

	As ppm	Pb ppm	Mn Wt%	Fe Wt%	Sediment Type
Ouray Spring	4200	181	0.12	58.02	Fe-oxide
Chief Spring	3200	225	N.M.	N.M.	Fe-oxide
Iron Geyser	3400	170	N.M.	N.M.	Fe-oxide
Thieves Canyon 31.5 cm	7400	3900	1.14	72.87	Goe+Hem*
Thieves Canyon 39.0 cm	8800	6200	1.50	73.23	Goe+Hem

N.M.=Not Measured   * Goe = Goethite   Hem = Hematite

travel from Williams Canyon to the Gusher Spring. The high flow rate of the Creighton Spring indicates a subsurface input with a composition similar to the Cave of the Winds Spring, which, because of its low TDS, is also considered near-surface meteoric water (Table 2).

STOP 4: THE DOWNTOWN SPRINGS

The downtown springs include the Cheyenne, Navajo, Shoshone, Stratton, Twin, Wheeler, and Soda Springs. The composition of the downtown springs consists of a mixture of the near-surface waters, water with a composition similar to that of the Eastern 7-minute Spring and water with a composition similar to that of the Iron Geyser (Fig. 8). Manganese

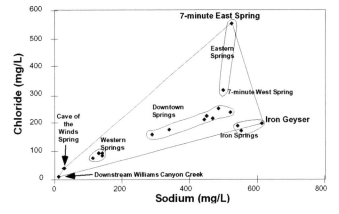

Figure 8. Plot of sodium versus chloride content of the springs of Manitou and low TDS waters. The polygon drawn on the plot contains almost all of the springs and streams in the Manitou Springs area. This suggests that the Western and Downtown Springs are a mixture of downstream Williams Canyon Creek and waters that have chemistry similar to the Iron Geyser, 7-minute East Spring, and Cave of the Winds Springs.

oxide layers intercalated with calcite that is precipitating from the water that issues from the Shoshone Spring is more evidence suggesting that manganese oxides are probably precipitating in the mixing zone. This correlates with the manganese sediments found in Cave of the Winds. The high manganese content of the Shoshone Springs has been attributed to the precipitation and dissolution of manganese near the boundary between the hypolimnion and epilimnion in the mixing zone.

STOP 5: THE EASTERN SPRINGS

One of the four mixing end members appears to have a composition similar to the 7-minute East Spring. The elevated amounts of sulfate, chloride, and boron (Table 2), suggest that the 7-minute East Spring has a different evolution than the other end members. Sulfate, chloride and boron are anions that are commonly associated with marine sediments, suggesting that the Eastern Springs may have been modified by rock-water interaction with marine sediments. The low sulfur and chlorine content of the marine limestone of the Manitou and Leadville Formations removes these formations from consideration. Another rock with which water could be interacting is the Gleneyrie Member of the Fountain Formation, a marine shale that overlays the limestone beds. Another possibility is the Pierre Formation, a marine shale that abuts the Rampart Range Fault east of Manitou Springs. More study would be needed to enable the assignment of the marine influence to either of these marine shales or other unknown sources.

THE MIXING ZONE

All of the inputs and outputs of the Manitou Springs mixing zone have either been measured directly or have been calculated (Luiszer, 1997). The mass balance of the system indicates that calcium and magnesium are being added to the water in the mixing zone from mixing corrosion. The calculations indicate that ~71 tonnes of limestone are being removed every year from beneath the city of Manitou Springs. The mass balance also indicates that iron and manganese are being lost in the mixing zone. The Cave of the Winds and the sediments found within the cave are strong evidence that the processes taking place today under Manitou Springs are the same processes that took place at Cave of the Winds some 5 m.y. ago.

REFERENCES CITED

Bianchi, L., 1967, Geology of the Manitou-Cascade Area, El Paso County, Colorado with a study of the permeability of Its crystalline rocks (M.S. Thesis): Golden, Colorado School of Mines.
Bretz, J. H., 1942, Vadose and phreatic features of limestone Caverns: Journal of Geology, v. 50, no. 6, part 2, p. 675-811.
Harland, W. B., and others, 1982, A geologic time scale: Cambridge, Great Britain, Cambridge University Press, p. 66.
Hawley, C. C., and Wobus, R. A., 1977, General geology and petrology of the Precambrian crystalline rocks, Park and Jefferson Counties, Colorado: Geological Survey Professional Paper 608-B, 77 p.
Luiszer, F. G., 1997, Genesis of Cave of the Winds, Manitou Springs, Colorado (Ph.D. thesis): Boulder, Colorado, University of Colorado, 122 p.
Morgan, G. B., 1950, Geology of Williams Canyon area, north of Manitou Springs, El Paso County, Colorado (Masters thesis): Golden, Colorado School of Mines, 80 p.
Strieby, W., 1893, The origin and use of the natural gas at Manitou, Colorado: Colorado College Studies, v. 4, p. 14-36.

Geological Society of America
Field Guide 1
1999

200,000 years of climate change recorded in eolian sediments of the High Plains of eastern Colorado and western Nebraska

Daniel R. Muhs
U.S. Geological Survey, MS 980, Box 25046, Federal Center, Denver, Colorado 80225, United States
James B. Swinehart
Conservation and Survey Division, Institute of Agriculture and Natural Resources, University of Nebraska, Lincoln, Nebraska 68588-0517, United States
David B. Loope
Department of Geosciences, University of Nebraska, Lincoln, Nebraska 68588-0304, United States
John N. Aleinikoff and Josh Been
U.S. Geological Survey, MS 980, Box 25046, Federal Center, Denver, Colorado 80225, United States

INTRODUCTION

Loess and eolian sand cover vast areas of the western Great Plains of Nebraska, Kansas and Colorado (Fig. 1). In recent studies of Quaternary climate change, there has been a renewed interest in loess and eolian sand. Much of the attention now given to loess stems from new studies of long loess sequences that contain detailed records of Quaternary glacial-interglacial cycles, thought to be a terrestrial equivalent to the foraminiferal oxygen isotope record in deep-sea sediments (Fig. 2). Loess is also a direct record of atmospheric circulation, and identification of loess paleowinds in the geologic record can test atmospheric general circulation models. Until recently, eolian sand on the Great Plains had received little attention from Quaternary geologists. The past decade has seen a proliferation of studies of Great Plains dune sands, and many studies, summarized below, indicate that landscapes characterized by eolian sand have had dynamic histories.

On this field trip, we will visit some key eolian sand and loess localities in eastern Colorado and southwestern Nebraska (Fig. 1). Stratigraphic studies at some of these localities have been conducted for more than 50 years, but others have been systematically studied only in the past few years. Many of the data which appear in this guidebook have been derived from previous studies (Swinehart and Diffendal, 1990; Madole, 1994; Loope and others, 1995; Maat and Johnson, 1996; Muhs and others, 1996, 1997a, 1999; Mason and others, 1997; Aleinikoff and others, 1999), but some are presented here for the first time.

LOESS STRATIGRAPHY IN THE CENTRAL GREAT PLAINS

Four middle-to-late Quaternary loess units, from oldest to youngest, Loveland Loess, the Gilman Canyon Formation, Peoria Loess and Bignell Loess, have been identified and correlated on the Great Plains (Schultz and Stout, 1945; Frye and Leonard, 1951). Loveland Loess is usually no more than a few meters thick and is often the oldest loess unit exposed at many localities. In places, it appears to have an eolian sand facies. Loveland Loess is identifiable by the presence of the last interglacial Sangamon Soil in its upper part. This paleosol is usually relatively thick (1-2 m), frequently exhibits 7.5YR hues, and has well-developed prismatic or subangular blocky structure with clay films in the upper part of the B horizon and carbonate coatings or nodules in the lower part of the B horizon. The Gilman Canyon Formation is thin (usually <2 m) and typically has an organic-rich soil developed in it. Commonly, the soil developed in the Gilman Canyon Formation is welded to the upper part of the Sangamon Soil. In places, including one locality to be visited on this field trip, the Gilman Canyon Formation has two buried soils. The Gilman Canyon Formation is overlain by Peoria Loess, which is the thickest (up to ~48 m) and areally most extensive of the Great Plains loess units. A dark, organic-rich buried soil, referred to as the Brady soil, caps the upper part of the Peoria Loess, separating it from the overlying Bignell Loess. Bignell Loess is usually no more than ~2 m thick and has a patchy distribution. It has been found in Nebraska, Kansas and Colorado, but has not been reported east of the Missouri River.

Muhs, D. R., Swinehart, J. B., Loope, D. B., Aleinikoff, J. N., and Been, J., 1999, 200,000 years of climate change recorded in eolian sediments of the High Plains of eastern Colorado and western Nebraska, *in* Lageson, D. R., Lester, A. P., and Trudgill, B. D., eds., Colorado and Adjacent Areas: Boulder, Colorado, Geological Society of America Field Guide 1.

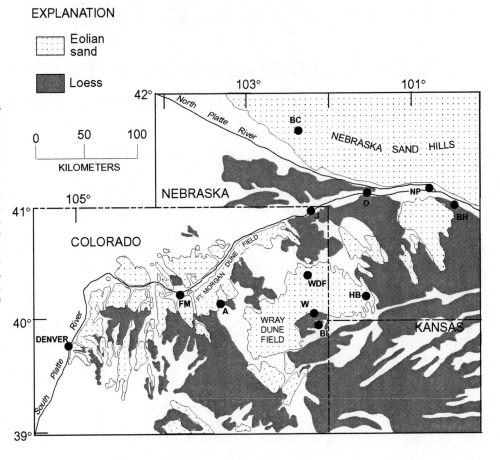

Figure 1. Map showing distribution of late Pleistocene loess and eolian sand in northeastern Colorado, southwestern Nebraska and northeastern Kansas, and field trip stops. Compiled from Swinehart (1990), Swinehart and others (1994), Kuzila and others (1990), Ross (1991), and Muhs and others (1996, 1999). FM, Fort Morgan (STOP 1); A, Akron; J, Julesburg; WDF, Wray dune field overlook (STOP 2); W, Wray (STOP 3); BI, Beecher Island (STOP 4); HB, Hoover blowout (STOP 5); BH, Bignell Hill (STOP 6); NP, North Platte; O, Ogallala; BC, Eldred Camp and Blue Creek (STOPS 7 and 8).

Geochronological studies indicate that the uppermost loess deposits on the Great Plains span the last interglacial-glacial cycle (Fig. 2). Most recent age estimates of Great Plains loesses have been from localities in Nebraska. At Eustis, Nebraska, there are numerous pre-Loveland loesses and intercalated paleosols, and a carbonate nodule from the Btk horizon of the youngest well-developed paleosol below the Loveland Loess gives a U-series age of 184,000 ± 5000 yr (analysis by B.J. Szabo, communicated to D.R. Muhs), which is a minimum-limiting age for this buried soil. Thermoluminescence (TL) ages that average about 163,100 yr have been made on the Loveland Loess itself, also at Eustis (Maat and Johnson, 1996). The TL ages and the underlying U-series age indicate that Loveland Loess could have been deposited during the penultimate glacial period, equivalent to deep-sea oxygen isotope stage 6, in good agreement with TL ages on Loveland Loess from the paratype locality in Iowa (Forman and others, 1992b). The Sangamon Soil, therefore, could have developed over a period from sometime after ~160,000 yr BP until deposition of the earliest Gilman Canyon Formation sediments, a timespan that may correspond to all of oxygen isotope stage 5 and perhaps part of stage 4 (Fig. 2). Based on radiocarbon ages of soil organic matter reported by Martin (1993), May and Holen (1993), Maat and Johnson (1996) and Muhs and others (1999), and TL analyses by Pye and others (1995) and

Maat and Johnson (1996), the age of the Gilman Canyon Formation is ~40,000 to ~22,000 ^{14}C yr BP. The Gilman Canyon Formation, therefore, corresponds in time to the mid-Wisconsin interstadial period and has its equivalent in the deep-sea record as oxygen isotope stage 3 (Fig. 2). Charcoal from spruce (Picea), as well as bone, snails, and detrital organic matter found within Peoria Loess give ages ranging from ~21,000 to ~10,000 ^{14}C yr BP (Wells and Stewart, 1987; Martin, 1993; May and Holen, 1993; Feng and others, 1994; Maat and Johnson, 1996), and TL dating of Peoria Loess in Nebraska gives ages ranging from ~24,000 to ~12,000 cal yr BP (Pye and others, 1995; Maat and Johnson, 1996). Direct dating of probable Peoria Loess at a locality in eastern Colorado using TL methods gives ages ranging from ~20,000 to ~15,000 cal yr BP (Forman and others, 1995). All these ages indicate that Peoria Loess found in the Great Plains, as with Peoria Loess east of the Missouri River, correlates to the late Wisconsin glacial period and deep-sea oxygen isotope stage 2 (Fig. 2). Maximum-limiting ages of Bignell Loess are based on radiocarbon ages of organic matter from the Brady soil, and range from ~11,800 to ~8,000 ^{14}C yr BP (Martin, 1993; Maat and Johnson, 1996; Muhs and others, 1999). Direct dating of Bignell Loess using TL gives ages ranging from ~9,000 to ~3,000 cal yr BP (Pye and others, 1995; Maat and Johnson, 1996).

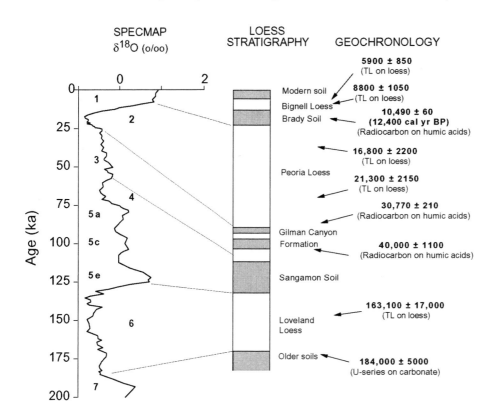

Figure 2. Generalized loess stratigraphy of the western Great Plains, with age estimates for deposits and soils and possible correlation to the deep-sea oxygen isotope record. Age data for western Great Plains loess from Maat and Johnson (1996) and Muhs and others (1999); oxygen isotope data from Martinson and others (1987).

EOLIAN SAND STRATIGRAPHY IN THE CENTRAL GREAT PLAINS

The most extensive eolian sands in North America are found in the central and southern Great Plains. Limited radiocarbon ages and degree of soil development suggest that eolian sand sheet and dune deposition took place during the last glacial period in Nebraska and Colorado. Late glacial eolian activity is supported by maximum-limiting ages of ~13,000 ^{14}C yr BP for some of the largest barchanoid ridges in the Nebraska Sand Hills (Swinehart and Diffendal, 1990) and evidence for eolian sand movement in the southwestern part of the dune field (Loope and others, 1995), seen on this trip. In Colorado, radiocarbon and soil evidence indicate that eolian sheet sands were deposited over large areas during the last glacial period (Madole, 1995; Muhs and others, 1996).

The mid-Holocene (~8000-5000 ^{14}C yr BP) has long been considered to be a dry period in central North America (Webb and others, 1993), and would seem to be an optimum time for eolian sand movement. However, there are actually few records of paleoclimatic conditions from the Great Plains itself during this period, and there is not widespread evidence for mid-Holocene eolian sand activity in the region. Radiocarbon ages from a few localities in the Nebraska Sand Hills indicate some probable mid-Holocene eolian activity (Loope and others, 1995; Stokes and Swinehart, 1997). In Colorado, Forman and co-workers (Forman and Maat, 1990; Forman and others, 1992a, 1995) infer that the mid-Holocene was an important

time of eolian sand movement, but only one locality really has definitive evidence for this. In Texas, mid-Holocene eolian sand is found in the stratigraphic record of dry valleys (Holliday, 1989; 1995b). Elsewhere in the Great Plains, the record for mid-Holocene eolian sand deposition is scanty, and is based on indirect lines of evidence such as dunes with a degree of soil development that is greater than that on late Holocene eolian sand, but less than that on late Pleistocene deposits (Madole, 1995; Muhs and others, 1996). Although it seems likely, from independent evidence, that conditions were optimal for eolian sand activity over the central Great Plains during the mid-Holocene, extensive late Holocene eolian activity may have removed much of the geomorphic record.

Both radiocarbon and luminescence methods demonstrate that eolian sands over much of the Great Plains have been active in the past 3,000 yr (Ahlbrandt and others, 1983; Swinehart and Diffendal, 1990; Madole, 1994, 1995; Holliday, 1995a, 1997a, 1997b; Forman and others, 1992a, 1995; Loope and others, 1995; Muhs and Holliday, 1995; Arbogast, 1996; Muhs and others, 1996, 1997a, 1997b; Wolfe and others, 1995; Stokes and Swinehart, 1997). In addition, most of these studies have stratigraphic data indicating multiple periods of eolian activity in the late Holocene. The number of radiocarbon ages and their analytical uncertainties do not yet make it possible to test the hypothesis of regional synchroneity of activity. However, these observations indicate that, contrary to earlier beliefs, eolian sands in this region can be active under an essentially modern climatic regime.

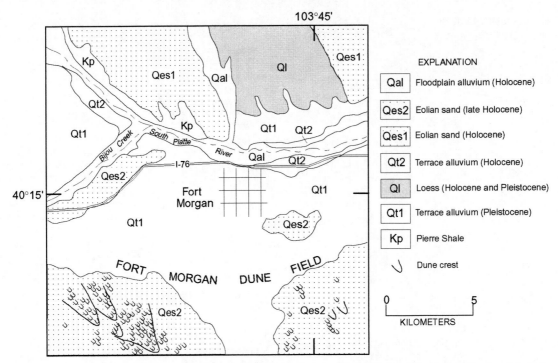

Figure 3. Simplified geologic map of the Fort Morgan, Colorado area (STOP 1), based on unpublished aerial photograph interpretation and field mapping by D.R. Muhs, and soil survey data of Spears and others (1968).

BETWEEN DENVER AND STOP 1

Eolian deposits can be seen shortly after passing the town of Hudson on I-76. East of Hudson, at the Kersey Road exit, there is a cut made for a railroad visible on the right. The sediments exposed in this cut have been studied by Forman and others (1992a, 1995) and Madole (1995). Between Hudson and Keenesburg, the landscape you see is covered with 2-3 m of late Wisconsin (Peoria) loess. A few miles east of Keenesburg, as you climb the hill, the Fort Morgan dune field will be visible, on both sides of I-76. The landforms here are low-relief parabolic dunes and eolian sand sheets, and can be easily distinguished by the abundance of *Artemisia* (sage) cover and frequent blowouts. The soils here belong mostly to the Valent series, and have simple A/AC/C profiles. Between mileposts 45 and 46, just before entering Roggen (exit 48), partially active dunes can be seen on both sides of the highway, but particularly on the north side. These dunes were active during the 1930s drought; just after passing through Roggen, there is a good view to the north of other dunes that were also active in the 1930s and are barely stable now. After passing by the town of Wiggins, I-76 traverses part of the Broadway-Kersey-Qt1 terrace (see discussion below); just outside of Wiggins, the terrace can be viewed to the north, eolian sand to the south.

STOP 1: SOUTH PLATTE RIVER NEAR FORT MORGAN

The major drainage for northeastern Colorado is the South Platte River, which heads in the Colorado Rocky Mountains near Colorado Springs and joins the North Platte River just east of the city of North Platte, Nebraska. Because the river figures prominently in the origin of both eolian sand and loess in the region, our first stop will be a short landscape view of the South Platte River near Fort Morgan. Fort Morgan itself and much of I-76 in this area are built on what has been called the "Broadway" (Scott, 1978), "Kersey" (Holliday, 1987) and "Qt1" (Muhs and others, 1996) terrace, of probable late-glacial age. It is as much as 8 km wide in places (Fig. 3). To the south, compound parabolic dunes of the Fort Morgan dune field, with their minimally developed soils (A/AC/C profiles) are visible and overlie this terrace. To the northwest and northeast, more subdued dunes and sand sheets of the Sterling dune field, with their relatively well developed soils (A/Bt/Bk/C profiles), can be seen. Immediately north of Fort Morgan, the highest part of the landscape is an alluvium-mantled upland overlain by loess.

Isotopic analyses (discussed at STOP 4) indicate that South Platte River sediments were a contributing, but not sole source of loess in eastern Colorado (Aleinikoff and others, 1999). Geochemical and isotopic analyses demonstrate that the most likely source of sediment for both late Pleistocene and Holocene eolian sand in the Fort Morgan and Wray dune fields (Fig. 1) was the South Platte River (Muhs and others, 1996).

Stratigraphic and radiocarbon studies by Madole (1994) show that the most recent episodes of eolian sand movement in the Fort Morgan dune field occurred in the past ~1500 yr (Fig. 4), and helped build the compound parabolic dunes

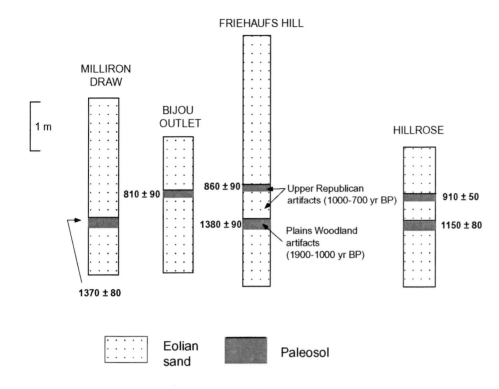

Figure 4. Stratigraphy and radiocarbon ages of late Holocene eolian sands from sections in the Fort Morgan dune field. Redrawn from Madole (1994).

shown in Fig. 3. Stratigraphic studies through augering show that the late Holocene parabolic dunes are underlain by eolian sheet sands that also occur at the surface in interdune areas and have well-developed soils (Muhs and others, 1996). These older sheet sands have maximum-limiting radiocarbon ages of ~27,000 yr BP, suggesting they were probably deposited during the last glacial period.

BETWEEN STOPS 1 AND 2

From the city of Brush (immediately east of Fort Morgan) to ~25 km southwest of Julesburg, on the Colorado/Nebraska state line, I-76 is built on late Holocene sands of the Fort Morgan dune field. This dune field parallels the South Platte River (which provided the source sediments) for more than 100 km northeast of Brush (Fig. 1). Near Julesburg, we turn south on U.S. 385. Note the exposure of carbonate-cemented sand and gravel in Miocene Ogallala Group rocks behind the gas station on the right immediately after turning onto highway 385. Lugn (1968) and Hunt (1986) considered sediments derived from Ogallala Group rocks to be the most important source of eolian sands on the Great Plains, a concept no longer supported, at least for dunes in northeastern Colorado (Muhs and others, 1996). About 10 km south of the turnoff onto 385, we ascend the High Plains surface on Ogallala Group rocks. Scott (1978) mapped loess on much of this surface. New field work by Muhs and others (1999) confirms the presence of this loess, but it is patchy and frequently only about a meter thick. About 9 km south of Holyoke, we enter the Wray dune field.

STOP 2: GEOMORPHOLOGY OF THE WRAY DUNE FIELD

This stop is an overview of landforms in the Wray dune field, the largest eolian sand body in Colorado and southwestern Nebraska. Sediments in this dune field range from medium to fine sand, and show a general northwest-to-southeast fining, consistent with dune orientations that indicate northwesterly winds at the time of formation (Fig. 5). At this stop, we are near the northern edge of the Wray dune field, but the landscape here has some of the finest examples of parabolic dune forms in eastern Colorado. At the intersection of the county line and highway 385, we are standing on the left arm (as you look downwind) of a small parabolic dune that is part of a much larger compound parabolic dune (Fig. 6). Other small dunes that are part of this megadune are visible to the southwest and U.S. Highway 385 itself cuts through two arms of such a dune. Simple parabolic dunes are also found in the area, and are of approximately the same dimensions as those that make up the compound dunes (Fig. 6). Soils on both the simple and compound parabolic dunes belong to the Valent series, an Ustic Torripsamment with a simple A/AC/C profile, indicating relatively young deposits, similar to those in the Fort Morgan area.

Interdune areas in this region are occupied by eolian sand sheets, such as the one with the center-pivot irrigation system visible to the southwest. These eolian sands have much better developed soils with A/Bt/Btk/C profiles, indicating a considerably greater age than late Holocene. Although stratigraphic studies have yet to confirm it, it is likely that, as with the Fort Morgan dune field, these older eolian sheet sands underlie the parabolic dunes.

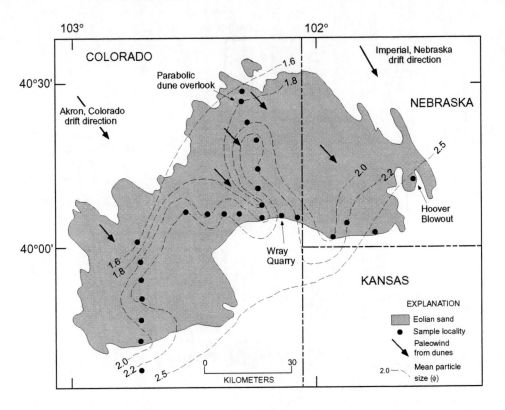

Figure 5. Map of the Wray dune field, Colorado and Nebraska, locations of STOPS 2, 3, and 5, and mean particle sizes for late Holocene eolian sands. Particle size data are plotted geographically for the first time here, but are from Muhs and others (1996).

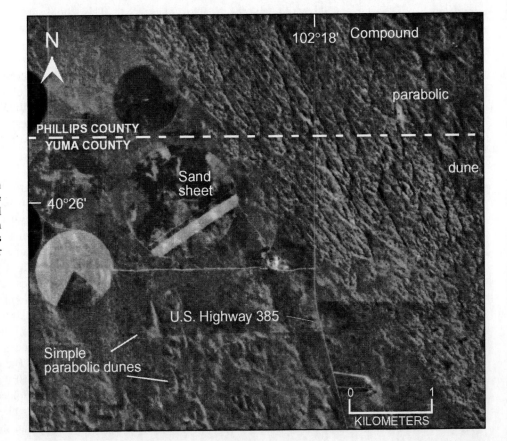

Figure 6. Aerial photograph of a portion of the northern part of the Wray dune field (STOP 2), showing simple and compound parabolic dunes and eolian sheet sands. Note erosion of sheet sands since last cultivation of area in circular field.

STOP 3: STRATIGRAPHY AND SEDIMENTOLOGY OF A PARABOLIC DUNE NEAR WRAY, COLORADO

Having seen the geomorphology of the Wray dune field at its northern end, at the next stop we will examine the stratigraphy and sedimentary structures in the left arm (looking downwind) of a parabolic dune in a quarry exposure immediately north of the North Fork of the Republican River, near the town of Wray, Colorado (Figs. 7, 8). The lowermost unit is a medium-to-fine sand that is thinly bedded and has apparent gentle (5-7°) dips to the east. About 3 m of this eolian sand is exposed and it is characterized by prominent clay lamellae (also called "clay bands," cf., Gile, 1985 and "dissipation structures," cf. Ahlbrandt and Fryberger, 1980) in the lower part of the exposure. At least nine bands, 0.5-1.5 cm thick, can be found over an ~80-cm depth zone. The lowermost eolian unit has a minimally developed A/AC/C profile, but one which is more than a meter thick. Organic matter content in this soil is extremely low and not much different from the underlying sand (Fig. 7), but it is easily identifiable by its darker colors (10YR 5/2, dry), the presence of roots, and abundant krotovina (infilled rodent-sized burrows). Its combination of minimal horizon development (i.e., lack of B horizon formation) but relatively great thickness suggests that the final stages of deposition of eolian sand were slow, such that pedogenesis kept pace with sedimentation. The contact between the buried soil and the younger eolian unit above it is unusually sharp, suggesting that the soil profile, thick as it is, has been truncated by later deflation.

The middle eolian unit is light brown or light yellowish brown (10YR 6/3, 6/4, dry) sand about 5 m thick and also has beds that dip gently (7-10°) to the east (Fig. 7). Particle size is extremely variable and ranges from coarse to very fine sand. "Pin stripe" or "wood grain" ripple strata 1-3 mm thick are common, particularly in the upper part of the unit. Secondary structures are also common. Root casts are visible in the upper beds

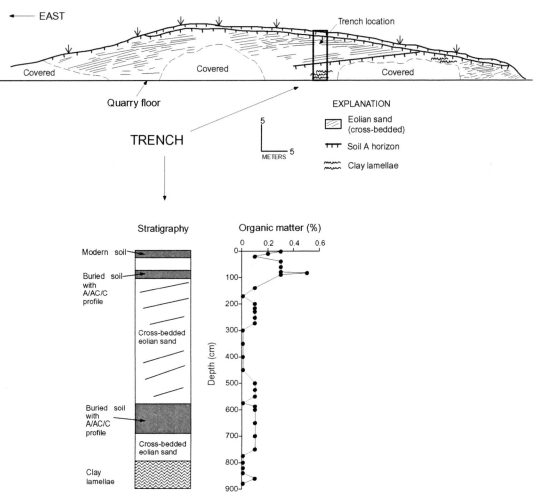

Figure 7. Upper: stratigraphy and structures of the left arm of a parabolic dune exposed in a quarry at Wray, Colorado (STOP 3), and location of trench shown in detail. Lower: detail of trench stratigraphy shown above, and organic matter content (done by the Walkley-Black method) of eolian sand and soils shown as a function of depth. Previously unpublished data of the authors.

Figure 8. Aerial photograph of the Wray, Colorado area with surficial geology superimposed and location of quarry (STOP 3). Previously unpublished geologic mapping by D.R. Muhs.

and suggest that some vegetation was present at the time of sedimentation. In addition, possible bison hoof prints (cf. Loope,
1986) can be seen about 0.5 m above the lower paleosol and
possibly in the some of the finer-grained strata upsection. The
middle eolian unit is capped by a thin (10-15 cm), dark (10YR
5/2, dry) paleosol with an A/AC/C profile that has higher
organic matter content than the underlying sand (Fig. 7). A
meter or less of very young eolian sand occurs above the thin
paleosol and could be historic, because only the simplest of A
horizons has developed in the upper 5 cm or so.

The stratigraphy and degree of soil development in this

dune suggests that all three eolian units were probably
deposited in the late Holocene. The lowermost paleosol closely
resembles those in a similar stratigraphic position in the Fort
Morgan dune field that have radiocarbon ages of ~800 to 1400
yr BP (Madole, 1994, 1995). The thick, lower paleosol, and the
presence of root casts in primary structures above it, suggest
that vegetation colonization with soil development and sedimentation were competing processes. Despite this, the volume
of sand that was transported in late Holocene time is considerable, given the thickness seen in section here and the size of the
dune itself.

South of Wray, the North Fork of the Republican River has cut magnificent cliffs into Ogallala Group rocks, underlain by White River Group rocks (Fig. 8). These cliffs, which are visible to the south of STOP 3, are mantled with loess but little or no eolian sand. Dune sand of the Wray dune field apparently did not cross the North Fork of the Republican River. However, loess thickness here is considerable, and is the subject of the next stop.

STOP 4: LOESS STRATIGRAPHY AT BEECHER ISLAND, COLORADO

Loess in northeastern Colorado is the westernmost part of an almost continuous loess blanket in the North American mid-continent. Loess is distributed widely but discontinuously to the southeast of the South Platte River, with only isolated occurrences to the north of this major drainage (Figs. 1 and 9). The thickest loess we have observed in eastern Colorado is about 12 m. Thicknesses of 2-5 m are more typical, and in the northeasternmost part of the state, thin loess occurs in a patchy distribution on the surface of Ogallala Group bedrock.

Mean particle sizes of loess in northeastern Colorado vary from fine silt in the western part of the area to coarse silt in the eastern part of the area (Fig. 9). This eastward coarsening is in part a function of decreasing clay content to the east, which is as high as 30-40% in the western part of the region and less than 10% near Julesburg (Muhs and others, 1999). The latter workers suggested that much of the clay in eastern Colorado loess may have been eroded from high-clay bedrock units such as the Pierre Shale, which are exposed more widely in the west than they are in the east, and where deflation hollows have been mapped (Colton, 1978). This clay may have been transported as silt-sized clay aggregates. Silt-sized grains in eastern Colorado loess require other sources, which we discuss below.

A roadcut near Beecher Island, Colorado (Fig. 9), has one of the thickest exposures of loess yet reported for eastern Colorado. Buried soils can be found in this section and are identifiable on the basis of morphology, organic matter maxima, CaCO$_3$ minima, and clay maxima (Fig. 10). At this locality, grey-green calcareous clays of unknown origin (not visible in the roadcut, but accessible by augering) are overlain by eolian (?) sands and silts in which a strongly expressed buried soil

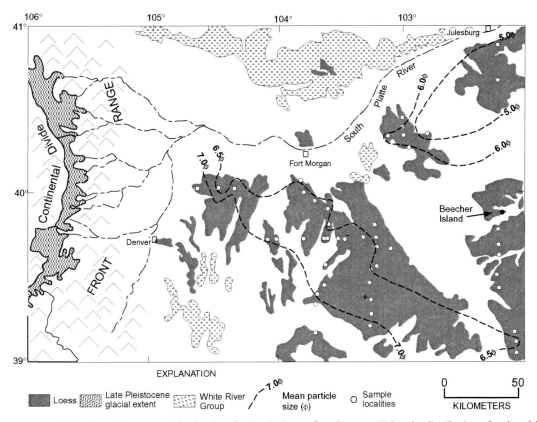

Figure 9. Distribution and mean particle size data for Peoria loess of northeastern Colorado, distribution of rocks of the White River Group, and extent of last-glacial (Pinedale) glaciers on the east side of the Continental Divide. Particle size data are from Muhs and others (1999), but are plotted geographically for the first time here; plus/minus symbols indicate anomalous samples. Distribution of White River Group rocks from Scott (1978); extent of Pinedale glaciers from Madole and others (1998).

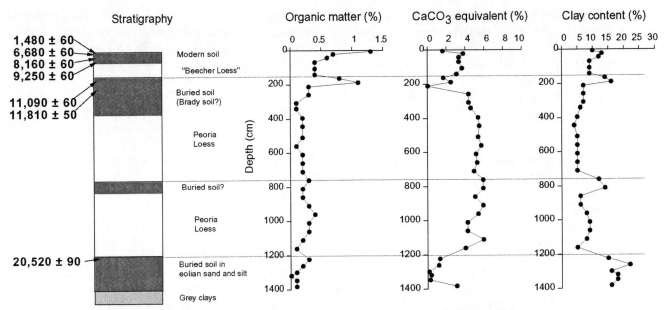

Figure 10. Stratigraphy, AMS radiocarbon ages, organic matter content, CaCO₃-equivalent content, and clay content as a function of depth in the loess section at Beecher Island, Colorado (STOP 4). From Muhs and others (1999).

developed. This buried soil has a subangular-to-angular blocky structure with well-expressed clay films, and up to 22% clay in the Bt horizon. Approximately 10 m of what is interpreted to be Peoria Loess overlies this buried soil, although a possible thin buried soil, with strong, coarse prismatic structure and 12-14% clay, is found at a depth of ~8 m. Between a depth of ~1.5 m and 3.5 m, there is a thick, buried soil with an A1/A2/AB/Bw1/Bw2/C profile (Fig. 10) that may be equivalent to the Brady soil of Nebraska and Kansas. This buried soil is in turn overlain by stratified eolian silt and sand with a modern soil, characterized by an A/Bw1/Bw2/C profile, in its upper part. Land snails (*Succinea grosvenori* Lea) can be found in both the upper part of the Peoria Loess and in the younger loess above the uppermost buried soil.

AMS radiocarbon dating of carefully extracted humic acids from paleosols, following the methods given in Abbott and Stafford (1996) and reported by Muhs and others (1999), provides a chronology of loess deposition at Beecher Island (Fig. 10). Humic acids from the upper part of the lowermost buried soil give a radiocarbon age of 20,520 ± 90 ^{14}C yr BP, and those from the A1 and A2 horizons of the buried soil (between 1.5 and 3.5 m depth) give ages of 11,090 ± 60 and 11,810 ± 50 ^{14}C yr BP (~13,000 and ~13,700 cal yr BP), respectively. Collectively, the radiocarbon ages indicate that Peoria loess deposition occurred between about 20,000 and 12,000 ^{14}C yr BP. AMS radiocarbon age determinations were also made on humic acids from the A, Bw1, Bw2, and upper C horizons of the modern soil, in order to provide a minimum age for the youngest, sandy loess at this locality. These ages are consistently younger up through the modern soil profile, and suggest that the youngest loess was deposited between about 11,000 and 9,000 ^{14}C yr BP

(~13,000 and 10,000 cal yr BP). We had originally assumed that this younger loess was the Bignell Loess of Nebraska and Kansas (seen at the type locality, STOP 6), but the radiocarbon age of the deepest part of the modern soil suggests either that it is slightly older or that deposition of Bignell Loess was time-transgressive.

The complex origin of silts in loess of eastern Colorado, alluded to earlier, can be demonstrated by examination of Pb-isotopic data from Peoria Loess at Beecher Island. One likely source of silt-sized particles during the last glacial period in Colorado is glaciogenic silt derived from alpine glaciers of Pinedale age in the Front Range. Rock flour from Front Range glaciers likely would have been carried by the South Platte River and its tributaries to the Great Plains and deposited in what are now sediments of the Qt1-Broadway-Kersey terrace. Such an origin was proposed for western Great Plains loess by Bryan (1945), Frye and Leonard (1951), Swineford and Frye (1951), and Pye and others (1995). A less obvious source of loess is the White River Group of Eocene-Oligocene age (Fig. 9), which is rich in silt-sized particles. These two sources can be distinguished from one another because the Pb-isotopic compositions of K-feldspars are distinctly different (Aleinikoff and others, 1999). K-feldspars in modern South Platte River silts, derived from Precambrian crystalline rocks of the Front Range (ages of 1.0, 1.4, and 1.7 Ga), have [206]Pb/[204]Pb values of 17.0 to 17.7, whereas silt-sized K-feldspars of the White River Group (age of volcanism is ~34 Ma) have [206]Pb/[204]Pb values that are much more radiogenic, and range from 18.1 to 19.6. In addition, U-Pb ages of silt-sized zircons from these two sediment groups are distinctly different (Proterozoic vs. Tertiary).

At Beecher Island, K-feldspars from Peoria Loess have Pb-

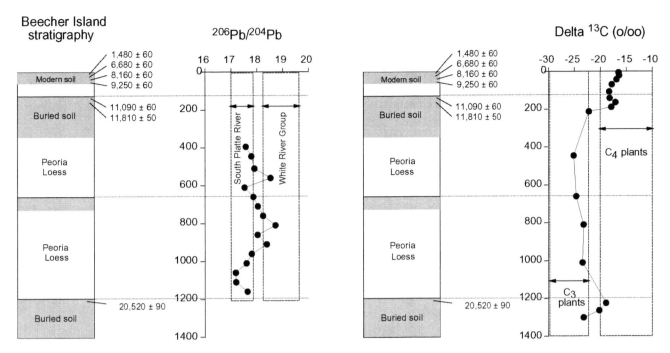

Figure 11. Pb isotopic composition of K-feldspars, ranges of these values in modern South Platte River sediments and sediments of the White River Group, and carbon isotopic composition of soil organic matter and detrital organic matter in loess at Beecher Island. Pb isotope data from Aleinikoff and others (1999); carbon isotope data from Muhs and others (1999).

isotopic compositions that span the entire range of ratios measured in both possible sources (Fig. 11). This indicates that loess was derived from both silt from the South Platte River (and therefore could be glaciogenic) and silt from the White River Group. The isotopic ratios vary systematically within the loess section at Beecher Island and within another section found farther west, near Last Chance, Colorado (Aleinikoff and others, 1999). In both sections, loess just above the ~20,000 yr old paleosol has Pb isotopic compositions within the range of values measured in K-feldspars from the South Platte River. Upsection, the ratios increase (to values corresponding to those found in the White River Group) and decrease twice. Aleinikoff and others (1999) suggested that under relatively cold conditions within the last (Pinedale) glacial period, valley glaciers of the Front Range advanced and glaciogenic silt derived from Proterozoic crystalline rocks was entrained within the ice, with relatively little silt released to streams during short summer ablation periods. Concomitantly, vegetation, which even now is fairly sparse on rocks of the White River Group, decreased, thereby destabilizing the volcaniclastic sediments and making them more susceptible to erosion. Although there may have been some eolian erosion directly from sediments of the White River Group, it is more likely that reduced vegetation cover would allow greater fluvial erosion and delivery to tributaries of the South Platte River. As conditions became warmer during the Pinedale glacial period, vegetation cover increased in eastern Colorado, including that on sediments of the White River Group, and Pinedale glaciers of the Front Range receded, gen-

erating greater amounts of outwash. The Pb-isotopic data from Beecher Island suggest the occurrence of an earlier cycle of warming and cooling between peak late Pinedale glaciation and final deglaciation (Fig. 11).

Carbon isotopic composition of organic matter in buried soils and loess show changes in dominant vegetation types over the last glacial-interglacial cycle (Fig. 11). The lowermost buried soil, dated at ~20,000 ^{14}C yr BP, has $\delta^{13}C$ values of -19‰ to -23‰, indicating a probable mix of C_3 and C_4 vegetation. In the overlying Peoria Loess, detrital organic matter has $\delta^{13}C$ values ranging from about -23‰ to -25‰, indicating a dominance of C_3 vegetation at the time of loess fall. However, the A horizon of the ~11,000 ^{14}C yr BP buried soil and the loess and modern soil above it have $\delta^{13}C$ values of -16‰ to -17‰, indicating a dominance of C_4 vegetation. Overall, the carbon isotopic compositions indicate a dominance of C_3 vegetation from about 20,000–12,000 ^{14}C yr BP, and a dominance of C_4 vegetation after ~12,000 ^{14}C yr BP. Two vertebrate faunal localities near Beecher Island studied by Graham (1981) provide additional details of the environment at the time of loess deposition. Fossil ungulates and rodents recovered from Peoria Loess at these localities led Graham (1981) to conclude that grassland was the predominant vegetation in eastern Colorado during the time of Peoria Loess deposition. The combined faunal and carbon isotope data indicate that, during the last glacial period, eastern Colorado supported a cool grassland, perhaps similar to that found today in southern Canada, Montana, or the Dakotas.

STOP 5: HOOVER BLOWOUT, WRAY DUNE FIELD, NEBRASKA

In the eastern part of the Wray dune field a deep deflation hollow called the Hoover blowout (Madole, 1995; Muhs and others, 1997a) exposes ~8 m of eolian sand within the nose area of a northwest-trending compound parabolic dune (Fig. 12). The eolian sands contain two buried soils with A/AC/C profiles and are underlain by pond or lacustrine sands, silts, and clays. Fossil mollusk shells (*Stagnicola palustris*) from the paludal sediments gives a radiocarbon age of 13,130 ± 295 ^{14}C yr B.P. (DIC-2198) and humus from the lowermost buried soil gives a radiocarbon age of 7870 ± 240 ^{14}C yr B.P. (DIC-2270) (Madole, 1995). Impressions that we interpret as bison hoof marks (cf. Loope, 1986) can be found in the middle eolian unit and transect primary bedding structures. At a depth of ~4.5 m, within the middle eolian unit, we recovered a long bone fragment attributable to Bison. The bone gives a carboxyl age of 290 ± 60 ^{14}C yr B.P. (490-0 cal yr B.P.) and a collagen age of 360 ± 60 ^{14}C yr B.P. (515-290 cal yr B.P.) (Muhs and others, 1997a). From all of these radiocarbon ages, we infer that eolian sedimentation began sometime after ~13,000 ^{14}C yr B.P. and was episodic, based on the presence of the two buried soils. The two most recent episodes of eolian sedimentation began sometime after ~7870 ^{14}C yr B.P., and one episode apparently occurred within the past ~500 cal yr. Thus, the data indicate that the eastern Wray dune field, like the Fort Morgan dune field and the Nebraska Sand Hills, has been active in the past 1000 cal yr, and could even have been active in historic time.

STOP 6: BIGNELL HILL, NEBRASKA

Bignell Hill, Nebraska, is one of the most famous loess sections in the midcontinent of North America. The locality has been studied by Quaternary geologists for more than 50 years, and results of this work have been reported by Schultz and Stout (1945), Frye and Leonard (1951), Dreeszen (1970), Johnson (1993), Feng and others (1994), Maat and Johnson (1996) and Muhs and others (1999). It is the type locality for the Brady soil, developed in the uppermost Peoria Loess, and the Bignell Loess, which overlies the Brady soil (Schultz and Stout, 1945).

Bignell Hill contains what may be the thickest (>50 m) late Quaternary loess section in North America. At the northern end of the roadcut, a reddish-brown paleosol can sometimes be seen cropping out on the road surface. Above this, and visible in the roadcut itself (to about 4.5 m above road level), is a well developed brown (10YR 5/4, dry) paleosol with a Btk/Bt/C profile almost a meter thick that has developed in eolian (?) silty sands (Fig. 13). Both paleosols are undated, but they may correlate to some part of the Sangamon interglacial period. Overlying these two paleosols is the Gilman Canyon Formation, consisting of two loesses that both contain minimally developed but organic-rich soil A horizons that are darker (10YR 5/2 and 4/3, dry) than the underlying and overlying loess (10YR 6/3, 6/4, dry). The

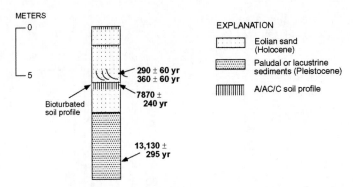

Figure 12. Stratigraphy and radiocarbon ages in eolian sands at the Hoover blowout (STOP 5). Data from Madole (1995) and Muhs and others (1997b).

Gilman Canyon Formation contains abundant evidence of burrowing in the form of krotovina. Overlying the Gilman Canyon Formation is one of the thickest (~48 m) exposures of Peoria Loess yet described from North America. This loess has distinct laminae, particularly in the upper 20 m of the unit, which Feng and others (1994) suggest may represent annual layers. At the top of the Peoria Loess, there is a distinctive paleosol (10YR 3/2 dry colors in the A horizon) with an A/Bw1/Bw2/BC/C profile called the Brady soil, first reported by Schultz and Stout (1945). The Brady soil is overlain by about 2 m of Bignell Loess (2.5Y 6/3, dry) and a modern soil.

There have been many geochronological studies of the loesses at Bignell Hill (Fig. 13). Maat and Johnson (1996) reported an age of 30,970 ± 780 ^{14}C yr BP on organic matter from the oldest of the two paleosols from the Gilman Canyon Formation. Muhs and others (1999) used the humic acid extraction method of Abbott and Stafford (1996) in an attempt to minimize contamination from both older (reworked) and younger carbon, and obtained an age of 40,600 ± 1100 ^{14}C yr BP for the same paleosol. Feng and others (1994), Maat and Johnson (1996) and Muhs and others (1999) also dated the uppermost Gilman Canyon paleosol. The earlier studies reported an age of 28,130 ± 610 ^{14}C yr BP; Muhs and others reported an age of 30,770 ± 210 ^{14}C yr BP. Maat and Johnson (1996) reported concordant total-bleach and partial-bleach TL ages of 28,300 ± 5100 cal yr BP and 28,200 ± 8400 cal yr BP, respectively, for the Gilman Canyon Formation. All age estimates for the uppermost paleosol are in good agreement with one another within analytical uncertainties and possible radiocarbon-to-calendar year calibration uncertainties (which are unknown for this time period). However, there is a significant difference (almost 10,000 yr) in the ages reported for the lowermost Gilman Canyon paleosol. Part of the difference may be due to the portion of the paleosol that was sampled in the two studies as well as the extraction methods themselves.

There are fewer ages of the Peoria Loess at Bignell Hill. Feng and others (1994) found a charcoalized twig of *Picea glauca* from the upper part of the Peoria Loess, ~3.5 m below the base of the Brady soil, and reported an age of 11,880 ± 90

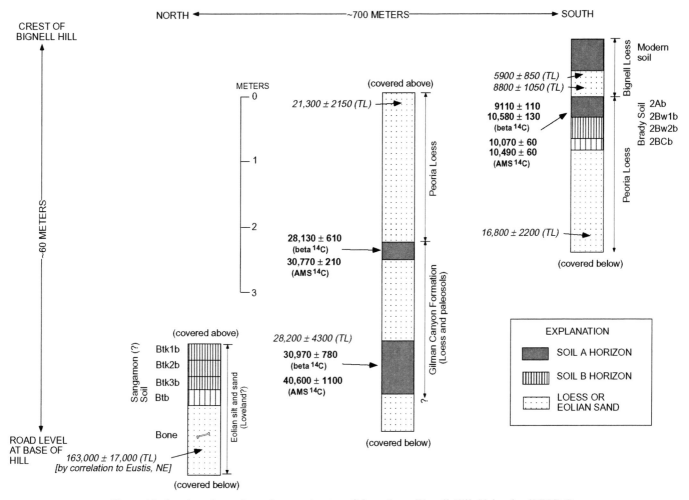

Figure 13. Stratigraphy, soils and age estimates of deposits at Bignell Hill, Nebraska (STOP 6). Stratigraphy from Maat and Johnson (1996), soils data from the present authors, and age data from Maat and Johnson (1996) and Muhs and others (1999).

[14]C yr BP. This radiocarbon age translates to a calendar-year age of ~13,600-14,200 yr BP. Maat and Johnson (1996) reported basal Peoria Loess TL ages of 21,700 ± 3200 (total bleach) and 20,900 ± 3000 cal yr BP (partial bleach). These same workers also reported TL ages for the uppermost Peoria Loess (~4 m above Feng and others' sample depth) of 17,900 ± 2500 (total bleach) and 12,400 ± 5000 (partial bleach) cal yr BP.

The Brady soil and overlying Bignell Loess have received the most attention for geochronological studies at Bignell Hill, starting with Dreeszen (1970). Johnson (1993) and Maat and Johnson (1996) redated the upper and lower parts of the Brady soil and obtained ages of 9110 ± 110 ^{14}C yr BP and 10,580 ± 130 ^{14}C yr BP, respectively. Muhs and others (1999) also redated the upper and lower Brady soil and obtained ages of 10,070 ± 60 and 10,490 ± 60 ^{14}C yr BP (11,007-12,054 and 12,176-12,590 cal yr BP), respectively. Maat and Johnson (1996) reported TL ages of 8300 ± 1200 (total bleach) and 10,400 ± 2300 (partial bleach) cal yr BP for the lower Bignell

Loess and 6100 ± 900 (total bleach) and 4500 ± 2200 cal yr BP (partial bleach) for the upper Bignell Loess. The TL ages obtained by both methods are concordant and stratigraphically consistent with the radiocarbon ages of the underlying Brady soil. Collectively, all the age estimates indicate that the Bignell Loess was deposited during the Holocene, in agreement with TL ages of this unit reported from elsewhere in Nebraska (Pye and others, 1995).

The radiocarbon and TL results from Bignell Hill indicate that while there is general agreement in the timing of last-glacial loess deposition in various parts of the midcontinent, there are differences in detail. Loess deposition in Iowa, Illinois, and areas eastward was, to a great extent, a function of source sediment availability from the Laurentide ice sheet via the Mississippi and Missouri Rivers, and the timing of loess deposition closely followed the history of movement of the ice sheet (see Grimley and others, 1998). Based on the ages from Bignell Hill, Peoria Loess deposition in western Nebraska could have begun earlier (sometime just after ~30,000 ^{14}C yr BP) and continued

84

D. R. Muhs et al.

until significantly later (~10,500 ^{14}C yr BP) than in areas to the east (Ruhe, 1983; Curry and Follmer, 1992; Grimley and others, 1998). Furthermore, eolian silt of Holocene age, such as the Bignell Loess, has not been reported from areas east of the Missouri River. These observations suggest that sources of loess in the central Great Plains are unrelated to the specific dynamics of the Laurentide ice sheet, a conclusion supported by recent isotopic data found in Aleinikoff and others (1998, 1999).

STOP 7: ELDRED CAMP, NEBRASKA SAND HILLS

The 50,000 km^2 Nebraska Sand Hills area (Fig. 1) is the largest dune field (active or stabilized) in North America. Diverse stabilized eolian landforms are found in the Nebraska Sand Hills (Smith, 1965; Ahlbrandt and Fryberger, 1980; Swinehart, 1990). Barchanoid-ridge dunes are as much as 50

km long and are easily visible on Landsat imagery. Barchans, linear dunes, parabolic dunes, dome-like dunes, and sand sheets all are found in the Nebraska Sand Hills. Measurement of slipfaces of stabilized dunes and high-angle foreset bed dip azimuths indicate that paleowinds originated from the northwest and north, similar to modern wind regimes (Warren, 1976; Ahlbrandt and Fryberger, 1980). In the next two stops we will examine the field evidence for at least two episodes of eolian sand movement that were dramatic enough to dam drainages within the western Nebraska Sand Hills (Figs. 1 and 14).

At Eldred Camp (STOP 7), the steep groundwater gradient between Crescent Lake and the springs at the head of Blue Creek (1:115) contrasts with the 1:450 gradient of Blue Creek and the 1:1100 slope of the groundwater table north of the lake (Fig. 14). Crescent Lake, the southernmost of hundreds of lakes in the western Sand Hills, lies 3 km north of and 25 m higher

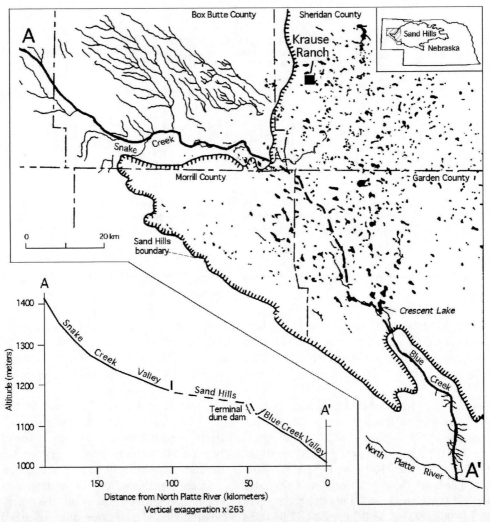

Figure 14. Western Nebraska Sand Hills showing lakes and present Snake Creek and Blue Creek drainages. Section A-A' was drawn down the primary Snake Creek and Blue Creek valleys and across the Sand Hills (dashed line) along a probable trace of the dune dammed valley. The position of the buried valley is well known only in the area of Crescent and Swan lakes (Fig. 15).

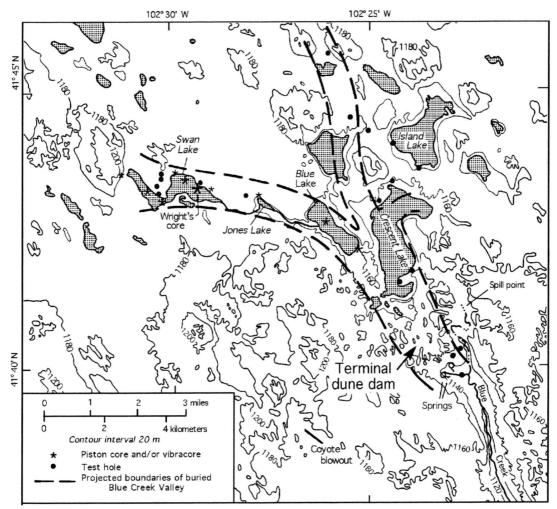

Figure 15. At the southwest margin of the Sand Hills, springs at the head of Blue Creek (lower right) emerge from a sand dam, the southernmost of numerous sand bodies that block this major paleovalley. Configuration of the buried valley system is based on vibracores, rotary drilled test holes, outcrops, long axes of lakes, and the dimensions of Blue Creek valley. The high water table behind the dam creates lakes in interdune positions. None of the lakes presently has natural surface water inlets or outlets. Note position of the intermittent stream course just east of the terminal dune dam that acted as a spillway when the level of Crescent Lake was about 2 m higher.

than the springs at the head of Blue Creek (Figs. 14 and 15). This portion of the valley of Blue Creek was not formed by headward sapping; dune sand clearly dammed a through-going extension of Snake Creek that occupied this valley (Loope and others, 1995; Mason and others, 1997).

Although dune sand is presently mounded across the position of the paleovalley, an abandoned spill point for the system lies on Ogallala Group bedrock about 1 km east of the sand-filled paleovalley, at an elevation 2 m higher than the adjacent lake surface (Fig. 15). A sinuous channel below this spill point that is cut into Ogallala Group rocks testifies to overflow during a former lake high stand. Catastrophic drainage of the lake immediately behind the dam did not take place because the spillover is floored by partially lithified material, not dune sand. Seepage through the dune dam must have taken place at a sufficient rate to prevent massive overflow and deep entrenchment.

Re-establishment of Blue Creek as a through-going stream would require that many dune dams be removed. If the southernmost dune dam were overtopped, then the southernmost lake or cluster of a lakes would drain, and the head of Blue Creek would migrate upstream several kilometers to the base of the next dam. A counterintuitive conclusion of our work is that a positive change in the water budget could cause a drastic drop in the water table throughout the catchment area due to overtopping and removal of the dune dams.

STOP 8: BLUE CREEK DUNE DAM, NEBRASKA SAND HILLS

Across broad areas of the Sand Hills, interdune surfaces intersect the groundwater table, forming extensive wetlands. Paradoxically, the part of the sand sea with the least precipita-

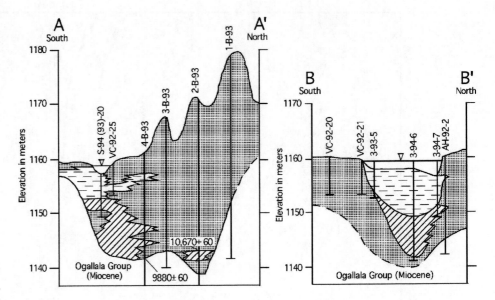

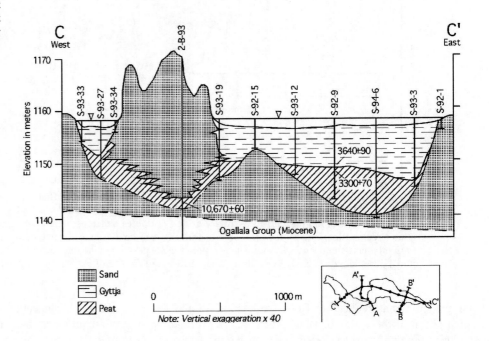

Figure 16. Swan Lake basin cross-sections constructed from lake piston cores and probes, vibracores (VC), rotary test holes (1 to 4-B-93), and an auger test hole (AH), and radiocarbon ages. After Mason and others (1997).

tion—the western Sand Hills—contains the greatest number of lakes. Estimates of the total number of interdune lakes in the region vary from 1500 to 2500.

Eastward-flowing water courses on the dune- and lake-free tableland west of the Sand Hills disappear when they reach the western margin of the sand sea (Fig. 14). At the southern edge of the Sand Hills, Blue Creek, a perennial, spring-fed stream that occupies a valley cut into Miocene bedrock, emerges from dune sand (Figs. 14 and 15). This "terminal dune dam" is the southernmost of scores of valley-blocking sand bodies; modern lakes and Holocene lacustrine and wetland sediments occupy the parts of the paleovalleys not filled by dune sand.

Based on rotary drilling, vibracores, piston cores, outcrops, the orientation of the long axes of lakes, and the dimensions of Blue Creek's valley, we project a 1200 to 2200-m-wide, 30-m-deep buried fluvial paleovalley between Swan Lake and the head of Blue Creek, 8 km southeastward (Fig. 15). Sixteen radiocarbon ages on various materials in cores from three different lake basins (Figs. 16, 17, 18) establish a chronological framework for correlation and interpretation (Loope and others, 1995; Mason and others, 1997).

At least two distinct episodes of blockage are required to explain the history of sedimentation in Swan, Blue and Crescent Lakes. Recently, we obtained five radiocarbon ages from

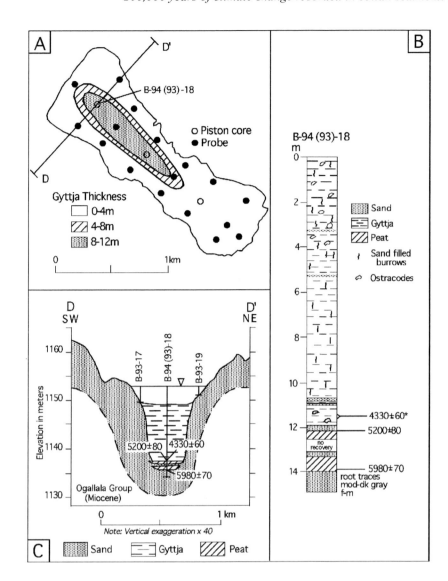

Figure 17. Blue Lake basin gyttja thickness and stratigraphy. (A) Gyttja thickness. (B) Composite section of two piston cores taken at the same location. The AMS radiocarbon age of 4330 yr BP (marked by asterisk) was obtained from plant fragments in gyttja. (C) Cross-section D-D'. After Mason and others (1997).

basal lacustrine/wetland sediments in cores taken near Krause ranch (Fig. 14), about 60 km north-northwest of Crescent Lake (Sweeney and others, 1998). These sites were part of the ancestral Blue Creek drainage basin prior to the latest Wisconsin blockage. The radiocarbon ages range from 12,160 to 12,360 ^{14}C yr BP and suggest dune blockage of the drainage just prior to 12,000 radiocarbon years ago. It appears that a significant arid interval well after the last glacial maximum led to a major episode of dune blockage in the Blue Creek drainage. We postulate that the blockage at Swan Lake formed prior to 10,600 ^{14}C yr BP and possibly as early as 12,000 ^{14}C yr BP since the 10,600 ^{14}C age comes from the upper part of a 2-m-thick peat. The dune dams that created Blue and Crescent basins were emplaced during remobilization of this part of the dune field during the mid-Holocene (about 6000 ^{14}C yr BP).

The second blockage event most likely reflects the middle Holocene period of minimum effective moisture in the Great Basin and Colorado Plateau summarized by Thompson and others (1993). Lake and pollen data from the north-central U.S. also support a middle Holocene period of minimum effective moisture (Webb and others, 1993). Holliday (1989) gave evidence for a prolonged drought and widespread eolian activity on the Southern High Plains between 6500 and 4500 ^{14}C yr BP, while Stokes and Swinehart (1997) presented direct evidence of middle Holocene eolian activity in the northern Sand Hills based on an optically stimulated luminescence age of ~5700 cal yr BP.

We postulate that as the Sand Hills area became increasingly arid and vegetation became sparse, dune sand would have covered a larger and larger proportion of the interfluves. Infiltration of precipitation into the permeable dune sand would have reduced the magnitude of runoff events, and as rainfall diminished further, surface flow would have ceased when subsurface flow through the unconsolidated, highly per-

88   D. R. Muhs et al.

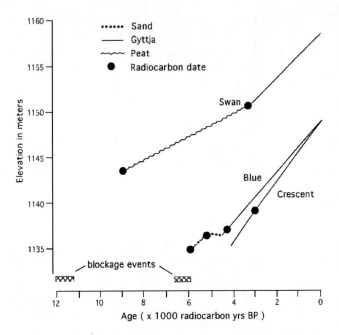

Figure 18. Estimated sedimentation rates and water level changes in the three lake basins based on radiocarbon ages and sedimentary facies distribution from piston cores. After emplacement of the initial dune dam 10,500 to 12,000 ^{14}C yr BP, peat accumulation in Swan Lake basin kept pace with the water table rise. Following a second episode of dune blockage south of Blue and Crescent Lakes, the water table gradient was lowered through the Swan basin dam and led to the formation of an open-water lake. Lacustrine deposition in Swan Lake basin lagged about 1000 years behind that of Blue and Crescent basins.

locally controlled changes in the rate of groundwater flow and requires two arid episodes during which dunes blocked dry stream courses.

Marshes initially formed in a dune-blocked (western) arm of the paleovalley system prior to about 12,000 ^{14}C yr ago and thereafter steadily aggraded for more than 6000 years. Ingram (1982) showed that, in areas of much higher precipitation, peat accumulation impedes the drainage of rain water, resulting in the growth of groundwater mounds and domed mires. We suggest that in the Sand Hills, peat accumulation did not passively keep pace with the rise of the water table, but rather that the deposition of this impermeable material actively contributed to the rise of the water table by progressively impeding the down-gradient movement of water through the valley-blocking dune sand. About 6000 ^{14}C yr BP a second blockage event took place down flow from Swan Lake and led to emplacement of the dune dams southeast of the present Blue and Crescent Lakes. Crescent and Blue Lakes formed about 4000 years ago (Figs. 18 and 19) and Swan Lake about 1000 years later. The lag in lake sedimentation reflects the time needed for the regional water table to rise about 11 m between the Blue and Swan Lake basins.

After 11,000 years of sedimentation and water table rise, Swan Lake is now nearly full of impermeable sediment. The difference in elevation between its surface and the water table beneath the dunes to the north and south (transverse to the paleovalley) is very small. This situation, combined with the low elevation of the now-abandoned spill point on the east side of Crescent Lake (Fig. 15), indicates that the water table rise cannot be sustained very far into the future. The present extent of wetlands in the study area is therefore near both the maximum for Holocene time and the maximum possible.

We postulate that the paleovalleys control the wide variation of lake water chemistry in the western Sand Hills: the fresh-water lakes at the southern margin of the sand sea, where the gradient of the groundwater table is steep (Winter, 1976), are flow-through lakes that lose salts to the springs at the head of Blue Creek. Only short segments of thick, sand-filled paleovalleys lie between these lakes and the discharge point. We interpret the alkaline, saline lakes to the north, however, as discharge points for closed, local groundwater flow systems (Toth, 1962; Gosselin and others, 1994): these lakes cannot lose salts to the regional aquifer because they occupy an area with a low hydraulic gradient and because groundwater flow through the thinner valley fills in the north is impeded by the thick, impermeable mud deposited in lakes to the south.

The capillary fringe in sand is thin and sedimentary structures show that upland vegetation was very sparse when dunes were active; the diminished stream flow that allowed blockage cannot be explained by an increase in evapotranspiration. The blockages of valleys by dune sand—like the giant bedforms themselves—testify to long periods during which precipitation was much less than at present.

meable alluvium on the floor of the channel (derived in large part from the Broadwater Formation) and the underlying poorly consolidated sediments of the High Plains aquifer could accommodate all the water input to the drainage basin. Given the high hydraulic conductivity of the materials that underlie the Sand Hills, it seems likely that prolonged drought would have eventually eliminated surface flow in many reaches of the streams of the region. The sand-carrying capacity of the wind is reduced in the lee of obstacles to air flow such as cliffs and stream banks (Greeley and Iversen, 1985). This tendency probably led to preferential deposition of eolian sand in valleys. Large masses of dune sand then moved into blocking positions on the dry floors of streams that lacked the potential to generate flash floods.

Thick peat deposits beneath the floor of Swan Lake (Fig. 16) indicate that a slow, steady rise of the water level in the basin began in early Holocene time and continued for over 6000 years. Gyttja above the peat indicates that this slow rise was followed by a more rapid rise that, at about 3700 years ago, created the open-water conditions present today (Fig. 18). Rather than interpreting the 17-m rise of the wetland surface during Holocene time as a result of a long-term trend toward a wetter regional climate (Wright and others, 1985), our working hypothesis for the history of the Swan Lake basin calls upon

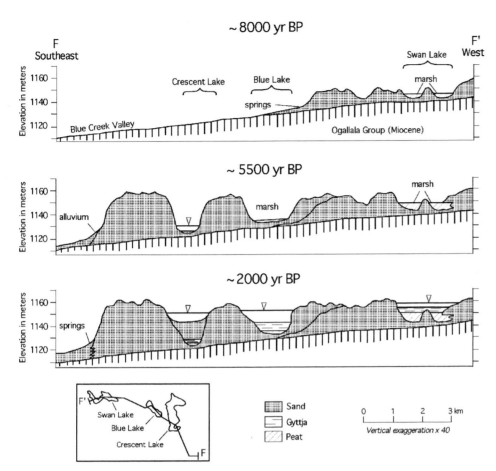

~ 8000 yr BP

~ 5500 yr BP

~ 2000 yr BP

Figure 19. Schematic cross sections (F-F') along the west axis of Blue Creek Valley from Swan Lake through Crescent Lake at three times in the Holocene. Dune sand blocked the valley between the present Swan Lake and Blue Lakes sometime during 10,500 to 12,000 years ago. (8000 yr BP)—The regional water table was high enough for about 2 m of peat to accumulate in two separate areas of Swan Lake basin. Downstream from the dune dam, Blue Creek Valley is unblocked. A second episode of blockage probably occurred prior to 5500 years ago and created Blue and Crescent basins. Some dune sand advanced into parts of Swan Lake basin but was stopped from reaching the central parts of the basin by riparian vegetation. (5500 yr BP)—Between 5000 and 6000 years ago less than two m of peat accumulated in marshes in Blue basin while lacustrine sediments accumulated in the deeper of the two Crescent Lake basins. Lacustrine sedimentation of gyttja began about 4000 years ago in Blue and Crescent basins followed by the partial collapse of the deep, southern Crescent basin about 1000 years later. Open water conditions lagged almost 1000 years behind the two lakes down gradient. (2000 yr BP)—All three lake basins reach their present configuration (after Mason and others, 1997).

ACKNOWLEDGMENTS

Most of this work was supported by the Earth Surface Dynamics Program of the U.S. Geological Survey (USGS authors) and the National Science Foundation (University of Nebraska authors). We thank Vic and Martha Eldred, owners of Crescent and Swan Lakes, for providing hospitality, encouragement and unlimited access to their ranch, and Sonny and Meryl Ritchey for access to the parabolic dune quarry at Wray. Thanks also go to Barney Szabo (USGS, retired) for the U-series analysis of carbonates at Eustis, Nebraska and Ruth Ann Lucas and Jeannine Honey (USGS) for last-minute assistance. Dave Lageson, Tom Ager and Walt Dean read an earlier version of this guidebook and made helpful comments for its improvement.

REFERENCES CITED

Abbott, M.B., and Stafford, T.W., Jr., 1996, Radiocarbon geochemistry of modern and ancient Arctic lake systems, Baffin Island, Canada: Quaternary Research, v. 45, p. 300-311.

Ahlbrandt, T.S., and Fryberger, S.G., 1980, Eolian deposits in the Nebraska Sand Hills: U.S. Geological Survey Professional Paper 1120-A, 24 p.

Ahlbrandt, T.S., Swinehart, J.B., and Maroney, D.G., 1983, The dynamic Holocene dune fields of the Great Plains and Rocky Mountain basins, U.S.A., in M.E. Brookfield and T.S. Ahlbrandt, eds., Eolian sediments

and processes: New York, Elsevier, p. 379-406.

Aleinikoff, J.N., Muhs, D.R., and Fanning, C.M., 1998, Isotopic evidence for the sources of late Wisconsin (Peoria) loess, Colorado and Nebraska: Implications for paleoclimate, in Busacca, A.J. (ed.), Dust Aerosols, Loess Soils and Global Change. Washington State University College of Agriculture and Home Economics, Miscellaneous Publication No. MISC0190, Pullman, WA, p. 124-127.

Aleinikoff, J.N., Muhs, D.R., Sauer, R.R., and Fanning, C.M., 1999, Late Quaternary loess in northeastern Colorado, II: Pb isotopic evidence for the variability of loess sources: Geological Society of America Bulletin, in press.

Arbogast, A.F., 1996, Stratigraphic evidence for late-Holocene aeolian sand mobilization and soil formation in south-central Kansas, U.S.A.: Journal of Arid Environments, v. 34, p. 403-414.

Bryan, K., 1945, Glacial versus desert origin of loess: American Journal of Science, v. 243, p. 245-248.

Colton, R.B., 1978, Geologic map of the Boulder-Fort Collins-Greeley area, Colorado: U.S. Geological Survey Miscellaneous Investigations Series Map I-855-G, scale 1:100,000.

Curry, B.B., and Follmer, L.R., 1992, The last interglacial-glacial transition in Illinois: 123-25 ka, in The last interglacial-glacial transition in North America (P.U. Clark and P.D. Clark, eds.): Geological Society of America Special Paper 270, p. 71-88.

Dreeszen, V.H., 1970, The stratigraphic framework of Pleistocene glacial and periglacial deposits in the Central Plains, in Dort, W., Jr., and Jones, J.K., Jr., Pleistocene and recent environments of the Central Great Plains: Lawrence, University of Kansas Press, p. 9-22.

Feng, Z., Johnson, W.C., Lu, Y., and Ward, P.A., III., 1994, Climatic signals from loess-soil sequences in the central Great Plains, USA: Palaeogeography, Palaeoclimatology, Palaeoecology, v. 110, p. 345-358.

Forman, S. L., and P. Maat, 1990, Stratigraphic evidence for late Quaternary dune activity near Hudson on the piedmont of northern Colorado: Geology v. 18, p. 745-748.

Forman, S.L., Goetz, A.F.H., and Yuhas, R.H., 1992a, Large-scale stabilized dunes on the High Plains of Colorado: Understanding the landscape response to Holocene climates with the aid of images from space: Geology, v. 20, p. 145-148.

Forman, S.L., Bettis, E.A., III, Kemmis, T.J., and Miller, B.B., 1992b, Chronologic evidence for multiple periods of loess deposition during the late Pleistocene in the Missouri and Mississippi River valley, United States: Implications for the activity of the Laurentide Ice Sheet: Palaeogeography, Palaeoclimatology, Palaeoecology, v. 93, p. 71-83.

Forman, S.L., Oglesby, R., Markgraf, V., and Stafford, T., 1995, Paleoclimatic significance of Late Quaternary eolian deposition on the Piedmont and High Plains, Central United States: Global and Planetary Change, v. 11, p. 35-55.

Frye, J.C., and Leonard, A.B., 1951, Stratigraphy of the late Pleistocene loesses of Kansas: Journal of Geology v. 59, p. 287-305.

Gile, L.H., 1985. The Sandhills project soil monograph: Las Cruces, Rio Grande Historical Collections, New Mexico State University.

Gosselin, D.C., Sibray, S., and Ayers, J., 1994, Geochemistry of K-rich alkaline lakes, western Sand Hills, Nebraska, USA: Geochimica et Cosmochimica Acta, v. 58, p. 1403-1418.

Graham, R.W., 1981, Preliminary report on late Pleistocene vertebrates from the Selby and Dutton archeological/paleontological sites, Yuma County, Colorado: University of Wyoming Contributions to Geology, v. 20, p. 33-56.

Greeley, R., and Iversen, J.D., 1985, Wind as a geological process on Earth, Mars, Venus, and Titan: Cambridge University Press, 333 p.

Grimley, D.A., Follmer, L.R., and McKay, E.D., 1998, Magnetic susceptibility and mineral zonations controlled by provenance in loess along the Illinois and central Mississippi River valleys, Quaternary Research, v. 49, p. 24-36.

Holliday, V.T., 1987, Geoarchaeology and late Quaternary geomorphology of the middle South Platte River, northeastern Colorado: Geoarchaeology, v. 2, p. 317-329.

Holliday, V.T., 1989, Middle Holocene drought on the southern High Plains: Quaternary Research, v. 31, p. 74-82

Holliday, V.T., 1995a, Late Quaternary stratigraphy of the Southern High Plains, in E. Johnson, ed., Ancient peoples and landscapes: Lubbock, TX, Museum of Texas Tech University, p. 289-313.

Holliday, V.T., 1995b, Stratigraphy and paleoenvironments of late Quaternary valley fills on the Southern High Plains: Geological Society of America Memoir 186, 136 p.

Holliday, V.T. 1997a, Origin and evolution of lunettes on the High Plains of Texas and New Mexico: Quaternary Research, v. 47, p. 54-69.

Holliday, V. T., 1997b, Paleoindian geoarchaeology of the Southern High Plains: Austin, University of Texas Press, Austin.

Hunt, C.B., 1986, Surficial deposits of the United States. New York, Van Nostrand Reinhold Company, 189 p.

Ingram, H.A.P., 1982, Size and shape in raised mire ecosystems: a geophysical model: Nature, v. 297, p. 300-303.

Johnson, W.C., 1993, Surficial geology and stratigraphy of Phillips County, Kansas, with emphasis on the Quaternary Period: Kansas Geological Survey Technical Series 1, 66 p.

Kuzila, M.S., Mack, A.M., Culver, J.R., and Schaefer, S.J., 1990, General soil map of Nebraska: Conservation and Survey Division, Institute of Agriculture and Natural Resources, University of Nebraska, Lincoln and U.S. Department of Agriculture Soil Conservation Service, scale 1:1,000,000.

Loope, D.B., 1986, Recognizing and utilizing vertebrate tracks in cross-section: Cenozoic hoofprints from Nebraska: Palaios, v. 1, p. 141-151.

Loope, D.B., Swinehart, J.B., and Mason, J.P., 1995, Dune-dammed paleovalleys of the Nebraska Sand Hills: Intrinsic versus climatic controls on the

accumulation of lake and marsh sediments: Geological Society of America Bulletin, v. 107, p. 396-406.

Lugn, A.L., 1968, The origin of loesses and their relation to the Great Plains in North America, in C.B. Schultz and J.C. Frye, eds., Loess and related eolian deposits of the world: Lincoln, NE, University of Nebraska Press, p. 139-182.

Maat, P.B., and Johnson, W.C., 1996, Thermoluminescence and new ^{14}C age estimates for late Quaternary loesses in southwestern Nebraska: Geomorphology, v. 17, p. 115-128.

Madole, R.F., 1994, Stratigraphic evidence of desertification in the west-central Great Plains within the past 1000 yr: Geology, v. 22, p. 483-486.

Madole, R.F., 1995, Spatial and temporal patterns of late Quaternary eolian deposition, eastern Colorado, U.S.A.: Quaternary Science Reviews, v. 14, p. 155-177.

Madole, R.F., VanSistine, D., and Michael, J.A., 1998, Glaciation in the upper Platte River drainage basin, Colorado: U.S. Geological Survey Geologic Investigations Series I-2644, scale 1:300,000.

Martin, C.W., 1993, Radiocarbon ages on late Pleistocene loess stratigraphy of Nebraska and Kansas, central Great Plains, U.S.A.: Quaternary Science Reviews. v. 12, p. 179-188.

Martinson, D.G., Pisias, N.G., Hays, J.D., Imbrie, J., Moore, T.C., Jr., and Shackleton, N.J., 1987, Age dating and the orbital theory of the ice ages: Development of a high-resolution 0 to 300,000-year chronostratigraphy: Quaternary Research, v. 27, p. 1-29.

Mason, J.P., Swinehart, J.B., and Loope, D.B., 1997, Holocene history of lacustrine and marsh sediments in a dune-blocked drainage, southwestern Nebraska Sand Hills, U.S.A.: Journal of Paleolimnology, v. 17, p. 67-83.

May, D.W., and Holen, S.R., 1993, Radiocarbon ages of soils and charcoal in late Wisconsinan loess, south-central Nebraska: Quaternary Research, v. 39, p.55-58.

Muhs, D.R., and Holliday, V.T., 1995, Evidence of active dune sand on the Great Plains in the 19th century from accounts of early explorers: Quaternary Research v. 43, p. 198-208.

Muhs, D.R., Stafford, T.W., Jr., Cowherd, S.D., Mahan, S.A., Kihl, R., Maat, P.B., Bush, C.A., and Nehring J., 1996, Origin of the late Quaternary dune fields of northeastern Colorado: Geomorphology v. 17, p. 129-149.

Muhs, D. R., Stafford, T. W., Jr., Swinehart, J.B., Cowherd, S.D., Mahan, S.A., Bush, C.A., Madole, R.F., Maat, P.B., 1997a, Late Holocene eolian activity in the mineralogically mature Nebraska Sand Hills: Quaternary Research, v. 48, p. 162-176.

Muhs, D. R., Stafford, T. W., Jr., Been, J., Mahan, S.A., Burdett, J., Skipp, G., and Rowland, Z.M., 1997b, Holocene eolian activity in the Minot dune field, North Dakota: Canadian Journal of Earth Sciences, v. 34, p. 1442-1459.

Muhs, D.R., Aleinikoff, J.N., Stafford, T.W., Jr., Kihl, R., Been, J., Mahan, S.A., and Cowherd, S.D., 1999, Late Quaternary loess in northeastern Colorado, I: Age and paleoclimatic significance: Geological Society of America Bulletin, in press.

Pye, K., Winspear, N.R., and Zhou, L.P., 1995, Thermoluminescence ages of loess and associated sediments in central Nebraska, USA.: Palaeogeography, Palaeoclimatology, and Palaeoecology, v. 118, p. 73-87.

Ross, J.A., 1991, Geologic map of Kansas: Kansas Geological Survey, University of Kansas Map M-23, scale 1:500,000.

Ruhe, R.V., 1983, Depositional environment of late Wisconsin loess in the midcontinental United States, in Wright, H.E., Jr., and Porter, S.C., eds., Late-Quaternary environments of the United States Volume 1, The late Pleistocene: Minneapolis, University of Minnesota Press, p. 130-137.

Schultz, C.B., and Stout, T.M., 1945, Pleistocene loess deposits of Nebraska: American Journal of Science, v. 234, p. 231-244.

Scott, G.R., 1978, Map showing geology, structure and oil and gas fields in the Sterling 1° x 2° quadrangle, Colorado, Nebraska and Kansas: U.S. Geological Survey Miscellaneous Investigations Series Map I-1092, scale, 1:250,000.

Smith, H.T.U., 1965, Dune morphology and chronology in central and western Nebraska: Journal of Geology, v . 73, p. 557-578.

Spears, C.F., Amen, A.E., Fletcher, L.A., Healey, L.R., 1968, Soil survey of Morgan County, Colorado: Washington, D.C., U.S. Government Printing Office, 102 p.

Stokes, S., and J.B. Swinehart, 1997. Middle- and late-Holocene dune reactivation in the Nebraska Sand Hills, USA: The Holocene, v. 7, p. 263-272.

Sweeney, M.R., Swinehart, J.B., and Loope, D.B., 1998, Testing the hypothesis for latest Wisconsin blockage of streams at the west margin of the Nebraska Sand Hills [abs.]: Proceedings of the Nebraska Academy of Sciences, 118th Annual Meeting, Lincoln, NE, p. 50.

Swineford, A., and Frye, J.C., 1951, Petrography of the Peoria Loess in Kansas: Journal of Geology, v. 59, p. 306-322.

Swinehart, J.B., 1990, Wind-blown Deposits, in A. Bleed and C. Flowerday, eds., An atlas of the Sand Hills: Resource Atlas No. 5a, University of Nebraska-Lincoln, p. 43-56.

Swinehart, J.B., and Diffendal, R.F., Jr., 1990, Geology of the pre-dune strata, in A. Bleed and C. Flowerday, eds., An atlas of the Sand Hills: Resource Atlas No. 5a, University of Nebraska-Lincoln, p. 29-42.

Swinehart, J.B., Dreeszen, V.H., Richmond, G.M., Tipton, M.J., Bretz, R., Steece, F.V., Hallberg, G.R., and Goebel, J.E., 1994, Quaternary geologic map of the Platte River 4° x 6° quadrangle, United States: U.S. Geological Survey Miscellaneous Investigations Series Map I-1420 (NK-14), scale 1: 1,000,000.

Thompson, R.S., Whitlock, C., Bartlein, P.J., Harrison, S.P. and Spaulding, W.G., 1993, Climatic changes in the western United States since 18,000 yr B.P., in Wright, H.E., Jr., Kutzbach, J.E., Webb, T. III, Ruddiman, W.F., Street-Perrott, F.A., and Bartlein, P.J., eds., Global climates since the last glacial maximum: Minneapolis, University of Minnesota Press, p. 468-513.

Toth, J., 1962, A theoretical analysis of ground-water flow in small drainage basins: Proceedings Hydrology Symposium 3, Groundwater, Queens Printers, Ottawa, p. 75-96.

Warren, A., 1976, Morphology and sediments of the Nebraska Sand Hills in relation to Pleistocene winds and the development of eolian bedforms: Journal of Geology, v. 84, p. 685-700.

Webb, T., III, Bartlein, P.J., Harrison, S.P., and Anderson, K.H., 1993, Vegetation, lake levels, and climate in eastern North America for the past 18,000 years, in Wright, H.E., Jr., Kutzbach, J.E., Webb, T., III, Ruddiman, W.F., Street-Perrott, F.A., and Bartlein, P.J., eds., Global climates since the last glacial maximum: Minneapolis, University of Minnesota Press, p. 415-467.

Wells, P.V., and Stewart, J.D., 1987, Spruce charcoal, conifer macrofossils, and landsnail and small-vertebrate faunas in Wisconsinan sediments on the High Plains of Kansas, in Johnson, W.C., ed., Quaternary environments of Kansas, Kansas Geological Survey Guidebook Series 5, p. 129-140

Winter, T.C., 1976, Numerical simulation analysis of the interaction of lakes and groundwater: U.S. Geological Survey Professional Paper 1001, 45 p.

Wolfe, S.A., Huntley, D.J., and Ollerhead, J., 1995, Recent and late Holocene sand dune activity in southwestern Saskatchewan: Current Research 1995B; Geological Survey of Canada, p. 131-140.

Wright, H.E., Jr., Almendinger, J.C., and Gruger, J., 1985, Pollen diagram from the Nebraska Sandhills and the age of the dunes: Quaternary Research, v. 24, p. 115-120.

Printed in U.S.A.

Geological Society of America
Field Guide 1
1999

Walking tour of paleontologist George G. Simpson's boyhood neighborhood

Léo F. Laporte

Professor Emeritus, Earth Sciences Department, University of California, Santa Cruz, California 95064

ABSTRACT

George Gaylord Simpson (1902-1984) dominated American paleontology for some five decades spanning the middle of the twentieth century. This dominance was both quantitative and qualitative, for Simpson not only published hundreds of articles, monographs, and books (his bibliography includes more than 750 entries), but his work had major impact on contemporary views of the origin, classification, and evolution of mammals; historical biogeography; principles of taxonomy and systematics; biostatistical methods; and most significantly, the formulation of the modern evolutionary synthesis. The Capitol Heights neighborhood of Denver where Simpson spent his youth includes two of the houses where he lived; the home of his childhood playmate and future wife Anne Roe; the church where he pumped the organ and whose teachings he eventually disavowed; the elementary school where he excelled as a student and suffered for it; the firehouse where the fire horse died; and the corner where he sold lemonade for spending money.

BIOGRAPHICAL BACKGROUND

Simpson was born in Chicago on 16 June 1902. He was the third and last child of Helen J. (Kinney) and Joseph A. Simpson, having been preceded in the world by his sisters, Margaret (1895-1991) and Martha (1898-1984). His father was an attorney who handled railroad claims, then became involved in land speculation and mining in the West, which resulted in the family's resettlement in Denver while Simpson was still an infant. His mother was born in Iowa and, owing to the premature death of her mother, had been brought up in Hawaii by her grandparents who were lay missionaries. Simpson's Scots ancestry and missionary background led to a strict fundamental Presbyterian upbringing, which he turned his back on by his early teens. As a boy, Simpson was curious about everything. He talked his parents into subsidizing his purchase of the now classic 11th edition of the *Encyclopaedia Britannica*, which he then read straight through. It became the foundation of what was to become a huge personal research library, and he was still using it at the end of his long life. As a boy he also kept a notebook in which he recorded random facts that he learned, including such dubiously useful information as the densities of various materials.

Simpson had just a few close friends in childhood, chiefly a neighborhood chum, Bob Roe (Fig. 1) and his sister, Anne, whom he would marry years later (Fig. 2). In old age, Simpson reminisced about his childhood and noted that being more intelligent, shorter, and redheaded had guaranteed antagonism from his peers. He was also afflicted with an eye condition that made it difficult to follow the flight of a ball—a serious handicap for virtually all sports. His father, sister Martha, and Bob Roe all enjoyed the outdoors, so Simpson preferred to spend much of his recreational time exploring the Rocky Mountain landscape, which undoubtedly fed his interest in natural history.

Simpson attended Denver elementary and high schools. Despite losing a year or so, because of eye ailments and appendicitis, he managed to skip grades and graduated from East Denver high school, close to his 16th birthday. In the fall of 1918 he entered the University of Colorado at Boulder. He called Denver his home from 1903 until 1923 when he married and began his graduate studies at Yale University.

Laporte, L. F., 1999, Walking tour of George G. Simpson's boyhood neighborhood, *in* Lageson, D. R., Lester, A. P., and Trudgill, B. D., eds., Colorado and Adjacent Areas: Boulder, Colorado, Geological Society of America Field Guide 1.

Figure 1. George G. Simpson (right) around age 10, just about when he decided he "didn't want to give up being naughty." He is with his boyhood companion Bob Roe on their homemade raft. It was Bob Roe who brought Simpson home to meet his sister Anne, when both were still very young.

Figure 2. Simpson and his wife Anne in their New York City apartment in the late 1940s. They both lived on Milwaukee Street as children. It was only years later, after both had divorced their first spouses, that they married and remained so for more than 46 years, until Simpson died in 1984.

WALKING TOUR

The Capitol Heights neighborhood is two miles east of the downtown Civic Center. A bus from downtown on the East Colfax Ave. line stops at Milwaukee St. Walk four and one-half blocks south to 1048 Milwaukee St.

STOP 1—1048 Milwaukee St. The 1908 telephone book indicates that the Simpson family was living here then, but the first home they had in the neighborhood was on Vine St. (nine blocks to the west), later torn down for the Botanical Park. Simpson lived several years here while in the lower grades of elementary school. A favorite past-time of his was to balance along the top of the fence that ran the length of the property on the north side. Two of his household chores were to shovel coal into the basement and empty the furnace ashes into the backyard ash pit. The family later moved into a larger house diagonally across the street at 1069 Milwaukee St. (Continue one-half block to the north, almost to the intersection with 11th St.)

STOP 2—1069 Milwaukee St. Here Simpson was allowed to claim the whole attic for himself where he built a model of Machu Picchu, discovered in 1911, that he copied from pictures in the *National Geographic*. From the lenses of a discarded stereopticon he also constructed a make-shift telescope that he put on the peak of the roof to look at the stars and moon. Here, too, he taught himself the international ("Morse") code and would signal with a few other amateurs in the neighborhood. Planning to go to sea, he earned a 2nd class radio operator's license as a teen and soon had an offer to ship out to South America but declined, deciding he would be better off finishing school first.

When Simpson was about 9 years old, he cajoled his parents into guaranteeing half the cost of the new 11th edition of the *Encyclopaedia Britannica* (1910-11) that he saw advertised in the local newspaper, if he in turn could come up with his half share. Before long, to his parent's surprise and financial dismay, he had

accumulated enough money through a series of odd jobs in the neighborhood—mowing lawns, selling pop, and putting coal into cellars. He then proceeded to read all 28 volumes straight through! "I think it gave me my first conception of the world of learning as a whole, my first definite feeling for organized facts, and my first inkling of how to go systematically about finding out such facts."[1]

According to his oldest sister Margaret, when the streets were torn up to put in electric lighting her brother picked up all sorts of rocks and put them in a make-shift museum in the attic. He charged a penny admission, and when he later had fossils he charged two-cents.[2] He also had a small menagerie that included lizards, horned toads, turtles, and ants—the latter raised in Mason jars and occasionally let loose to observe their behavior when liberated.

STOP 3—Northwest corner of Milwaukee & 11th St. Site of the home-made stand where Simpson and Bob Roe sold cold lemonade, pop, and buttermilk. Proceeds went to support Simpson's half-share of the purchase of the *Encyclopaedia Britannica*. (Continue north a short distance—less than one-half block— on Milwaukee St.)

STOP 4—1110 Milwaukee St. Home of his playmate Bob Roe and the latter's sister Anne (1904-1991), whom Simpson would marry in 1938 in New York City. Anne has said that she cannot remember a time when she didn't know Simpson, because Bob Roe brought him home when she was about four years old and sick in bed, and they both read to her. They were all "great chums." They had a group called "The Eight," chiefly organized by Anne that met regularly in high school and early college years, usually Sunday afternoons at the Roes or Simpsons, and they would sing with Anne at the piano. At one point Simpson asked Anne to marry him (he was 18, she 16). After first saying yes, she called the next day and said no. Later, they corresponded on and off when Simpson was at the University of Colorado. Correspondence dwindled when he was at Yale and she at Denver University, at least in part because Anne had a brief teen-age flirtation with religious fundamentalism of which Simpson strongly disapproved. Anne says her intellectual interest in those days was stimulated by Simpson's conversation about such ideas as the fourth dimension and non-Euclidean geometry. Simpson saw Anne whenever he came back to Denver, and he and her brother Bob always remained good friends.[3]

(Return to 11th St. and go one block west to the intersection with Fillmore St.)

STOP 5—Capitol Heights Presbyterian church. Simpson's mother was raised by her grandparents who were lay missionaries in Hawaii, after her own mother died when she was quite young. She was brought up in a strongly religious setting that carried over into her adult life, for Simpson claimed that as a child he "commonly attended three services on Sunday at the Capitol Heights Presbyterian Church as well as the midweek prayer meeting, and in addition had family prayers and psalm recitations"[4] Simpson's father had a Presbyterian background as well, for his own father was a Welsh Presbyterian minister. Simpson was made a formal member of the church at the age of nine, but soon after de-converted when he decided in a fit of childish peevishness that he "did not want to forsake forever being naughty." One outcome of this decision was Simpson's being put with half-a-dozen other boys in a special Sunday school class that met in the church tower, out of sight and hearing of the more devout. The local physician was put in charge of this unruly group and rather than follow the prescribed biblical lessons told of his adventures in South America. Simpson also pumped the mechanism on the church organ to maintain the air pressure when it was being played, while his father played the flute and one of his sisters sang in the choir.[5]

(Continue two blocks west on 11th St. to the intersection with Clayton St.)

STOP 6—Fire House. Simpson's route to and from school passed the fire house and he often stopped by to talk to the firemen, all of whose names he knew as well as those of the horses that pulled the fire wagons. "One of the horses was killed in the street one day, and Stevens [the school principal] tried to keep us in school until the body was removed so we could not see our friend lying dead, but I sneaked out. The wife of one of the firemen was found dead one day in her room, and although I of course knew that everyone died when they were old, I somehow had not really felt that someone I knew and was no older than my very active parents could also die. I had nightmares for a while over that."[6]

(Continue two blocks west on 11th St. to the intersection with Columbine St. Turn right—north—to Stevens Elementary School.)

STOP 7—Stevens Elementary School. The school is named after Edward Stevens who was principal during the years when Simpson attended. Simpson skipped several grades and after seventh grade, at age 11, he went directly to the East Denver Latin High School (long since torn down). One day Simpson was called in by Principal Stevens and asked what he did during recess. Simpson replied that it was marbles just then, but other times it might be trading cards or flying kites. Later, Simpson was approached by three brothers, two of whom held him while the third and youngest brother punched him. "I asked what the matter was, and they said 'the principal called them in for misbehaving and said why can't you be regular boys like Simpson, why can't you be more like him.'"[7]

Simpson claimed that "as large as school loomed in my life, my real mental growth was entirely outside it and I still feel that I never learned anything before college that I would not have learned just as quickly and well without ever entering a school....As a child I was inclined to be solitary, or to prefer the constant companionship of one or two close friends to a larger circle of acquaintances, and in fact this is still true of me. I never cared for the orthodox sports...partly due to a sense of physical inferiority, for even when as strong as my playmates I was not as skillful at games & was always a little small for my age. But it was due much more to my character...My favorite outdoor occupations were such things as roller skating, bicycling, or hiking. Because of my dislike for gangs and team play and my interest in books and handiwork, I suffered a great deal from a sense of being different from other boys, and therefore inferior to them....but real as this unhappiness was, it was only a small part of my life, even then, and my childhood was very happy on the whole. My abiding interests were always books and the childish equivalent of science, & from these my chief happiness & most important development came."[8]

NOTES

1. Autobiographical Notes, Simpson Papers, American Philosophical Society.

2. Interview with Margaret Peck Simpson, 6 June 1987, Glendale, Calif.

3. Interview with Anne Roe Simpson, 17-18 December 1985, Tucson, Ariz.

4. Simpson's autobiography, *Concession to the Improbable,* 1978, Yale Univ. Press, p. 20-21, 25-26.

5. Ibid.

6. Letter from G.G. Simpson to LFL, 16 July 1980.

7. Interview with G.G. Simpson, 2 February 1979, Tucson, Ariz.

8. Autobiographical Notes, Simpson Papers, American Philosophical Society.

Geological Society of America
Field Guide 1
1999

Active evaporite tectonics and collapse in the Eagle River valley and the southwestern flank of the White River uplift, Colorado

R. B. Scott, D. J. Lidke, M. R. Hudson, W. J. Perry, Jr., Bruce Bryant, M. J. Kunk, J. R. Budahn, and F. M. Byers, Jr.
U.S. Geological Survey, MS 913, Denver Federal Center, Denver, Colorado 80225, United States

ABSTRACT

This field trip presents field evidence for Neogene evaporite tectonism, dissolution of evaporites, and related collapse in Eagle River valley and along the southwestern flank of the White River uplift. In the Eagle collapse center, Pennsylvanian evaporite flowed to form anticlinal diapirs, dissolved, and disrupted a lower Miocene basaltic plateau originally at elevations as high as 3.35 km by tilting, faulting, and sagging to elevations as low as about 2.1 km. Also in the Eagle collapse center, the 30 x 10-km, homoclinal Hardscrabble Mountain sank into evaporite during Triassic and Permian collapse followed by Neogene(?) tilting and collapse, based on seismic reflection data. Along the southwestern flank of the White River uplift in the northwestern part of the Carbondale collapse center, parts of the Grand Hogback monocline have collapsed northeastward toward a series of strike-elongate extrusive diapirs. The volume of evaporite removed from the Eagle and Carbondale collapse centers during the Neogene (about 2,250 km^3 from an area of roughly 4,500 km^2) was calculated by measuring the departure of collapsed basalts from an assumed original basalt plateau. Regional Neogene uplift and incision of the Rocky Mountains, which locally began about 8-10 Ma, probably triggered dissolution and collapse. Presently the Colorado River removes a dissolved-solids load of about 1.4 x 10^9 kg per year from the two collapse centers.

INTRODUCTION

Structures that do not fit the fabric of contractional or extensional tectonism are commonly encountered in western Colorado between the towns of Minturn and Rifle. Previous workers have reported considerable difficulty in drawing cross sections in the area using concepts of conventional tectonism. Although evaporite tectonism has been recognized in the region for nearly half a century, the scale and character of evaporite tectonism has only recently been proposed (Scott and others, 1998; Kirkham and others, 1997b). When U.S. Geological Survey (USGS) geologists began mapping in the area in the fall of 1995, we entered into a collaborative effort with Colorado Geological Survey (CGS) geologists who had already mapped several quadrangles in the Glenwood Springs-Carbondale area. During discussions with our CGS colleagues about perplexing

structural and stratigraphic relationships they had found, both of our groups began to realize that evaporite tectonism probably played a major role in regional tectonism. The scale of evaporite tectonism became more apparent as USGS geologists began finding similar structures in the Eagle River valley and along the Grand Hogback on the southwestern flank of the White River uplift (Fig. 1).

The structures that formed during evaporite-driven tectonism, particularly in areas affected by regional contractional and extensional tectonism require fresh, innovative explanations. These are examples of seemingly baffling structures we found:

- an elliptical normal fault that dips inward toward a downdropped 30 x 10-km block of older strata;
- as much as 1.2-km changes in elevations of originally horizontal Miocene basalt flows;

Scott R. B., Lidke D. J., Hudson M. R., Perry, W. J., Jr., Bryant B., Kunk M. J., Budahn J. R., and Byers, F. M., Jr., 1999, Active evaporite tectonics and collapse in the Eagle River valley and the southwestern flank of the White River uplift, Colorado, *in* Lageson, D. R., Lester, A. P., and Trudgill, B. D., eds., Colorado and Adjacent Areas: Boulder, Colorado, Geological Society of America Field Guide 1.

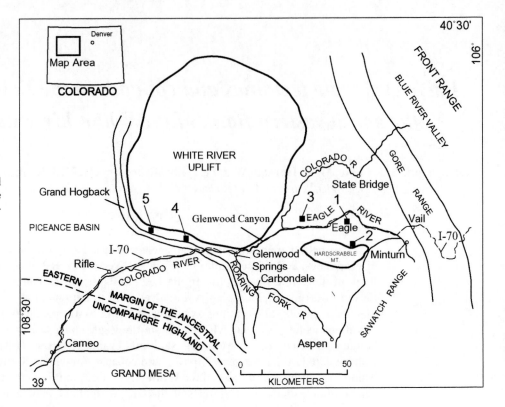

Figure 1. Map showing geologic and geographic features. Field trip stops are numbered. Modified from Scott and others (1998).

- hanging-wall strata of a large, south-dipping normal fault that dip steeply southward, whereas footwall strata that dip steeply north, forming a puzzling tepee-shaped structure across the fault;
- a tight syncline in a young conglomerate in contact with evaporite;
- local anticline-syncline pairs that deform the otherwise consistently steep-dipping Grand Hogback monocline;
- isolated low-angle normal faults along the southwest flank of the White River uplift.

Here are some of the factors and concepts we used to solve these seeming enigmas:

- Halite and gypsum have unique physical properties: they readily flow under gravitational forces, they are about 10% less dense than most clastic sedimentary rocks, and they are readily soluble. These properties allow a great degree of freedom for structural interpretation compared to interpretation of conventional sedimentary rocks.
- One of the most valuable tools to use in study of evaporite tectonism is that of a datum plane, from which the amount of deformation can be estimated. Fortunately, we have such a datum, a 10-25-Ma basaltic plateau emplaced before the latest phase of evaporite deformation (Scott and others, 1998; Kirkham and others, 1997b).
- Finally, nowhere does rock/water interaction play a more profound tectonic and hydrologic role than in evaporite tectonics

where dissolution and flow of evaporite control tectonism. Not only can evaporite dissolution control the morphology of the land surface, style of structures, rate of tectonism, and nature of geologic hazards, but it also strongly affects the chemistry of ground and surface waters indicating the rate of modern dissolution (Scott and others, 1998).

GEOLOGIC SETTING

Western Colorado geology provides evidence for a complex history that includes Early and Middle Proterozoic tectonic and magmatic events, Ancestral Rocky Mountain crustal shortening and uplift, Laramide crustal shortening, uplift, and magmatism, middle Tertiary extension and magmatism, and Neogene uplift, extension, and evaporite tectonism. The continental crust in Colorado was produced by the accretion of magmatic arcs and inter-arc basins 1.8-1.655 Ga, followed by 1.45-1.00-Ga Middle Proterozoic magmatism that is interpreted to have been produced mostly by anatectic melting of preexisting continental crust (Reed and others, 1993) (Figs. 2A and 2B). A thin Paleozoic shelf sequence covered the region during Cambrian through Early Mississippian time. In Colorado and northern New Mexico, these strata and underlying Proterozoic rocks were eroded from uplifts formed during shortening beginning in Early or Middle Pennsylvanian time. This deformation that formed the Ancestral Rocky Mountains and basins probably resulted from subduction along the southwest margin of the North American plate (Ye and others, 1996).

During the Pennsylvanian and Permian, clastic sediments shed from the Ancestral Rockies accumulated in flanking basins to thicknesses of more than 5 km locally (Fig. 2). Coarse arkosic sediments ringed the margins of fine-grained basin deposits. Locally thick sections of Middle Pennsylvanian evaporites were deposited in sub-basins, which were at least in part controlled by intrabasin deformation (De Voto and others, 1986), probably related to continued Ancestral Rocky Mountain shortening. The original stratigraphic distribution and thickness of evaporites is uncertain because flow of evaporites began shortly after deposition; at present active diapirs contain the greatest thicknesses of evaporites.

Although hundreds of meters of halite have been encountered in deep wells, only gypsum, anhydrite, and gypsiferous siltstone crop out at the surface. The principal late Paleozoic uplifts in central Colorado were the Front Range Highland, which occupied much of the area of the present Front Range and Gore Range, and the Uncompahgre Highland that lay southwest and west of the present Sawatch Range (Fig. 1). The Central Colorado trough lies between those two uplifts and includes two deep sub-basins, the Eagle basin to the northeast and a basin centered near Carbondale to the southwest.

Permian, Triassic, and Jurassic fluvial, eolian, and shoreline deposits covered the area as the Ancestral Rockies gradually eroded. Intertonguing marine and continental rocks of Cretaceous age, locally nearly 3 km thick, blanketed the Southern Rocky Mountain region. The onset of Laramide crustal shortening in the Late Cretaceous was accompanied by calc-alkaline intrusions and volcanism (Tweto, 1975). In the Piceance basin southwest of the White River uplift, the late Paleocene and Eocene fluvial Wasatch Formation accumulated as erosion of Laramide highlands continued (Johnson and May, 1980) (Fig. 1). By middle Eocene the lacustrine Green River Formation accumulated in the Unita and Piceance basins. During the late Eocene and Oligocene, widespread calc-alkaline volcanic and magmatic activity spread across western Colorado, concentrated in a belt from the San Juan Mountains in the southwest to the Front Range in the northeast. Lower to Upper Miocene basalts flowed out onto an erosion surface that extended from Grand Mesa to as far east as the Gore Range (Larson and others, 1975). During this period, extensional deformation displaced the old surfaces and produced new basins (Tweto, 1975, 1979), most notably along the trend of the Rio Grande rift. Much of the present topographic relief in Colorado resulted from differential regional uplift and dissection during the Neogene (Steven and others, 1997; Steven and others, 1995; T.A. Steven, USGS emeritus, oral commun., 1997).

PREVIOUS RESEARCH

For nearly a half century, evidence of salt tectonism has been studied in western Colorado. Hubert (1954) assigned some 300 m of relief on Tertiary basaltic flows near Eagle to "adjustment of the underlying gypsum," and Wanek (1953) recognized

900 m of synclinal subsidence of Tertiary basaltic flows at State Bridge (Fig. 1). Benson and Bass (1955) reported a young diapiric salt anticline along the Eagle River that tilted Tertiary basaltic flows, and Bass and Northrop (1963) attributed a tight syncline in conglomerate in a canyon northwest of Glenwood Springs to "subsidence into caverns in the underlying thick gypsum beds." Mallory (1966, 1971), Freeman (1971, 1972), and Piety (1981) recognized that evaporite tectonism persisted from Late Pennsylvanian to the Quaternary from investigations of diapiric evaporite anticlines, sink holes, back-tilted Pleistocene terraces, and changes in thicknesses of post-evaporite units in the Carbondale area. Ogden Tweto (1977) called for voluminous flow of evaporite that began soon after denser Permian-Pennsylvanian arkosic sediments covered the evaporite. He also attributed young geomorphic features near the Eagle River to evaporite tectonics. Post-10 Ma collapse by evaporite dissolution and flow modified the Laramide Grand Hogback (Murray 1966, 1969; Soule and Stover, 1985; Stover, 1986; and Unruh and others, 1993).

CURRENT RESEARCH

Since 1993, CGS geologists have mapped a series of 7.5' quadrangles near and southeast of the junction of the Roaring Fork and Colorado Rivers. Clear evidence of salt dissolution, flow, and resulting collapse has been confirmed by detailed mapping in the Glenwood Springs-Carbondale area (Kirkham and others, 1995, 1997a; Kirkham and others, 1996a, 1996b; Carroll and others, 1996; Streufert and others, 1997a, 1997b; and Kirkham and Widmann, 1997). In addition to many previously recognized evaporite-tectonic features, these more recent studies documented abundant evidence of more subtle Neogene deformation including: complex patterns of faults ranging from orthogonal to parallel sets that cut Tertiary basaltic rocks, synclinal sags that range from linear to arcuate shapes, structural depressions that are circular, elliptical, rectangular, trough-like, and half-graben in shape, local elongate closed basins that contained lakes and lacustrine sediments, chaotically brecciated strata above areas of suspected collapse, and Tertiary basaltic rocks folded to form monoclines and broad warps.

In late 1995, USGS geologists began our mapping project along the Eagle and Colorado Rivers near Interstate 70 (Fig. 1), and cooperative research began on topical issues recognized by both USGS and CGS mapping. Because USGS geologists also found areas of suspected collapse, we placed an emphasis upon supporting research into the timing and chemistry of basalts and study of seismic reflection lines in the vicinity of suspected collapse.

CONCEPTUAL MODEL FOR NEOGENE EVAPORITE TECTONISM IN WESTERN COLORADO

We have created a conceptual model of evaporite tectonism that involves the recognition of a datum to quantify Neogene

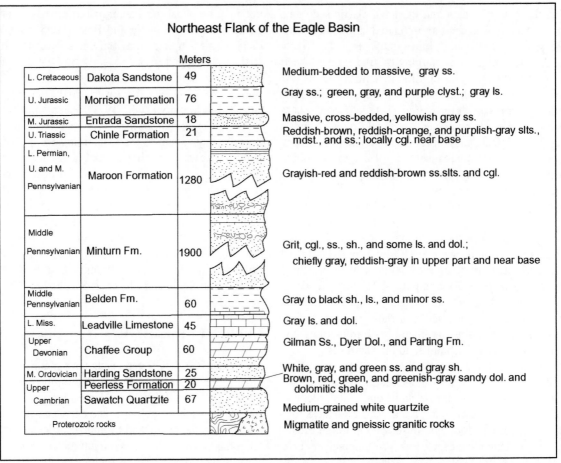

Figure 2 (this and following page). Stratigraphic relations in field trip area. A. Simplified stratigraphic column from the northeast flank of Eagle basin, modified after Tweto and Lovering (1977). B. Simplified stratigraphic column from White River uplift and southeastern Piceance basin, modified after Johnson and others (1990), Kirkham and others (1995, 1997), Bryant and others (1998), Scott and Shroba (1997), Scott and others (1999), and Shroba and Scott (1997, 1999). Thicknesses are in meters.

tectonism, the timing of uplift, and the unloading of evaporite by river incision. The resulting processes of diapirism, flow, and dissolution of evaporite cause large-scale geologic collapse (Scott and others, 1998; Kirkham and others, 1997). Both surface and ground waters play important roles in this tectonism.

Basaltic plateau datum

During the early phase of research, CGS and USGS geologists were trying to explain the presence of an isolated remnant of a 24-Ma basalt near the level of the Colorado River. Geomorphic concepts dictate that older basalts, such as this 24-Ma basalt, should be at higher levels and only younger basalts should be able to flow to lower elevations along young river valleys. In an unrelated conversation, Tom Steven (USGS emeritus, oral commun., 1996) pointed out that many of the 10-25 Ma basalts in the region stand at remarkably constant elevations between 3.0-3.4 km, well over a kilometer above the Colorado River. USGS geologists realized that these elevated basalts were

probably remnants of a basaltic plateau. If so, then removal of hundreds of meters of evaporite by dissolution could have allowed older basalt to collapse to near modern river levels.

Larson and others (1975) have established that much of the region between Grand Mesa on the southwest, the western part of the Gore Range north of Vail on the east, the White River uplift on the north, and areas southeast of Carbondale along the Roaring Fork River was covered by basaltic flows that erupted between about 25 and 10 Ma (Fig. 1). Although relatively small volumes of basaltic volcanism continued to 4 ka (Giegengack, 1962), volcanism younger than about 8 Ma appears to postdate significant stream incision of the plateau.

We made the assumption that the plateau provides a datum from which to measure the amount of basalt collapse during Neogene evaporite tectonism (Scott and others, 1998). Areas between these plateau remnants contain 25-10-Ma flows that are commonly at elevations significantly lower than 3.0 km, and a growing body of independent detailed structural evidence suggested that these areas had undergone collapse. For example, 9

White River Uplift and Southeastern Piceance Basin

Era	Epoch	Formation	Meters	Description
Tertiary	Pliocene or Miocene	Conglomerate of Canyon Creek	250	Bouldery cobble and pebble cgl.
	Miocene	Basaltic rocks	75	Olivine basalt and trachybasalt
	Eocene and Paleocene	Wasatch Formation	1960	Shire Mbr: variegated clyst.,:mdst., slts.,and minor ss.. Molina Mbr.: cyst., mdst. and prominent ss.. Atwell Gulch Mbr.: clyst., mdst., slts., ss. and cgl. rich in porphyritic andesite; underlain by a unit of ss., chert pebble cgl., slts., and clyst. of Late Cretaceous or Paleocene age
CRETACEOUS	Upper	Mesaverde Grp.	1450	Williams Fork Fm. : ss., slst., mdst., sh.,coal and clinker Iles Fm.: ss.,sh.,and slst.. Rollins Ss. Mbr. at top; Cozzette Ss Mbr. in lower part; Corcoran Sandstone Mbr. at base
	Cretaceous	Mancos Shale	1395	Upper mbr.: dark-gray, locally bentonitic, carbonaceous sh.; ss. beds near top, calcareous sh. at base. Niobrara Mbr.: shaly ls.; ls.beds in lower part. Lower mbr: gray, very pale gray-weathering, calcareous sh., and dark gray sh., silty sh., calcareous ss., slts, and ls.
	L. Cret.	Dakota Ss.	50	Yellowish-gray weathering very pale gray ss
		Morrison Fm.	150	.Greenish-gray to grayish-red slts., clyst., white ss., and gray ls.near base
	M. Jurassic	Entrada Ss.	30	Gray to pale orangish-gray medium to very fine-grained cross-bedded ss.
	Upper Triassic	Chinle Fm.	90	Pale to reddish-brown calcareous slts., silty ss.ls. and ls.. pebble cgl.
	L. Trias., U. Perm.	State Bridge Fm.	48	Pale red, grayish-red, and reddish-brown slts. and minor ss.
	L. Permian	Maroon Formation	1160	Schoolhouse Mbr.; gray ss. Main body: grayish-red and reddish-brown arkosic, ss., congl. ss., and mdst. Lower mbr.; grayish-red and reddish-brown ss.,slts., mdst., white ss., and a few beds of ls., silty ls., gypsum, and anhydrite
	Middle Pennsylvanian	Eagle Valley Formation	1000	Ss., slts., ls., and silty ls. in various shades of gray; some red, brown, and orange. beds;a few beds of gypsum and anhydrite
		Eagle Valley Evaporite	?	Pale gray to white gypsum, anhydrite, halite and gray, partly gypsiferous slts., sh., ss.. and fossliferous ls.
	M. and L. Pennsylvanian	Belden Fm.	250	Medium-gray to black and dark brown carbonaceous sh. and gray fossliferous ls.
	Mississippian	Leadville Ls.	60	Bluish gray, coarse- to fine-grained ls. and dol.
	Upper Devonian	Chaffee Group	70	Dolomitic ss., dol.,ls., gray fossiliferous ls., ss.and dol. sh.
	L. Ordovician	Manitou Fm.	40	Dol., ls. cgl., calcareous sh.,ss. and ls.
	Upper Cambrian	Dotsero Fm.	30	Thinly bedded dol., dolomitic ss., dolomitic sh., dolomitic cgl., and algal ls.
		Sawatch Quartzite	165	White to yellowish-gray, qzt.; beds of brown sandy dol. in upper part; local arkosic quartz pebble cgl. near base
Proterozoic Rocks				Gneissic granitic rocks, sillimanite mica gneiss, amphibolite, and felsic gneiss

km southeast of Eagle, a 22.3 ± 0.1-Ma flow has been traced from an elevation of about 2.7 km to 2.3 km (Kunk and Snee, 1998; Lidke, 1998).

We assumed that areas containing remnants of basalts below the datum have undergone collapse. Where evidence of a basalt datum is absent, evaporite dissolution, flow, and collapse probably occurred in areas that either exposure evaporite or are underlain by evaporite, particularly where structures are similar to those in areas of known collapse. A conservative estimate of the present datum elevation of 3.0 km was used, although parts of the uncollapsed plateau reach 3.4 km. Using these criteria, an estimated area affected by collapse was calculated to be about 4,500 km^2 (Fig. 2). Where sub-datum basalts are reasonably abundant, as in the Carbondale collapse center and in the

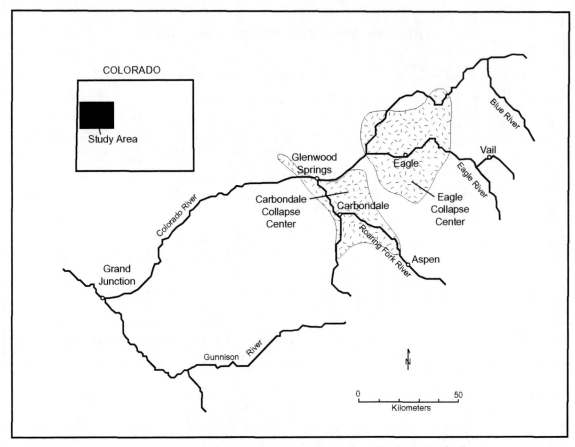

Figure 3. Map showing areas affected by evaporite flow, dissolution, and collapse. The two major collapse centers near Carbondale and Eagle coincide roughly with the two areas underlain by the greatest thickness of evaporite (De Voto and others, 1986).

central and northern part of the Eagle collapse center (Fig. 3), the basalts could be contoured. In areas of suspected collapse in the absence of basalts, contours were extrapolated following patterns of collapse. About 2,250 km³ of collapse was calculated from these contours.

Where basalt flows drape over significant topographic relief, a question needs to be answered: Were basalts erupted on a nearly horizontal surface, or were they erupted on an irregular topography and simply flowed downhill? K.A. Hon (USGS, oral commun., 1996) confirmed that the pahoehoe-type flow structures in pre-10-Ma basaltic flows are indicative of flow on relatively horizontal surfaces, not on significant slopes. Also, initial paleomagnetic study of the basalts indicates that the basalts were deposited on a nearly horizontal surface that was subsequently tilted about a horizontal axis.

How far did the basaltic lavas flow? Are there only a few major vents or large numbers of eruptive vents on the plateau? Accurate correlation of basalts, using high-field-strength trace-elemental signatures, $^{40}Ar/^{39}Ar$ geochronology, petrography, and paleomagnetic data, has been used to establish both the stratigraphy and the continuity of basalts on the plateau. About

165 basalt samples from the area have been analyzed for over 40 major, minor, and trace elements by x-ray diffraction and instrumental neutron activation analysis, and over 90 $^{40}Ar/^{39}Ar$ basalt samples have been dated. About 8 separate major groups of basalts have been recognized from data using chondrite-normalized La/Yb and Hf/Ta ratios. Basalts >20 Ma have La/Yb ratios of 16-20, basalts 10-11 Ma have La/Yb ratios of 5-8, basalts 8-9 Ma have La/Yb ratios of 8-10, and basalts 3-4 Ma have La/Yb ratios of 10-13. Two or three subgroups of each major group have been identified by differences in Ba, Sr, Rb, and Th. The differences in basalt compositions are assumed to be the results of differences in partial melting, crystal fractionation and, in particular, crustal assimilation. Few flows can be correlated farther than tens of kilometers. Although remnants of basalt flows of statistically identical ages are found within 10-20 km of one another, trace-element data do not support correlation. This requires that fairly closely spaced volcanic centers erupted magmas with different degrees of partial melting, types of fractionation, or levels of crustal contamination. Thus, the plateau probably was partly covered with discontinuous flows from scattered vents.

Initiation of Neogene evaporite tectonism

Both USGS and CGS geologists consider that regional Neogene uplift triggered more rapid downcutting in the Rocky Mountains that, in turn, prompted renewed evaporite tectonism in the Eagle and Carbondale areas. Basaltic flows older than about 10 Ma are stacked upon one another in normal stratigraphic order, suggesting that the plateau basalts erupted on low-relief topography (Larson and others, 1975) and predate significant renewed uplift in the White River uplift area. Timing of uplift in the Eagle and White River uplift area probably is slightly older than the 7.8-Ma age of a local basaltic flow collected 7.5 km south-southwest of the junction of the Colorado and Eagle Rivers (Kunk and others, 1997; Kunk and Snee, 1998) because the flow rests on river gravels 200 m below older basalts (Streufert and others, 1997b; Kirkham and others, 1997). Also, young (~5-Ma) apatite fission track ages have recently been determined in the Gore Range (C.W. Naeser, USGS, written commun., 1997), suggesting late Miocene uplift in the area. Finally, Steven and others (1997) proposed late Miocene uplift in the Front Range of the Rocky Mountains. Although local variations in the timing and degree of uplift probably exist, they are all part of a regional Neogene uplift event in the southern Rocky Mountains.

Increased stream gradients would have increased the rate of downcutting as uplift proceeded; this, in turn, removed sufficient overburden to triggered local diapiric upwelling of evaporite in deeper river valleys. Once sufficient overburden had been removed by stream erosion, either pre-existing diapirs or diapirs initiated by that unloading would be activated. When the evaporite encountered shallow ground water or surface water, dissolution removed material from the upper part of the diapir and the rate of removal of material increased significantly. Therefore, the rate of new exposure of evaporite along the crest of anticlinal diapirs probably rapidly increased, "unzipping" their tops. In both the Eagle and the Carbondale collapse centers, the major gypsum-cored diapiric anticlines coincide with stream valleys. Replacement of dissolved evaporitic material required progressive flow of ductile evaporite from adjacent highlands toward river valley diapirs, causing progressive collapse of these highlands as material was removed. Irregular and relatively rapid dissolution of exposed or shallowly buried evaporite formed local sinkholes.

Geologic significance of dissolved-solids load of streams

Study of the dissolved-solids load of the upper Colorado River drainage provides a critical component of the evaporite collapse story, allowing us to estimate the length of time the process has been active. About 90 km downstream from the collapse centers (Fig. 1), the mean annual dissolved-solids load at Cameo is 1.3×10^9 kg (Fig. 1). The mean annual dissolved-solids loads from individual streams upstream from the collapse centers is less than 5×10^7 kg. However, within the collapse centers,

the annual load, including discharge from saline hot springs near the town of Glenwood Springs (Cappa and Hemborg, 1995; Barrett and Pearl, 1978), is nearly 1.2×10^9 kg (Bauch and Spahr, 1998; Butler, 1996; Liebermann and others, 1989; U.S. Department of the Interior, 1997). These dissolved-solids loads consist largely of Na^+, Cl^-, Ca^{+2}, and SO_4^{-2} ions. These rates do not include unmeasured saline ground water discharge into the Colorado River between Glenwood Springs and Cameo and unmeasured loads from several streams, some of which drain the relatively saline Green River Formation.

These data clearly indicate that evaporite dissolution is active, and therefore, evaporite-induced collapse must also be active. If the rate of dissolution has been constant, if the annual load at Cameo represents the total load added exclusively from the collapse centers, and if the volume of evaporite removed is 2,250 km^3, then the length of time during which the process has operated would be about 3.7 Ma. However, Neogene downcutting probably began about 8-10 Ma, so presumably the rate of dissolution has increased during this period rather than remain constant.

GEOLOGIC COLLAPSE IN THE FIELD TRIP AREA

This discussion will address broader observations of collapse in the general areas to be visited on this trip. In the following Abbreviated Road Log, more site-specific information is provided.

Eagle collapse center

Throughout much of the Eagle collapse center, the Eagle Valley Evaporite is widely exposed or present at relatively shallow depths (Fig. 4). In the central part of the collapse center along I-70, between the towns of Wolcott and Gypsum, evidence for Neogene salt tectonics includes large amplitude, trough-like collapse features and linear to arcuate, discontinuous, high-angle faults that locally form horsts and grabens, both of which deform early Miocene basaltic flows. Broad, east- to northeast-trending anticlines that follow the Eagle River probably reflect Neogene upwelling of the evaporite at depth, related to unloading from downcutting of the Eagle River (Lidke, 1998, 1999; unpublished mapping, Mark Hudson and Richard Moore, Gypsum 7.5' quadrangle, 1998). The westernmost anticline is clearly younger than the deformed 22.3-Ma basalts (Tweto, 1977), and it is probably younger than about 8 Ma based on timing of initial downcutting of the Colorado River.

The timing of initial uplift of the basalt plateau in the Eagle Valley area can be estimated by dividing the elevation of assumed uncollapsed basalt above the river level by the rate average of incision of the rivers. The elevations of basalts at Castle Peak are about 1.44 and 1.49 km above the Eagle and Colorado Rivers, respectively. The rate of incision based on the Lava Creek B volcanic ash just below the junction of the Eagle and Colorado Rivers is 0.14 m x 10^{-3} yr. (Izett and Wilcox,

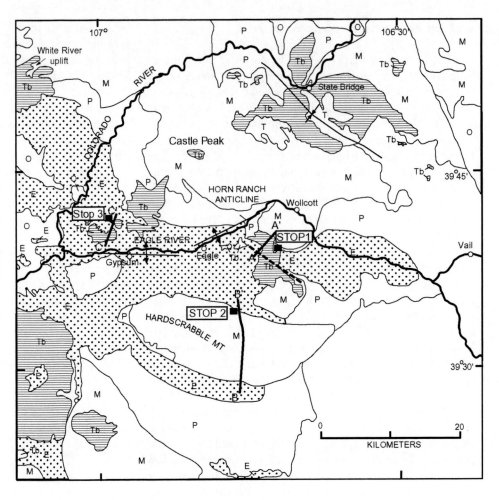

Figure 4. Map of the area of Stops 1, 2, and 3 showing the generalized bedrock geology of the Eagle Valley area modified from Tweto and others (1978). T = Tertiary lacustrine strata; Tb = Tertiary basaltic flows; M = Mesozoic strata; P = Permian and Pennsylvanian strata; E = Eagle Valley Formation and Pennsylvanian Eagle Valley Evaporite; O = Paleozoic and Precambrian rocks older than Eagle Valley Evaporite. Two large gypsum-cored anticlines are shown along the Eagle River and a trough-like feature is shown east of Eagle by a dashed line. Modified from Scott and others (1998).

1982). The rate of tectonic uplift of about 0.18 m x 10^{-3} yr. was based on the Derby Peak fauna in the Flat Tops area of the White River uplift (Colman, 1985). The timing of uplift based on the incision rate is 8.0-8.3 Ma and the timing based on uplift rate is 10.3-10.6 Ma, both of which are close to the 7.8 Ma age of nearby basalts deposited on incised river gravels 7.5 km southwest of the junction of the rivers.

A northwest-trending sag, or trough-like Neogene collapse feature (Lidke, 1998), expressed in 22.3-Ma basaltic flows 5 km southwest of Wolcott (Fig. 5) will be visited at Stop 1. Evidence for pre-Neogene deformation of the pre-basalt strata is seen in south-dipping flows that overlie northeast-dipping Pennsylvanian to Jurassic strata. Also, the evaporite has been intensely deformed beneath the less deformed younger Paleozoic and Mesozoic strata. Clearly the evaporite decoupled structurally from the overlying strata, but the genesis and timing of this deformation is problematic.

A few kilometers east of Eagle, linear to arcuate, northwest- to north-trending faults displace basalt flows and clastic basin facies rocks, and younger strata have either fault or intrusive contacts with the Pennsylvanian evaporite (Lidke, 1999) (Fig. 4). Many of these Neogene faults are attributed to evaporite dissolution and collapse because they cannot be traced later-

ally within the evaporite. The isolated northeast-trending Horn Ranch anticline is interpreted to be a Neogene structure related to diapiric upwelling of the evaporite in response to unloading during Neogene downcutting by the Eagle River.

Southwest of Eagle, Tweto and others (1978) interpreted Hardscrabble Mountain to be a 30 x 10-km block of gently north-dipping Triassic to Cretaceous strata that has been downdropped into Eagle Valley Evaporite (Fig. 4). At Stop 2 we observe the northern fault contact of Hardscrabble Mountain with evaporite. Tweto (1977) proposed that the evaporite flowed diapirically around the mountain. Seismic reflection data (Fig. 6) confirm his conclusion and add significantly to it. Faults that drop the mountain into the Pennsylvanian evaporite do not offset the older Paleozoic strata. Only a small, unrelated Laramide or Permian-Pennsylvanian thrust fault offsets the older strata below the center of the mountain. Evaporite deformation probably occurred at least twice, once during deposition of the unusually thick sequence of Triassic-Permian State Bridge Formation and again sometime afterward, possibly during the Neogene. The unusual thickness of the State Bridge reflects the likelihood that the block was subsiding within the evaporite as State Bridge deposition occurred. The tilt of the Mesozoic homoclinal block to the north requires post-State Bridge deformation, probably Neogene.

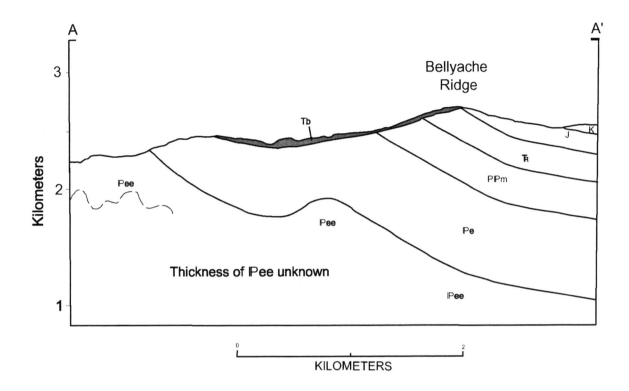

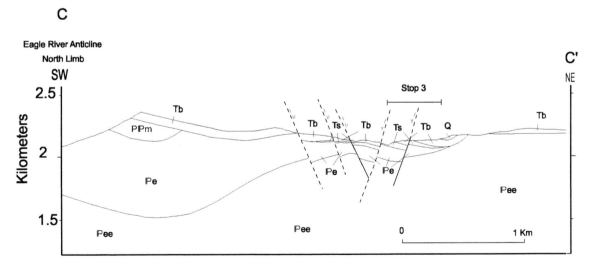

Figure 5. Cross sections shown in the Eagle collapse center. A. Section A-A' at Stop 1 trends northeast across a trough-like sag developed in basaltic flows and underlying strata, which reflects removal of evaporite. Tb = Tertiary basalt flows; K = Cretaceous Dakota Sandstone; J = Jurrasic Morrison Formation and Entrada Sandstone; ΤͰP Triassic and Permian Chinle and State Bridge Formations; PⱣm = Pennsylvanian and Permian Maroon Formation; Ⱥe = Eagle Valley Formation; Ⱥee = Eagle Valley Evaporite. B. Section C-C' at Stop 3 trends northeast across the collapsed north limb of the Eagle River anticline. Symbols are the same as for A except Q = Quaternary deposit, Ts = Tertiary sediment.

Brush Creek/Hardscrabble Mountain Area

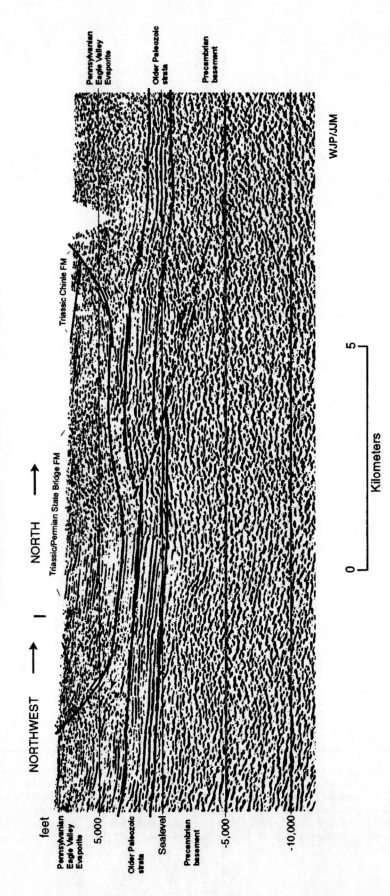

Figure 6. Seismic depth section through Hardscrabble Mountain along section B-B' visited at Stop 2. Strata exposed along section in the structural block are the Triassic-Permian State Bridge Formation and Triassic Chinle Formation. Note that the nominal vertical scale is in feet.

The evaporite-cored Eagle Valley anticline follows the Eagle River where it bends to a westward course between Eagle and Gypsum. The north limb of that diapiric anticline displays a 10° northward dip on 22.7-Ma basalts (Larson and others, 1975), clearly recording Neogene evaporite tectonism. In the vicinity of Gypsum, flow remnants are folded into a series of large synclines and smaller anticlines and are cut by minor southwest-dipping normal faults (unpublished mapping, Mark Hudson and Richard Moore, Gypsum 7.5' quadrangle, 1998). We will visit this deformation at Stop 3.

About 17 km northeast of Gypsum, Castle Peak is capped by basalt at an elevation of 3,437 m and near Gypsum the basalt is at an elevation of 2,135 m, recording an apparent collapse of about 1,300 m. Several faults in this area appear to terminate laterally in Eagle Valley Evaporite. A channel filled with Neogene(?) river gravels dips gently southwest on a limb of the highest syncline, in the same direction as the basaltic rocks that appear to underlie the gravels. In this area, underlying Mesozoic and Paleozoic strata dip consistently to the northeast, suggesting that deformation of the basalt also unfolded the preexisting dip of these older rocks. These relationships suggest that diapirism may have already been operating to some degree before the basalts were emplaced.

Volcanic ash identified by Izett and Wilcox (1982) as the 0.66-Ma Lava Creek B (redated by Izett and others, 1992), the youngest of three (0.66, 1.27, and 2.02 Ma) far-traveled ashes from the Yellowstone caldera complex (Izett and Wilcox, 1982), overlies terrace gravel about 85 m above the Colorado River about 4.5 km southwest of its junction with the Eagle River (Fig. 4) (Streufert and others, 1997a). The ash elevation is consistent with that of other reported occurrences of the Lava Creek B at 85-90 m above stream level in northwestern Colorado (Izett and Wilcox, 1982; Scott and Shroba, 1997). An ash on or within main-stream terrace deposits along the Eagle River, about 26 km east of its junction with the Colorado River, (Fig. 4) has been identified as Lava Creek B by correlation based on its signature of high field-strength trace elements. We propose that 25-30 m of the underlying Eagle Valley Evaporite dissolved since 0.66 Ma collapsing Lava Creek B ash that amount.

At State Bridge in the northern part of Figure 4, a large synclinal sag of basalt between Castle Peak and the mountain north of State Bridge poses a structural puzzle. Unlike other structures in the collapsed areas, much of this structure does not overlie known evaporite deposits. Tweto and others (1978) placed the boundary between the basin evaporitic facies and more proximal coarser clastic facies largely south of this structure. However, the presence of a saline warm spring in the area suggests the presence of evaporite at depth. Also the absence of contractional structures in the basalts leaves the probability of withdrawal of evaporite by flow and dissolution as the most likely explanation for the sag.

Collapse structures along the southwestern flank of the White River uplift

Northwest of Glenwood Springs, a string of elliptical evaporite diapirs intrudes Permian to Devonian rocks along the southwestern flank of the White River uplift (Fig. 7). South of the diapirs, the adjacent Maroon and Eagle Valley Formations form bench-like sags that depart significantly from the relatively steep southwest dips of the Grand Hogback monocline. North of the diapirs, older Paleozoic strata and even Proterozoic rocks are also deformed by evaporite tectonism.

North of the Colorado River at Stop 4, Canyon Creek cuts through a tight syncline in a Tertiary conglomerate (Bryant and others, 1998) (Fig. 8). The upright southern limb of the syncline is in contact with a narrow zone of Pennsylvanian evaporite

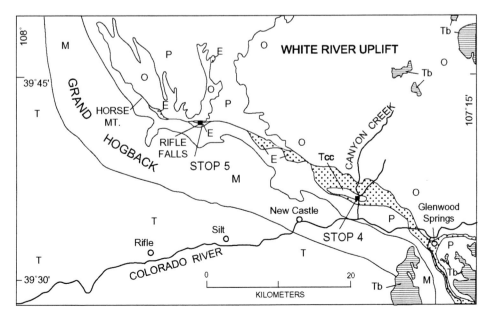

Figure 7. Map showing generalized geology in the area of Stops 4 and 5 northwest of Glenwood Springs modified from Tweto and others (1978). Tcc = Conglomerate of Canyon Creek; Tb = Tertiary basalt flows; T = Tertiary sedimentary strata; P = Permian and Pennsylvanian strata; E = Pennsylvanian Eagle Valley Evaporite; O = Paleozoic and Precambrian rocks older than Eagle Valley Evaporite. Clusters of and individual diapiric pod-shaped intrusions are highly generalized for simplicity. Modified from Scott and others (1998).

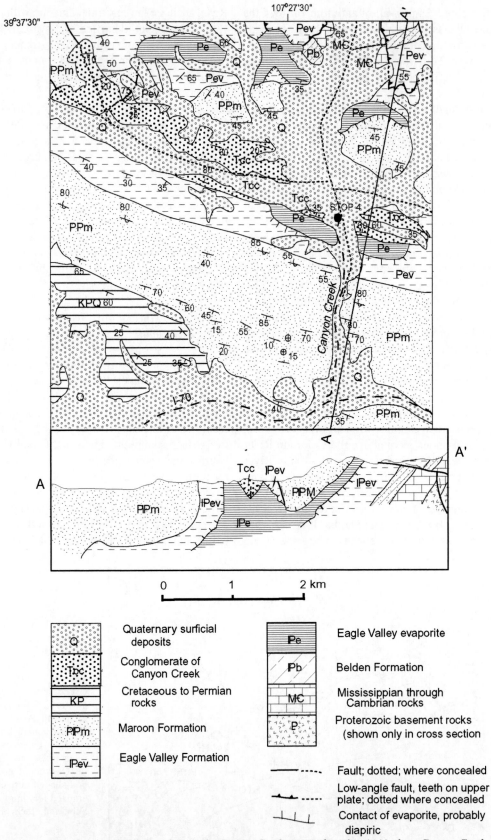

Figure 8. Map showing detailed geology in the Canyon Creek area and section A-A' along Canyon Creek near Stop 4. Modified from Bryant and others (1998).

that, in turn, is in contact with overturned Pennsylvanian clastic basin facies rocks that dip steeply northward. The clast composition of the conglomerate suggests that the deposit is locally derived and young, probably Pliocene. Deposition of the Tertiary conglomerate on Pennsylvanian strata presently beneath the conglomerate would require the highly unlikely geometry of nearly completely overturned Pennsylvanian rocks. Based on locally nonconformable contacts and pod-shaped bodies of evaporite in the area, we conclude that the bodies of evaporite have intrusive, not depositional, boundaries and that both flow- and dissolution-induced collapse occurred (Bryant and others, 1998). The bench-like anticlinal and synclinal structures in the Grand Hogback monocline south of the conglomerate near the mouth of Canyon Creek resulted from collapse as flow and dissolution of material undermined the monocline and allowed tight synclinal sagging of the conglomerate as it collapsed into dissolving evaporite. This explanation is similar to that proposed by Bass and Northrop (1963). A 10° low-angle normal fault that places Mississippian Leadville Limestone above Cambrian Sawatch Quartzite with 250 m of dip separation is probably related to the diapir complex exposed only 300 m to the south.

About 24 km northwest of the Canyon Creek stop, another example of collapse along the Hogback monocline occurs at Stop 5 near Rifle Falls (Scott and others, 1999). A bench-like structure similar to that at Canyon Creek is well exposed south of a pod-shaped diapir (Fig. 9). In this case, the diapir is 0.5 km south of a major south-dipping normal fault displaying several hundreds of meters of dip separation. Not only does the footwall of the fault dip steeply southward, but also the footwall strata locally dip steeply northward, forming a perplexing tepee-like structure. Because the north-dipping segment of the footwall is locally restricted to areas adjacent to the diapir, we conclude that the origin of the "tepee" is related to intrusion of evaporite across the fault into lower strata in the footwall.

In the adjacent Horse Mountain quadrangle to the west of Rifle Falls, similar bench-like structures are found (unpublished mapping, W.J. Perry, 1998). However about 20 km to the southeast of Rifle Falls, in the northeastern part of the New Castle quadrangle (Scott and Shroba, 1997), a different type of structure seems to be associated with evaporite diapirism. In this case, a 2.5-km-wide overturned, northeast-dipping sequence of the Maroon and the Eagle Valley Formations occurs south of a large evaporite diapir that is exposed about 1 km north of the New Castle quadrangle.

CONCLUSIONS

We assume that 25-10-Ma plateau basalts formed a datum from which the area and volume of collapse can be calculated. The area affected by collapse in the Carbondale and Eagle collapse centers is estimated to be 4,500 km^2 and the volume of evaporite removed is estimated to be 2,250 km^3.

Uplift of the basalt plateau and downcutting of rivers in the area around the Eagle collapse center is estimated to have begun between 7.8 and 10.6 Ma.

On the southwest flank of the White River uplift, the monoclinal limb of the hogback collapsed adjacent to diapiric intrusions of evaporite, diapiric intrusion is associated with normal faults, and a low-angle normal fault in lower Paleozoic and Proterozoic rocks was caused by diapiric intrusion.

In the Eagle collapse center, 22-23-Ma basalt flows are draped and faulted over an elevation change of at least 790 m, form graben-like depressions, and are tilted away from anticlinal crests along the Eagle River. The Lava Creek B volcanic ash from Yellowstone has apparently collapsed about 25 m in the last 0.66 m.y. Hardscrabble Mountain sank along an elliptical fault during Triassic-Permian deposition and probably still sinks into the Pennsylvanian Eagle Valley Evaporite.

ABBREVIATED ROAD LOG

On our drive to the Eagle Basin, we cross the Front Range, a major Laramide uplift modified by younger events, then the Blue River valley, an en echelon fault-valley extension of the Rio Grande rift, and finally the Gore and Tenmile Ranges that show fission-track and geomorphologic evidence of renewed uplifted during Neogene time. Numerous road logs are available for the route along I-70 to the Eagle Basin (such as Kirkham and others, 1996a; Reed and others, 1988). Our highly abbreviated road log starts at Vail Pass; a detailed log will be available for attendees and for those who request it from the first author.

Cumulative miles

0.0   Mileage begins on I-70 at exit 190 (Vail Pass rest area). Continue westward.

20.2   Here we see an abrupt facies change from massive, resistant, coarse-grained arkosic rocks of the proximal facies of the Pennsylvanian Minturn Formation into siltstone of the Eagle Valley Formation that contains gypsiferous evaporite of the distal facies of the Eagle Valley Evaporite.

32.9   On the left is the landslide complex of Bellyache Ridge, expressed by hummocky topography.

33.4   At exit 157 to Wolcott, turn off I-70 and turn left under I-70. Proceed up Bellyache Ridge Road to enter the landslide complex, which formed in the dip slope of northerly-dipping Cretaceous Benton Shale and the underlying Cretaceous Dakota Sandstone and Jurassic Morrison Formation.

37.6   At the intersection, stay on Bellyache Ridge Road.

40.0   At the electronic gate, enter a gated community.

40.7   **Stop 1.** South-dipping basalt caps northeast-dipping Mesozoic strata. The northwest view in foreground shows dark-colored, south-dipping lower Miocene basaltic flows that overlie reddish-colored, northeast-

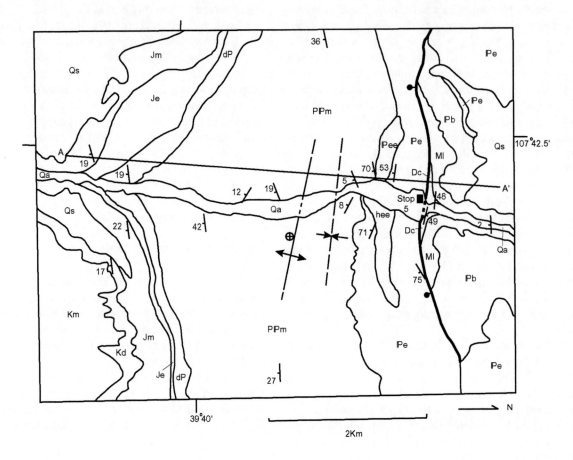

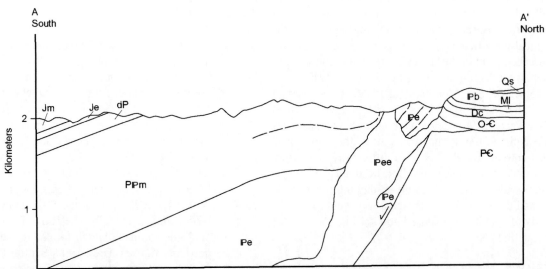

Figure 9. Map showing detailed geology in the Rifle Falls area and section B-B' at Stop 5. Qs = Landslide deposit, Km = Mancos Shale, Jm = Morrison Formation, Je = Entrada Sandstone, TʀP = Chinle Formation and State Bridge Formation, PIPm = Maroon Formation, IPe = Eagle Valley Formation, IPee = Eagle Valley Evaporite, IPb = Belden Formation, Ml = Leadville Limestone, Dc = Chaffee Group, OC = Ordovician and Cambrian strata, PC = Proterozoic rocks.

dipping Mesozoic sedimentary rocks along the crest of Bellyache Ridge. The south-dipping flows define the northern limb of a west-northwest-trending, trough-like sag defined by the basaltic flows (Figs. 4 and 5). Below, and to the south-southwest, basaltic flows in the central part of the sag can be seen at lower elevations. Basalts in the central part of the sag overlie the lower part of the Eagle Valley Formation and the Eagle Valley Evaporite. These basaltic flows yielded $^{40}Ar/^{39}Ar$ isotopic ages of about 22-23 Ma (Kunk and others, 1997; Kunk and Snee, 1998; Lidke, 1998). The south-dipping flows on Bellyache Ridge are at an elevation of about 2,750 m, whereas continuations of these flows in the trough of the sag are at about 2,320 m, providing evidence for at least 430 m of collapse of the basalt and underlying strata. If the original elevation of the basalt was that of Castle Peak and the White River Uplift to the northwest at about 3,350 m, then over 1 km of collapse occurred here.

40.9   Turn around at the 6903 sign and return to I-70.
48.3   Turn west on I-70
51.0   Enter Red Canyon and begin driving through red beds of the Triassic-Permian State Bridge and underlying Permian-Pennsylvanian Maroon Formations. The axis of the broad, northeast-plunging, Horn Ranch anticline coincides with the canyon here. This anticline is probably related to diapiric rise of the underlying Eagle Valley Evaporite in response to rapid Neogene downcutting by the Eagle River.
52.5   The valley opens ahead and is underlain by the Eagle Valley Evaporite.
58.2   Take exit 147 to Eagle, turn left at stop sign, proceed south across I-70 and the Eagle River to Eby Creek Road (U.S. 6) just beyond bridge, and turn right.
58.5   Turn left on Broadway and proceed through downtown Eagle.
58.8   Turn left at 5th Street, go one block, and turn right on Capitol Street which becomes Brush Creek Road about 1 mile ahead.
67.0   **Stop 2.** Hillside to left exposes spires of intensely fractured Jurassic Entrada Sandstone just south of major south-dipping normal fault revealed by reflection seismic line (Figs. 4 and 6). This line reveals the Hardscrabble Mountain homocline to be a collapse feature bounded on south and north sides by normal faults dipping toward and beneath the homocline and vanishing in the Eagle Valley evaporite. The thickness of the Eagle Valley Evaporite interval beneath the homocline, revealed by the seismic data, is relatively thin. The extraordinary thickness (~1.5 km) of Eagle Valley Formation and Evaporite to the north and south, indicates major flow of evaporite out from beneath the homocline, flow which probably continues. The extraordinary thickness of State Bridge Formation on

Hardscrabble Mountain mapped by Tweto and others (1978) suggests that initial collapse and subsidence of this structure may have occurred during the Permian. Return to I-70.

75.8   Turn west on I-70.
80.3   On the right, folds are well displayed within Eagle Valley Evaporite. Disharmonic folding is common within evaporites of the Eagle basin and reflects plastic flow within the thick deposits. Widmann (1997) demonstrated that fold axes are parallel to the long axes of evaporite upwellings and to coincident major drainages within the western Eagle Basin.
83.0   At Gypsum exit 140, turn right on good dirt road.
84.3   At the intersection, go left.
86.9   At road intersection, go left.
87.1   At road intersection, turn left on BLM road 8460.
87.2   Note the overview on the right of a 1 x 0.5 km closed basin that is interpreted to reflect near-surface dissolution of the underlying evaporite.
87.3   Turn around at jeep road to left.
87.6   Turn right on jeep road just past fence.
88.1   Turn right through fence
88.3   **Stop 3.** We will take a short walk (about 500 m) to examine basalt and underlying basal Tertiary sediments in area of maximum collapse (Figs. 4 and 5). Park and walk right from road down to wash to north. At wash turn left and walk southwest downstream. The top basalt flow is well exposed at sharp left turn in wash. This locality illustrates the faulted and tilted character of basalt flows at their lowest elevation of 2070 m within the Gypsum 7.5' quadrangle. To the northeast the basalts rise through a series of faults and folds to elevations as high as 2865 m. To the southwest the basalts climb the north limb of the Eagle River anticline to a maximum elevation of 2380 m. This wash exposes at least two basalt flows repeated by a fault. The basalt in the top of the highest flow is relatively fresh and contains cooling joints and elongate vesicles that indicate a dip of about 27° NE. Walking farther southwest down the wash, note a lower altered flow with an oxidized base that overlies northeast-dipping siltstone. Continuing down the wash, the basalt sequence is repeated across a poorly exposed north-northwest striking fault. The local correlation of the basalt flows across the fault is supported by paleomagnetic data, which reveal that the top flow in each fault block, as well as on the north limb of the Eagle River anticline 2.5 km to the southwest, carries reversed polarity remanent magnetization. All lower flows at these localities carry normal polarity remanent magnetization. A 22-Ma K-Ar date reported by Larson and others (1975) for the basalt on the west edge of the Gypsum quadrangle also carries reversed polarity magnetization and thus is probably correlative with the

top basalt flow at this stop. Northeast-dipping Tertiary sedimentary rocks are present below the basal basalt in the second fault block, and they are relatively well exposed along the wash and at its confluence with a larger southeast-trending wash. This sequence is about 40 m thick and consists of arkosic pebbly sandstone, an interval of well-rounded river cobbles, and basal sandstone and interbedded claystone that overlie Eagle Valley Formation. Southwest of the larger wash, basalt is downdropped across another northwest striking fault. Retrace path back to vehicles and return to I-70.

92.8   Turn west on I-70

101.9   Entrance to Glenwood Canyon.

111.3   Take exit 121 to Grizzly Creek rest stop and have lunch.

112.1   Return to I-70 and head west.

116.4   At Glenwood Springs, note saline hot springs that flow from the Leadville limestone on the right.

119.7   Enter a narrow canyon where Maroon Formation dips 40-60° SW.

123.9   At exit 109, turn right onto frontage road.

124.2   Turn left up Canyon Creek Road. The locally unexposed Maroon Formation is folded into a structural terrace with irregular and gentle dips in marked contrast to 40-60° SW dips in the canyon of the Colorado River (Figs. 7 and 8). At the first exposures we see on the right, the beds are vertical to overturned, and we see a transition between the Maroon and Eagle Valley Formations. Farther north we cross a poorly exposed contact with a probable evaporite diapir.

125.9   **Stop 4.** The tight syncline in the well-indurated conglomerate of Canyon Creek, exposed in the valley of Canyon Creek, creates a baffling structure (Fig. 8). The steeply dipping beds of the conglomerate are right-side-up, yet they overlie steeply dipping overturned beds of Eagle Valley Evaporite. To the west, angular discord between the attitudes of Pennsylvanian rocks and trends of faults and the attitudes of the overlying conglomerate is marked. The evaporite probably forms an intrusive rather than depositional contact with the base of the conglomerate. The conglomerate was probably deposited close to its sources by high-gradient streams because here this unit contains subangular to round clasts of sandstone and limestone from the Maroon and Eagle Valley Formations, and 2-3 km west of here, many of the clasts are limestone, dolomite, sandstone, and quartzite from Cambrian through Mississippian units, which form the surface of the southern part of the White River plateau. The syncline becomes shallower 5.5 km west of here. The east end of the syncline is on the slope across the valley to the east.

We conclude that after 10 Ma, collapse and diapiric movement of evaporite resulted in three unusual structures. 1) The most obvious is a structural terrace on the south flank of Storm King Mountain. The length of this terrace is similar to the known extent of the conglomerate of Canyon Creek. 2) In the vicinity of the stop, a local basin formed in the Pennsylvanian rocks, in which the conglomerate of Canyon Creek was deposited. Continued collapse associated with continuing dissolution and diapirism deformed the conglomerate into a syncline. This interpretation is practically the same as that given by Bass and Northrup (1963), who first recognized and mapped the conglomerate of Canyon Creek. 3) To the north, not visible from here, a low-angle normal fault places Leadville Limestone over Sawatch Quartzite. Either upwelling or removal of material related to the evaporite diapir probably caused this rare occurrence of low-angle normal faulting. Return to I-70.

127.7   Turn west on I-70. Half a mile west at 2-3 o'clock is a hill with horizontal beds of light-gray sandstone of the Schoolhouse Member of the Maroon Formation near the west end of the structural terrace we crossed at Stop 4, but did not see.

145.9   At Exit 90, turn right on State Highway 13 and drive north through Rifle.

149.9   Turn right on State Highway 325 toward Rifle Gap.

154.0   Pass through Rifle Gap and turn right at Rifle Gap Reservoir dam on route 325.

159.5   Rifle Falls campground is on the right; dips of the Maroon Formation approach horizontal.

159.8   The Maroon has a slight northward dip. As we approach exposures of the gypsum, notice that we drive through a shallow syncline and the Maroon dips steeply to the south close to the contact with the evaporite. Because the Eagle Valley Formation, which overlies the evaporite at most localities, is absent here, the pod-shaped gypsum body is assumed to be a diapiric intrusion. We consider the structural terrace we just drove through to the south to be analogous to the structure described at Canyon Creek (Fig. 9).

161.1   **Stop 5.** At the entrance to the narrow canyon cut by East Rifle Creek, the Leadville Limestone forms the light bluish-gray strata at the top of the cliffs. The darker exposures below consist of the Upper Devonian Chaffee Group that dips 45 degrees northward. Mapping in the quadrangle (Scott and others, 1999) indicates that a south-dipping normal fault with an offset of 400-650 m occurs at the base of the cliffs (Fig 9). If so, then why are middle Paleozoic strata dipping northward? Note that upstream, the attitude of middle Paleozoic strata shallows abruptly to dip 2° southward. Note also that the north-dipping strata on the upthrown block occur only adjacent to the diapir of evaporite. We conclude that at depth some of the evaporite injected across the normal fault into the upthrown block, making this local tepee-like structural feature. We also conclude that the local structural terrace, which departs from the Grand Hogback geometry, is related to evaporite flow, dissolution, and resulting collapse.

ACKNOWLEDGMENTS

When we began our work, Colorado Geological Survey colleagues Bob Kirkham, Randy Streufert, and Chris Carroll led field trips to familiarize us with local stratigraphic details and evaporite-related collapse structures. A series of informal seminars with CGS geologists encouraged a valuable exchange of ideas and data. Nancy Driver and Nancy Bauch, USGS Water Resources Division hydrologists with the Upper Colorado River Basin project of the National Water-Quality Assessment Program, provided us with critical dissolved-solids load data that provided us with the modern rate of dissolution. USGS emeritus geologist Tom Steven inspired us with his fresh ideas and depth of knowledge of Colorado geology. Anne Harding provided valuable editorial assistance, and Tom Cooper provided cartographic help. Ren Thompson, Dave Lageson, and Bruce Trudgill provided constructive reviews.

REFERENCES CITED

Barrett, J.K., and Pearl, R.H., 1978, An appraisal of Colorado's geothermal resources: Colorado Geological Survey Bulletin 39, 224 p.

Bass, N.W., and Northrop, S.A., 1963, Geology of the Glenwood Springs quadrangle and vicinity, northwestern Colorado: U.S. Geological Survey Bulletin, 1142-J, 74 p.

Bauch, N.J., and Spahr, N.E., 1998, Salinity trends in surface waters of the Upper Colorado River Basin, Colorado: Journal of Environmental Quality, v. 27, p. 640-655.

Benson, J.C., and Bass, N.W., 1955, Eagle River anticline, Eagle County, Colorado: American Association of Petroleum Geologists Bulletin, v. 39, no. 1, p. 103-106.

Bryant, Bruce, Shroba, R.R., and Harding, A.E., 1998, Revised preliminary geologic map of the Storm King Mountain quadrangle: U.S. Geological Survey Open-File Report 98-472, scale 1:24,000.

Butler, D.L., 1996, Trend analysis of selected water-quality data associated with salinity-control projects in the Grand Valley, in the lower Gunnison River basin, and at Meeker Dome, western Colorado: U.S. Geological Survey Water-Resources Investigations Report 95-4274, 38 p.

Cappa, J.A., and Hemborg, H.T., 1995, 1992-1993 low-temperature geothermal assessment program, Colorado: Colorado Geological Survey Open-File Report 95-1, 19 p.

Carroll, C.J., Kirkham, R.M., and Stelling, P. W., 1996, Geologic map of the Center Mountain quadrangle, Garfield County, Colorado: Colorado Geological Survey Open-File Report 96-2, scale 1:24,000.

Colman, S.M., 1985, Map showing tectonic features of late Cenozoic origin in Colorado: U.S. Geological Survey Miscellaneous Geologic Investigations Series Map I-1556, scale 1:1,000,000.

De Voto, R.H., Bartleson, B.L., Schenk, C.J., and Waechter, N.B., 1986, Late Paleozoic stratigraphy and syndepositional tectonism, northwestern Colorado, in Stone, D.S., ed., New interpretations of northwest Colorado geology: Rocky Mountain Association of Geologists-1986 Symposium, p. 37-49.

Freeman, V.L., 1971, Permian deformation in the Eagle basin, Colorado, in Geological Survey Research 1971: U.S. Geological Survey Professional Paper 750-D, p. D80-D83.

Freeman, V.L., 1972, Geologic map of the Ruedi quadrangle, Pitkin and Eagle Counties, Colorado: U.S. Geological Survey Quadrangle Map GQ-1004, scale 1:24,000.

Giegengack, R.F., Jr., 1962, Recent volcanism near Dotsero, Colorado: Boulder, Colorado, University of Colorado, unpublished M.S. thesis, 69 p.

Hubert, J.F., 1954, Structure and stratigraphy of an area east of Brush Creek, Eagle County, Colorado: Cambridge, Mass., Harvard University, unpublished M. S. thesis, 104 p.

Izett, G.A., and Wilcox, R.E., 1982, Map showing localities and inferred distributions of the Huckleberry Ridge, Mesa Falls, and Lava Creek ash beds (Pearlette family ash beds) of Pliocene and Pleistocene age in the western United States and southern Canada: U.S. Geological Survey Miscellaneous Investigations Series Map I-1325, scale 1:4,000,000.

Izett, G.A., Pierce, K.L., Naeser, N.D., and Jaworowski, C., 1992, Isotopic dating of Lava Creek B tephra in terrace deposits along the Wind River, Wyoming: Implications for post 0.6 Ma uplift of the Yellowstone hotspot: Geological Society of America Abstracts with Programs, v. 24, no. 7, p. 102.

Johnson, R.C., and May, Fred, 1980, A study of the Cretaceous-Tertiary unconformity in the Piceance Basin, Colorado: The underlying Ohio Creek Formation (Upper Cretaceous) redefined as a member of the Hunter Canyon or Mesaverde Formation: U.S. Geological Survey Bulletin 1482-B, 27 p.

Johnson, S.Y., Schenk C.J., and Karachewski, J.A., 1988, Pennsylvanian and Permian depositional systems in the Eagle Basin, northwest Colorado: in Holden, G.S., ed., Geological Society of America 1888-1988 Centennial Meeting Field Trip Guidebook: Colorado School of Mines Professional Contributions 12, p. 156-175.

Johnson, S.Y., Ander, D.L., and Tuttle, M.L., 1990, Sedimentology and petroleum occurrence, Schoolhouse Member, Maroon Formation (Lower Permian), northwestern Colorado: American Association of Petroleum Geologists Bulletin, v. 72, no. 2, p. 135-150.

Kirkham, R. M., and Widmann, B.L., 1997, Geologic map of the Carbondale quadrangle, Garfield County, Colorado: Colorado Geological Survey Open-File Report 97-3, scale 1:24,000.

Kirkham, R. M., Streufert, R.K., and Cappa, J.A., 1995, Geologic map of the Glenwood Springs quadrangle, Garfield County, Colorado: Colorado Geological Survey Open-File Report 95-3, scale 1:24,000.

Kirkham, R.M., Bryant, Bruce, Streufert, R.K., and Shroba, R.R., 1996a, Field-trip Guidebook on the geology and geologic hazards of the Glenwood Springs area, Colorado, in Thompson, R.A., Hudson, M.R., and Pillmore, C.L., eds., Geologic excursions to the Rocky Mountains and beyond: Colorado Geological Survey Special Publication 44, CD-ROM, 38 p.

Kirkham, R. M., Streufert, R.K., Hemborg, T.H., and Stelling, P.L., 1996b, Geologic map of the Cattle Creek quadrangle, Garfield County, Colorado: Colorado Geological Survey Open-File Report 96-1, scale 1:24,000.

Kirkham, R.M., Streufert, R.K., and Cappa, J.A., 1997a, Geologic map of the Glenwood Springs quadrangle, Garfield County, Colorado: Colorado Geological Survey Map Series 31, 22 p., scale 1:24,000.

Kirkham, R.M., Streufert, R.K., Scott, R.B., Lidke, D.J., Bryant, Bruce, Perry, W.J., Jr., Kunk, M.J., Driver, N.E., and Bauch, N.J., 1997b, Active salt dissolution and resulting geologic collapse in the Glenwood Springs region of west-central Colorado: Geological Society of America Abstracts with Programs, v. 29, no. 6, p. A-416.

Kunk, M.J., and Snee, L.W., 1998, $^{40}Ar/^{39}Ar$ age-spectrum data of Neogene and younger basalts in west-central Colorado: U.S. Geological Survey Open-File Report 98-243, p. 112.

Kunk, M.J., Kirkham, Robert, Streufert, R.K., Scott, R.B., Lidke, D.J., Bryant, Bruce, and Perry, W.J., 1997, Preliminary constraints on the timing of salt tectonism and geologic collapse in the Carbondale and Eagle collapse centers, west-central Colorado: Geological Society of America Abstracts with Programs, v. 29, no. 6, p. A-416.

Larson, E.E., Ozima, Minoru, and Bradley, W.C., 1975, Late Cenozoic basic volcanism in northwestern Colorado and its implications concerning tectonism and the origin of the Colorado River system, in Curtis, B.F., ed., Cenozoic history of the southern Rocky Mountains: Geological Society of America Memoir 155, p. 155-187.

Lidke, D.J., 1998, Geologic map of the Wolcott quadrangle, Eagle County, Colorado: U.S. Geological Survey Miscellaneous Investigations Series Map I-2656, scale 1:24,000.

Lidke, D.J., 1999, Geologic map of the Eagle quadrangle, Eagle County, Colorado: U.S. Geological Survey Miscellaneous Field Studies Map MF-in press, scale 1:24,000.

Liebermann, T.D., Mueller, D.K., Kircher, J.E., and Choquette, A.F., 1989, Characteristics and trends of streamflow and dissolved solids in the Upper Colorado River Basin, Arizona, Colorado, New Mexico, Utah, and Wyoming: U.S. Geological Survey Water Supply Paper 2358, 64 p.

Mallory, W.W., 1966, Cattle Creek anticline, a salt diapir near Glenwood Springs, Colorado, in Geological Survey Research, 1966: U.S. Geological Survey Professional Paper 550-B, p. B12-B15.

Mallory, W.W., 1971, The Eagle Valley Evaporite, northwest Colorado-A regional synthesis: U. S. Geological Survey Bulletin 1311-E, p. E1-E37.

Murray, F.N., 1966, Stratigraphy and structural geology of the Grand Hogback monocline, Colorado: Boulder, Colo., University of Colorado, unpublished Ph.D. dissertation, 219 p.

Murray, F.N., 1969, Flexural slip as indicated by faulted lava flows along the Grand Hogback Monocline: Journal of Geology, v. 77, p. 333-339.

Piety, L.A., 1981, Relative dating of terrace deposits and tills in the Roaring Fork valley, Colorado: Boulder, Colo., University of Colorado, unpublished M.S. thesis, 209 p.

Reed, J.C., Jr., Bryant, Bruce, Sims, P.K., Beaty, D.W., Bookstrom, A.A., Grose, T.L., Mallory, W.W., Wallace, S.R., and Thompson, T.B. 1988, Geology and mineral resources of central Colorado, in Holden, G.S., ed., Geological Society of America 1888-1988 Centennial Meeting, Denver, Colorado: Colorado School of Mines Professional Contributions 12, p. 68-121.

Reed, J.C., Jr., Bickford, M.E., and Tweto, Ogden, 1993, Proterozoic accretionary terranes of Colorado and southern Wyoming, in Van Schmus, R.R., and Bickford, M.E., eds., Transcontinental provinces, in Reed, J.C., Jr., Bickford, M.E., Houston, R.S., Link, P.K., Rankin, D.W., Sims, P.K., and Van Schmus, W.R., eds., Precambrian of the conterminous United States, v. C-2 of Geology of North America: Geological Society of America Decade of North American Geology, p. 211-228.

Scott, R.B., and Shroba, R.R., 1997, Revised preliminary geologic map of the New Castle quadrangle, Garfield County, Colorado: U.S. Geological Survey Open-File Report 97-737, scale 1:24,000.

Scott, R.B., Lidke, D.J., Shroba, R.R., Hudson, M.R., Kunk, M.J., Perry, W.J., Jr., and Bryant, Bruce, 1998, Large-scale active collapse in western Colorado: Interaction of salt tectonism and dissolution, in Brahana, J.V., Eckstein, Yoram, Ongley, L.K., Schneider, Robert, and Moore, J.E., eds., Gambling with Groundwater—Physical, chemical, and biological aspects of aquifer-stream relations: Proceedings of the joint meeting of the XXVIII Congress of the International Association of Hydrologists and the annual meeting of the American Institute of Hydrologists, Las Vegas, Nevada, USA, 28 September-2 October, 1998, p. 195-204.

Scott, R.B., Shroba, R.R., and Egger, A.E. 1999, Geologic map of the Rifle Falls quadrangle, Garfield County, Colorado: U.S. Geological Survey Miscellaneous Investigations Series, scale 1:24,000, in press.

Shroba, R.R., and Scott, R.B., 1997, Geology of the Rifle quadrangle, Garfield County, Colorado: U.S. Geological Survey Open File Report 97-852, scale 1:24,000.

Shroba, R.R., and Scott, R.B., 1999, Geology of the Silt quadrangle, Garfield

County, Colorado: U.S. Geological Survey Miscellaneous Investigations Series, scale 1:24,000, in press.

Soule, J.M., and Stover, B.K., 1985, Surficial geology, geomorphology, and general engineering geology of parts of the Colorado River Valley, Roaring Fork River Valley, and adjacent areas, Garfield County, Colorado: Colorado Geological Survey Open-File Report 85-1.

Steven, T.A., Hon, K., and Lanphere, M.A., 1995, Neogene geomorphic evolution of the central San Juan Mountains near Creede, Colorado: U.S. Geological Survey Miscellaneous Investigations Series Map I-2504.

Steven, T.A., Evanoff, Emmett, and Yuhas, R.H., 1997, Middle and Late Cenozoic tectonic and geomorphic development of the Front Range of Colorado, in Bolyard, D.W., and Sonnenberg, S.A., eds., Geologic history of the Colorado Front Range: Rocky Mountain Association of Geologists 1997 RMS-AAPG Field Trip #7, p. 115-124.

Stover, B.K., 1986, Geologic evidence of Quaternary faulting near Carbondale, Colorado, with possible associations to the 1984 Carbondale earthquake swarm, in Rogers, W.P., and Kirkham, R.M., eds., Contributions to Colorado seismicity and tectonics-A, 1986 update: Colorado Geological Survey Special Publication 28, p. 295-301.

Streufert, R.K., Kirkham, R.M., Shroeder, T.S., and Widmann, B.L., 1997a, Geologic map of the Dotsero quadrangle, Eagle and Garfield Counties, Colorado: Colorado Geological Survey Open-File Report 97-2, scale 1:24,000.

Streufert, R.K., Kirkham, R. M., Widmann, B.L., and Shroeder, T.S., 1997b, Geologic map of the Cottonwood Pass quadrangle, Eagle and Garfield Counties, Colorado: Colorado Geological Survey Open-File Report 97-4, scale 1:24,000.

Tweto, Ogden, 1975, Laramide (Late Cretaceous-early Tertiary) orogeny in the southern Rocky Mountains, in Curtis, B.F., ed., Cenozoic history of the Southern Rocky Mountains: Geological Society of America Memoir 144, p. 1-44.

Tweto, Ogden, 1977, Tectonic history of west-central Colorado, in Veal, H.K., ed., Exploration frontiers of the central and southern Rockies: Rocky Mountain Association of Geologists-1977 Symposium, p. 11-22.

Tweto, Ogden, 1979, Geologic Map of Colorado: U.S. Geological Survey, 1:500,000 scale.

Tweto, Ogden, and Lovering, T.S., 1977, Geology of the Minturn 15-minute quadrangle, Eagle and Summit Counties, Colorado: U.S. Geological Survey Professional Paper 956, 96 p.

Tweto, Ogden, Moench, R. H., and Reed, J.C., Jr., 1978, Geologic map of the Leadville 1o x 2o quadrangle, northwestern Colorado: U.S. Geological Survey Miscellaneous Investigations Series Map I-999, 1:250,000 scale.

Unruh, J.R., Wong, I.G., Bott, J.D., Silva, W.J., and Lettis, W.R., 1993, Seismotectonic evaluation, Rifle Gap Dam, Silt Project, Ruedi Dam, Fryingpan-Arkansas Project, northwestern Colorado: unpublished report prepared by William R. Lettis and Associates and Woodward-Clyde Consultants for U.S. Bureau of Reclamation, 154 p.

U.S. Department of the Interior, 1997, Quality of water, Colorado River Basin, Progress Report no. 18, Bureau of Reclamation, Salt Lake City, Utah.

Wanek, L.J., 1953, Geology of an area east of Wolcott, Eagle County, Colorado: Boulder, Colo., University of Colorado, unpublished M.S. thesis, 62 p.

Widmann, B.L., 1997, Evaporite deformation in the Dotsero, Gypsum, and Cottonwood Pass 7.5' quadrangles, Eagle County, Colorado: Golden, Colo., Colorado School of Mines, unpublished M.S. thesis, 93 p.

Ye, Hongzhuan, Royden, Leigh, Burchfiel, Clark, and Schuepbach, Martin, 1996, Late Paleozoic deformation of interior North America: The greater ancestral Rocky Mountains: American Association of Petroleum Geologists Bulletin, v. 80, p. 1397-1432.

Geological Society of America
Field Guide 1
1999

Coal mining in the 21st century: Yampa coal field, northwest Colorado

Michael E. Brownfield, Edward A. Johnson, Ronald H. Affolter, and Charles E. Barker
U.S. Geological Survey, MS 939, Box 25046, Denver Federal Center, Denver, Colorado 80225, United States

INTRODUCTION

This two-day excursion will travel to the Yampa coal field located in parts of Moffat, Rio Blanco, and Routt Counties, northwestern Colorado. The excursion will visit classic regression/transgression successions in the Upper Cretaceous coal-bearing Mesaverde Group, which was deposited along the western edge of the Cretaceous seaway. It includes visits to inactive and active mine sites where past and current mining practices will be discussed. This guide summarizes the stratigraphy, sedimentology, and coal geology of the Yampa coal field. The trip will emphasize the depositional setting, sedimentology, and quality of the Upper Cretaceous coals and coal-bearing strata of the Mesaverde Group.

In an era when coal mining and utilization is becoming more environmental friendly, many people are unaware that more than 60 percent of our electric power is produced using coal. This trip will visit the Trapper Mine near Craig and discuss various environmental concerns related to the mining and utilization of coal, including reclamation, trace-element and mineral characterization, sulfur emissions, and effective waste management.

Location

The Yampa coal field (Fig. 1), covering about 1,700 square miles, is located in northwest Colorado in Routt, Moffat, and Rio Blanco Counties (Fig. 1). U.S. Highway 40 traverses the coal field on the north and connects the small communities of Milner and Lay, which roughly define the east-west limits of the coal field. Major cities in the area are Steamboat Springs just east of the coal field, Hayden in the central part of the field, and Craig in the western part of the coal field (Fig. 1). The coal field is also transversed by the west-flowing Yampa River. The Yampa coal field is characterized by southwestward-facing cliffs separated by steep canyons on the southern boundary, and north trending dip slopes along the northern boundary of the Williams Fork Mountains. Elevations in the coal field range from about 6,000 ft above sea level near Lay to just over 10,600 ft in the northeastern part of the coal field.

Previous Geologic Studies

Hayden (1877) first discussed the presence of coal in northwestern Colorado, and minor studies of the coal resources of the Yampa coal field began as early as the 1880's. Fenneman and Gale (1906) published the first notable investigation just after the turn of the twentieth century. Campbell (1923) investigated the coal deposits in the Twentymile district in the eastern part of the coal field and Hancock (1925) investigated the coal resources on the western side of the coal field. Bass and others (1955) compiled the results of fieldwork conducted from 1923 to 1949 on the eastern side of the coal field. These three publications contain the primary sources of information on the geology of the Yampa coal field. The selected references section at the end of this report contains most of the significant publications pertaining to this coal field.

Mining Activities

According to Boreck and Murray (1979), 192 mines have operated in the Yampa coal field since the 1860's. Small wagon mines provided coal for domestic use during the second half of the ninetieth century, but larger-scale mining was hampered by a lack of adequate transportation. Following the arrival of the Denver and Salt Lake Railroad (now the Denver and Rio Grande Western Railroad) in 1906 (Campbell, 1923), several underground mines were established near Oak Creek. As the railroad was extended from Steamboat Springs several mines were opened near Bear River, McGregor, and Mount Harris in the central part of the coal field (Fig. 1). Notable in the Mount Harris area was the Colorado and Utah Coal Company's Harris Mine south of the Yampa River, and the Victor-American Fuel Company Wadge Mine north of the river. At Bear River, a coal

Brownfield, M. E., Johnson, E. A., Affolter, R. H., and Barker, C. E., 1999, Coal mining in the 21st century: Yampa coal field, northwest Colorado, *in* Lageson, D. R., Lester, A. P., and Trudgill, B. D., eds., Colorado and Adjacent Areas: Boulder, Colorado, Geological Society of America Field Guide 1.

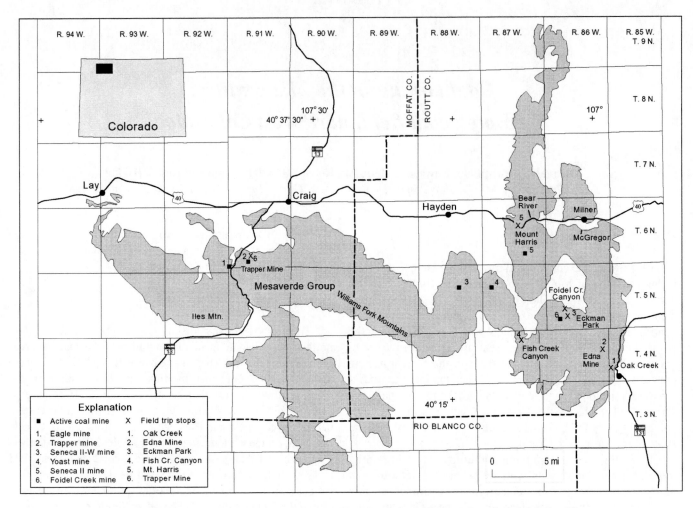

Figure 1. Location of the Yampa coal field, Moffat, Rio Blaco, and Routt Counties, Colorado. The outcrop of the Mesaverde Group (shaded) delineates the boundaries of the coal field. Approximate locations of existing mines and trip stops are shown.

bed about 600 feet below the top of the Trout Creek Sandstone was extensively mined. Two rail lines were constructed south of McGregor where several mines were developed in the Wadge and Wolf Creek coal beds. Later a strip mine was developed just south of McGregor that supplied coal for one of the early coal-fired power plants in northwest Colorado.

Large strip mines were developed during the 1950's in the eastern part of the coal field, notably the Edna Mine located about three miles north of Oak Creek, and continued into the 1970's with the Energy Fuels mines in the southern part of Twentymile Park. Both of these mines have ceased operations and are now in the reclamation stage.

Currently, only a few large mines operate in the Yampa coal field (Fig. 1). In the eastern part of the coal field, three major mines are the Foidel Creek Mine, located in the southern part of Twentymile Park, and the Seneca II-W, and Yoast Mines, located several miles south and southwest of Mount Harris. In the western part of the coal field, three notable mines are the Trapper Mine (surface), located about five miles south of Craig,

the Eagle No 5 Mine, and the Eagle No 9 Mine, both located about 12 miles southwest of Craig. The Eagle mines are temporarily closed.

Two mine-mouth power plants produce electricity in the coal field, the Craig plant, located about four miles south-southwest of Craig, and the Hayden plant, located about five miles east of Hayden. Tri State Generation and Transmission Association, Inc. own the Craig plant. Most coal burned at the plant comes from the Trapper Mine located about three miles to the south (Fig. 1). Public Service of Colorado, PaciCorp, and Salt River Project jointly own the Hayden plant. All of the coal burned at this plant comes from the Seneca II-W and Yoast mines located two miles to the south.

During 1997, the Yampa coal field accounted for about 40 percent of Colorado's total coal production (Resource Data International, 1998). The cumulative coal production for the Yampa coal field, from 1864 through 1997, is 266.19 million tons (Tremain and others, 1996; and, Resource Data International, 1998).

GEOLOGICAL SETTING

Cretaceous Paleogeography

During the Cretaceous Period, a large, north-trending epicontinental seaway, the Western Interior Seaway, occupied what is now central North America (Fig. 2). The seaway stretched from Mexico to Alaska in what is now the central part of the North America, the width of the seaway extended from central Utah to eastern Nebraska. A stable cratonic platform bordered the seaway on the east, and the tectonically active Sevier orogenic belt bordered the seaway on the west. Sediments moving

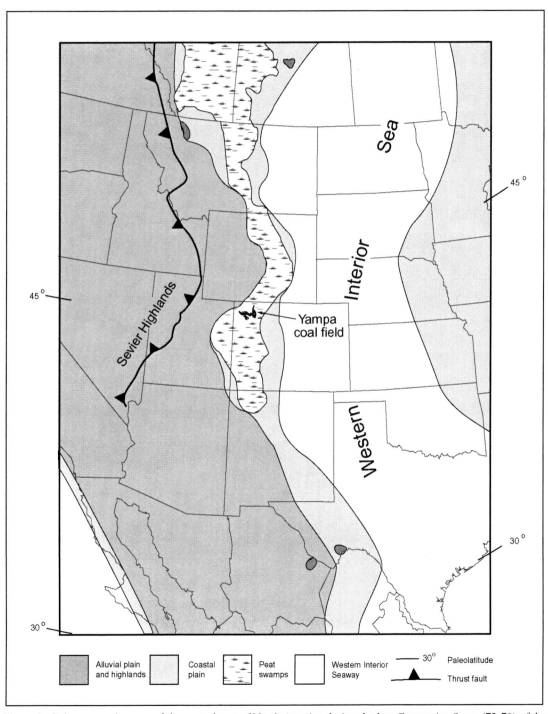

Figure 2. Paleogeography map of the central part of North America during the late Campanian Stage (72-79) of the Late Cretaceous Period. The Yampa coal field is shown in relation to the western shoreline, coastal plain, and peat swamps associated with the Western Interior Seaway. Modified from Roberts and Kirschbaum (1995).

eastward from the highlands were deposited along the fluctuating shoreline resulting in a complex package of Cretaceous sedimentary formations.

Stratigraphy of Upper Cretaceous Sedimentary Rocks in Northwestern Colorado

In ascending order, the Upper Cretaceous lithostratigraphic units in northwestern Colorado (Fig. 3) are Mancos Shale, Mesaverde Group (Iles Formation and Williams Fork Formation), Lewis Shale, Fox Hills Sandstone, and Lance Formation. Major unconformities separate the Lance Formation from the

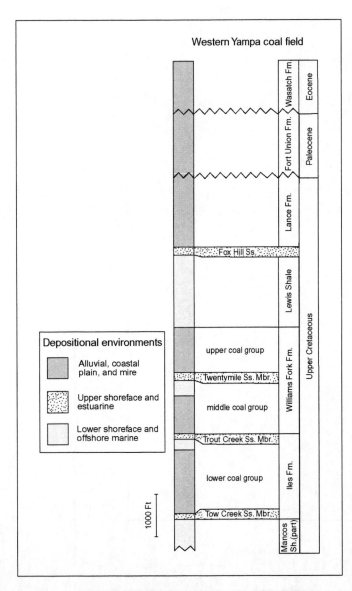

Figure 3. Stratigraphic column of upper Cretaceous and Tertiary rocks in the Yampa coal field showing depositional environments and coal groups. Modified after Bass and others (1955).

overlying Paleocene Fort Union Formation and the Fort Union from the overlying Wasatch Formation. The oldest Cretaceous unit exposed in the vicinity of the Yampa coal field is the Mancos Shale, which forms areas of low relief just south of the coal field. The youngest Cretaceous unit exposed in the vicinity of the coal field is the Lance Formation, which forms low hills just north of the coal field.

Mesaverde Group in the Yampa coal field

Holmes (1877) first applied the named Mesaverde Group to a sequence of sandstone, mudrock (siltstone, mudstone, shale, and claystone), and coal in southwestern Colorado. Fenneman and Gale (1906b), noting a similar sequence of rocks in the Yampa coal field, extended the name Mesaverde into northwestern Colorado. Fenneman and Gale did not subdivide the Mesaverde into formations, but they did describe two regional sandstones, which they named the Trout Creek and the Twentymile Sandstones. In addition, they noted that coal in the Mesaverde could be described as occurring in a lower, middle, or an upper coal group. The lower group contains all the coal below the Trout Creek, the middle group contains all the coal between the Trout Creek and the Twentymile, and the upper group contains all of the coal above the Twentymile.

Hancock (1925) later subdivided the Mesaverde Group into the Iles Formation and overlying Williams Fork Formation, and defined the Trout Creek Sandstone as a member at the top of the Iles and the Twentymile as a member more or less in the middle of the Williams Fork. Bass and others (1955) later extended these names into the eastern part of the coal field. Masters (1966) introduced the name Mount Harris member for the lower part of those rocks between the Trout Creek and the Twentymile, and reintroduced an earlier name, Holderness member, for those rocks in the Williams Fork above the Twentymile; neither of these names are used today.

In northwest Colorado, the Mesaverde (Fig. 4) comprises an eastward thinning, wedge-shaped package of marine and nonmarine rocks that overlies the Mancos Shale, which contains *Baculities perplexus* in its upper part, and underlies the Lewis Shale, which contains *Baculities clinolobatus* (Izett and others, 1971). Toward the east, the Mesaverde pinches out into marine rocks of the Pierre Shale, but evidence of this transition was lost to erosion when the Park Range was uplifted during the Tertiary. Toward the west, the Mesaverde becomes increasingly more fluvial to the point that in central Utah equivalent strata are composed almost exclusively of conglomerate.

DETAILED STRATIGRAPHY OF THE UPPER CRETACEOUS

Iles Formation

The Iles Formation (Fig. 3) of the Mesaverde Group was named by Hancock (1925) for coal-bearing rock exposed at Iles

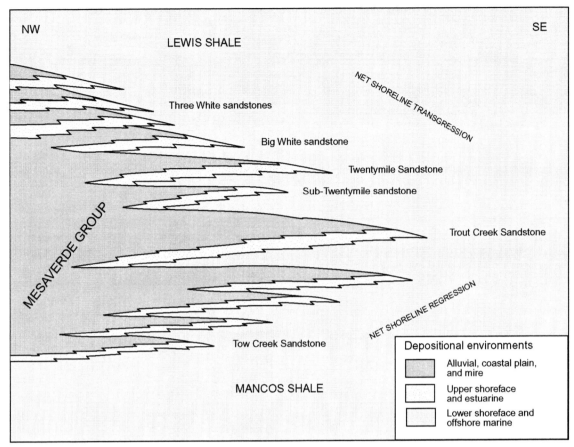

Figure 4. Generalized northwest-southeast cross section of the Mesaverde Group in the Yampa coal field showing the stratigraphic positions of major sandstone units and related major regressions and transgressions. Modified after Seipman (1985).

Mountain (Fig. 1) in the western part of the Yampa coal field. The Iles Formation is conformable with the Mancos Shale below and the Williams Fork Formation above. The Iles consists of sandstone interbedded with mudrock (siltstone, mudstone, shale, and claystone), carbonaceous shale, and coal and ranges in thickness from about 1,300 feet in the west (Hancock, 1925) to 1,500 in the eastern and central (Bass and others, 1955) parts of the coal field.

The lower two-thirds of the Iles Formation consists of massive ledge-forming beds of sandstone interbedded with mudrock, carbonaceous shale, and coal. This sequence of rocks form steep cliffs that rise above the broad lowland formed on the Mancos Shale along the southern and western boundaries of the Yampa coal field. Most of the coal beds within the lower coal group are in the upper part of this sequence and occur about 400 feet above the base of the Iles Formation (Bass and others, 1955). Three principal coal beds or zones were recognized by Fenneman and Gale (1906) in the lower coal group at Oak Creek and are referred to as No. 1, No. 2, and No. 3 coal zones (Bass and others, 1955). The upper part of the Iles Formation consists of a mudrock sequence capped by cliff forming sandstone. The mudrock sequence is a transgressive marine

tongue (Fig. 5) that ranges in thickness from about 400 feet just west of Oak Creek to about 100 feet in the western part of the coal field capped by the Trout Creek Sandstone.

Three persistent sandstone units in the Iles Formation deserve special mention as guides for correlation within the coal field. They are the Tow Creek Sandstone Member at the base, the double ledge sandstone unit about 400 feet above the base, and the Trout Creek Sandstone Member at the top of the formation.

Tow Creek Sandstone Member. The Tow Creek Sandstone Member of the Iles Formation (Figs. 3, 4, and 5) is the basal unit in the Iles in the eastern and central part of the Yampa coal field. Crawford and others (1920) named the Tow Creek for exposures between Milner and Bear River (Fig. 1). The Tow Creek consists of light-gray to white sandstone ranging in thickness from 35 to 125 feet (Bass and others, 1955). This unit is prominent west of Fish Creek canyon (Fig. 1) in the south central part of the coal field.

Double Ledge Sandstone. Bass and others (1955) informally named the double ledge sandstone. The double ledge sandstone unit consists of 1 to 3 beds of light-gray and white cliff-forming sandstone and interbedded siltstone unit about

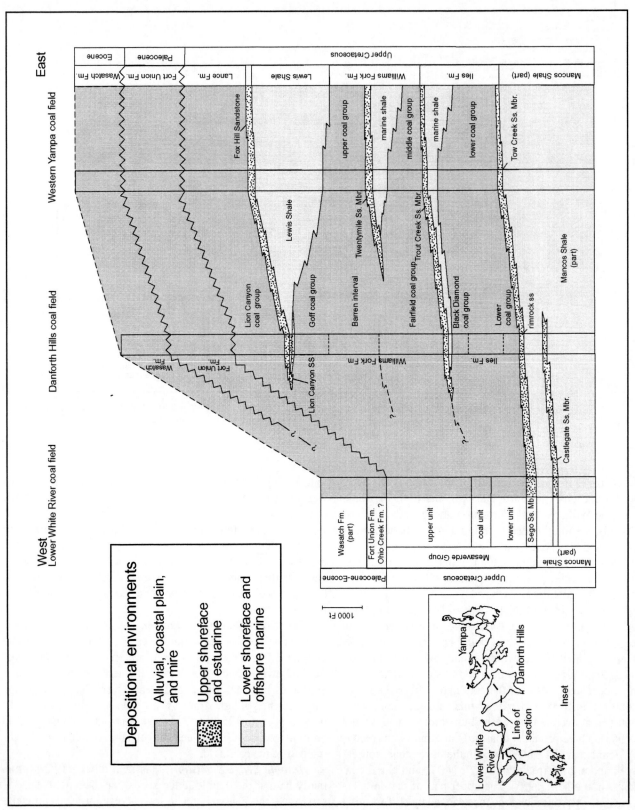

Figure 5. Generalized cross section showing part of the Upper Cretaceous and Tertiary rocks in the Lower White River, Danforth Hills, and Yampa coal fields with interpreted depositional environments.

400 to 460 feet above the base of the Iles Formation. The double ledge ranges in thickness from 130 to 250 feet. The unit forms prominent ledges visible along the steep south facing cliffs Williams Fork Mountains in the south central part of the Yampa coal field. North of Milner, the unit forms a single white ledge where it was useful in determining displacement on a fault.

Trout Creek Sandstone. Fenneman and Gale (1906) named the Trout Creek Sandstone Member of the Iles Formation (Figs. 3) for exposures along Trout Creek in the eastern part of the coal field. The Trout Creek is a cliff-forming unit of regional extent that transitionally overlies a sequence of marine shale (Figs. 4 and 5) containing *Exiteloceras jenneyi* (Izett and others, 1971) and is overlain by coal-bearing rocks. Fenneman and Gale (1906) and Bass and others (1955) reported the Trout Creek to be about 100 ft thick in the eastern part of the coal field and Johnson (1987) reported the unit as ranging from 67 to 79 ft thick in the western part of the coal field. Siepman (1985), in his regional study of the unit, determined that it ranges from 140 ft thick near Mount Harris to 31 ft thick just west of Oak Creek. The unit has an average thickness of about 67 ft. Toward the northwest, the Trout Creek pinches out in the subsurface of the Sand Wash Basin (Siepman, 1985; Roehler, 1987). Regionally, the Trout Creek (Fig. 5) is well exposed in the Danforth Hills (Hancock and Eby, 1930) and pinches out just east of the Lower White River coal field (Hail, 1978, Barnum and Garrigues, 1980). Toward the south, the Trout Creek is equivalent to the Rollins Sandstone in the southern Piceance Basin of west-central Colorado (Collins, 1976).

Williams Fork Formation

Hancock (1925) named the Williams Fork Formation for coal-bearing strata in the upper part of the Mesaverde Group. The Williams Fork rests conformably on the top of the Trout Creek Sandstone Member of the Iles Formation. The upper contact of the Williams Fork with the overlying Lewis Shale is also conformable. The Williams Fork Formation consists of mudrock interbedded with sandstone and lessor amounts of carbonaceous shale and coal. These rocks are well exposed in the Williams Fork Mountains (Fig. 1).

In the western part of the coal field, Hancock (1925) reported that the Williams Fork is approximately 1,600 ft thick; Johnson (1987) reported the formation to be about 1,880 ft thick in the Round Bottom quadrangle southwest of Craig. In the eastern part of the coal field, Bass and others (1955) reported the formation to range from 1,600 ft thick near Mount Harris to nearly 2,000 ft thick at the western margin of their study area. Four USGS coal exploration drill holes located in the western part of the coal field show an average thickness of 1,915 ft for the Williams Fork.

The lower two-thirds of the Williams Fork Formation consists of massive ledge-forming light gray and white sandstone interbedded with mudrock, carbonaceous shale, and coal of the middle coal group (Fig 3). Throughout the eastern part of the coal field the principal coal beds are, in ascending order, the Wolf Creek, Wadge, and Lennox (Bass and others, 1955). These beds occur in the lower 400 feet of the Williams Fork, immediately above the Trout Creek Sandstone. Data compiled by the authors suggest that the number of coal beds in the middle coal group increases westward in the Yampa coal field (Johnson and others, in press). The upper part of the middle coal group is characterized by mudrock capped by cliff-forming sandstone. The upper third of the Williams Fork contains the upper coal group and consists of sandstone, mudrock, carbonaceous shale, and coal.

A regional cross section by Roehler (1987) that extends from the vicinity of Mount Harris, Colorado northwest to Rock Springs, Wyoming shows that the Williams Fork is roughly equivalent to the upper part of the Ericson Sandstone and the Almond Formation. Southwest of the coal field, the Williams Fork is also used in the Danforth Hills (Fig. 5) coal field (Hancock, 1925, Hancock and Eby, 1930)

Five persistent stratigraphic units in the Williams Fork Formation have proven very useful as guides for correlation within the coal field. They are: 1) the Yampa bed, 2) sub-Twentymile sandstone in the middle coal group, 3) the Twentymile Sandstone Member, 4) the Big White sandstone in the upper coal group, and 4) the Three White sandstones at the top of the formation.

Yampa Bed. The term Yampa bed, introduced by Brownfield and Johnson (1986), is a regionally extensive tonstein (altered ash-fall tuff) in the lower part of the Williams Fork Formation. When exposed on the surface or observed in drill core, the unit is a white to grayish white, massisve, claystone. In the subsurface, the unit becomes an important regional marker bed that is easily identified on drill hole geophysical logs. The Yampa bed ranges in thickness from less than one foot to six ft. In the central and western parts of the coal field, the unit lies between 100 and 260 ft above the top of the Trout Creek Sandstone respectively. However, in the eastern part of the coal field, the stratigraphic separation is less than 20 ft, and on some geophysical logs, the Yampa bed appears to rest directly on the Trout Creek. The age of the Yampa bed, 72.5 mybp ± 5.1 my, was determined using K-Ar methods on andesine.

The Yampa bed has been identified in the Danforth Hills (Brownfield and others, in press) and in the subsurface of the Piceance and Sand Wash Basin (Brownfield and Johnson, 1986).

Sub-Twentymile Sandstone. The term sub-Twentymile sandstone was introduced by Kitely (1983) for a sandstone unit that lies about 150 ft below the base of the Twentymile Sandstone in the eastern-central part of the coal field (Fig. 4). Most of what is known about this unit comes from the subsurface, but it probably has many physical characteristics in common with the Trout Creek and Twentymile Sandstones. Based on information compiled by the authors, the unit averages about 30 ft thick and contains one to three sandstone bodies. Masters

(1966) defined a sandstone unit, the Hayden Gulch sandstone, at this same stratigraphic level, and this sandstone is undoubtedly equivalent to the sub-Twentymile. Siepman (1983) recognized the sub-Twentymile, and included it at the top of his sub-Twentymile unit. The sub-Twentymile pinches out toward the east, and toward the west, it looses its distinctiveness by splitting into a number of smaller sandstone bodies.

Twentymile Sandstone. The Twentymile Sandstone Member of the Williams Fork Formation was named by Fenneman and Gale (1906) for exposures in Twentymile Park (Fig. 1) in the eastern part of the coal field. The stratigraphic distance between the top of the Trout Creek Sandstone and the base of the Twentymile ranges from about 900 to 1,100 ft (Bass and others, 1955). The Twentymile typically forms distinctive cliffs of yellowish gray to white sandstone. Unlike the Trout Creek, the Twentymile often comprises two to three sandstone units separated by finer grained material. Bass and others (1955) reported that the Twentymile is about 100 to 200 ft thick in the eastern part of the coal field. Siepman (1985) reported that the unit ranges from 184 ft thick in Fish Creek Canyon in the eastern part of the coal field to 28 ft thick at Duffy Mountain in the western part of the coal field. The Twentymile Sandstone begins to lose its identity on the western edge of the coal field (Fig. 5), and pinches out toward the northwest in the subsurface of the Sand Wash Basin (Siepman, 1985) and to the southwest where it is not present in the Danforth Hills (Brownfield and others, in press).

Big White Sandstone. The term Big White sandstone was introduced by W. R. Grace and Company for a sandstone unit that lies about 200 ft above the Twentymile Sandstone in the central part of the coal field. In the western part of the coal field, a discontinuous sandstone exists at the same stratigraphic level as the Big White and is referred to by USGS geologists as the Fuhr Gulch sandstone of the Williams Fork Formation (Johnson, 1987). Johnson (1987) describes the unit at the mouth of Fuhr Gulch as a light-gray, very fine to fine grained sandstone with a thickness of 40 ft. In the eastern part of the coal field, Campbell (1923) named a sandstone unit at the same stratigraphic level of Big White (in Eckman Park, Figs. 1 and 8) the Fish Creek sandstone. With little doubt, all three sandstone units are equivalent, and represent a poorly exposed, regional unit.

Three White Sandstones. In the western part of the coal field, the upper part of the Williams Fork Formation is dominated by three, vertically stacked, thick sandstone units that are referred to informally by USGS geologists as the Three White sandstones (Fig. 4). Lithologically, the three sandstones resemble the Twentymile Sandstone. In addition, each sandstone unit overlies a thin sequence of marine rocks and is overlain by non-marine rocks. Each of the three sandstone units is 50 to 60 ft thick. Where best exposed along the Yampa River southwest of Craig, the base of the lower sandstone lies about 310 ft above the top of Big White sandstone. The stratigraphic distance between the base of the lower sandstone and the base of the

Lewis Shale is about 435 ft. Toward the west, the lower and middle sandstones pinch out, but the upper sandstone continues to the edge of the coal field. Toward the east, the three sandstones continue for several miles east of the Yampa River before pinching out.

DEPOSITIONAL SETTING OF THE MESAVERDE GROUP

During the Late Cretaceous, the western edge of the Western Interior Seaway was continually being modified by sediment influx from tectonically active areas to the west. According to Haun and Weimer (1960), as much as 11,000 vertical feet of sediment were deposited in the seaway during this time. Along the western margin of the Cretaceous Western Interior Seaway, the Mesaverde deposition was characterized by a series of westward transgressions and eastward regressions of the epicontinental seaway that resulted in the cyclic deposition of marine and non-marine lithofacies (Weimer, 1960; Zapp and Cobban, 1960). The sediment source area was located to the west in the Sevier orogenic belt.

The sediments deposited in northwestern Colorado during this time are now contained in the Mancos Shale and Iles and Williams Fork Formations of the Mesaverde Group (Fig. 3). Variations in the sedimentation rate and drainage patterns, basin subsidence, and eustatic changes in sea level contributed to the cyclic deposition of Upper Cretaceous lithofacies. Masters (1966) recognized three large-scale, regressive cycles in the Mesaverde: (1) from the base of the Iles to the base of the marine shale underlying the Trout Creek Sandstone, (2) from the base of the Trout Creek Sandstone to the base of the marine shale underlying the Twentymile Sandstone, and (3) from the base of the Twentymile Sandstone to the top of the Williams Fork (Fig. 4). In each cycle, the top of the coal bearing package and the base of the overlying marine shale are separated by a transgressive disconformity. In general, the Iles represents net shoreline regression, and the Williams Fork represents a net shoreline transgression. However in detail, the Williams Fork shows evidence of several shoreline fluctuations, as evidenced by the vertical juxtaposition of the formation's three major depositional settings: offshore marine, nearshore marine, and fluvial.

The westward-thinning tongues of mudrock that directly underlie the Trout Creek and Twentymile Sandstones were deposited on an open-marine shelf, as evidenced by the presence of marine body fossils and deep-marine trace fossils. Most likely, this same environment is represented in the strata that underlie the sub-Twentymile sandstone in the central part of the coal field. A shallower marine environment is represented in the strata that underlie the Three White sandstones in the western part of the coal field.

The thick sandstones, such as the Trout Creek, Twentymile Sandstone Members, and the sub Twentymile, Big White, and Three White sandstones were deposited in a progradational shoreface environment (Boyles and others, 1981). This conclusion is supported by an upward increase in grain size and an upward decrease in bed thickness, hummocky cross stratification followed upward by trough cross stratification, and the occurrence of shallow-marine trace fossils. Relatively complete stratigraphic sections of the Trout Creek and Twentymile reveal that the units are composed of a basal transitional part deposited below wave base, and an overlying shoreface part deposited above wave base with rare foreshore and backshore deposits. Many workers, most recently Siepman (1985), believe that these sandstone units were deposited along wave-dominated deltaic, strand plain, and barrier island systems, in a microtidal setting.

The strata that overlie the nearshore marine sandstones display characteristics suggesting fluvial, lagoonal, salt marsh, and freshwater swamp environments that accumulated on a coastal plain that sloped gently seaward with little topographic relief. The fluvial lithofacies contain channel sandstones and associated overbank deposits of sandstone, siltstone, shale, and lenticular coal beds. The lagoon and bay environments are represented by low-energy deposits of shale, siltstone, and minor fine-grained sandstone. The lagoonal sandstone deposits include washover fans, flood-tidal deltas, and crevasse splays. The salt marsh lithofacies contain evenly laminated, fine-grained deposits consisting of carbonaceous shale, siltstone, very fine-grained sandstone, and impure coal. The freshwater swamp lithofacies are represented by coal.

DEPOSITIONAL HISTORY OF THE MESAVERDE GROUP IN THE YAMPA COAL FIELD

Prior to the deposition of the Mesaverde Group, the shoreline of the Western Interior Seaway (Fig. 2) was positioned about 75 miles west of what is now northwestern Colorado (Zapp and Cobban, 1960), and in the area of the coal field, marine mud (Mancos Shale), was accumulating. The lowest unit of the Mesaverde, the Tow Creek Sandstone Member of the Iles Formation, was deposited when an eastward migration of the sea allowed nearshore sandstone to be deposited over the marine mud (Tow Creek regression). As this process continued, coal-bearing sediments of the Iles prograded eastward over the Tow Creek. Minor fluctuations are represented by the double-ledge and Oak Creek sandstones in the eastern part of the coal field. Toward the end of Iles time, the shoreline transgressed westward flooding the area with sea water, and marine mud was deposited over the nonmarine sediments. This event is represented by the marine rocks that underlie the Trout Creek Sandstone. At end of Iles time, the shoreline again moved eastward across the coal field and the nearshore marine sand of the Trout Creek was deposited (Siepman, 1985).

The lower Williams Fork sediments (middle coal group) prograded eastward over the Trout Creek and accumulated just behind the strandline or low on the coastal plain. On the eastern side of the coal field, a minor westward shift of the shoreline caused an interval of marine sediments to be deposited, fol-

lowed by an eastward shift of the shoreline that allowed nearshore marine sand to be deposited, now represented by the sub-Twentymile sandstone and its underlying marine shale. Next, nonmarine sediments prograded eastward over the sub-Twentymile followed by a westward shift of the shore line flooding the area, and marine mud was deposited over the nonmarine sediments represented by the marine shale underlying the Twentymile Sandstone.

The last phase of the Williams Fork deposition occurred when the shoreline again moved eastward across the coal field and nearshore marine sand and overlying nonmarine sediments now contained in the Twentymile Sandstone and uppermost part of the Williams Fork (upper coal group) were deposited. The depositional history at the close of Williams Fork time is represented by a major westward transgression of the shoreline flooding the area with sea water, and the deposition of the Lewis Shale. This signaled the end of Mesaverde deposition in northwestern Colorado. However, on the western side of the coal field, the Lewis transgression was interrupted at least three times by minor fluctuations in the shoreline during which time the Three White sandstones were deposited.

POST-CRETACEOUS DEFORMATIONAL HISTORY OF NORTHWESTERN COLORADO

Near the end of the Cretaceous, the Western Interior Seaway withdrew from what is now northwestern Colorado and marine and coastal plain deposition ceased. This event was followed closely by compressional tectonism associated with the onset of the Laramide orogeny. In northwestern Colorado, the Lance and older rocks were folded, mildly uplifted, and eroded. The peneplain that resulted is now represented by a regional unconformity between the Upper Cretaceous and Paleocene rocks. As the orogeny intensified in earliest Tertiary, the Park Range was uplifted and eroded, and a thin veneer of gravel spread westward as an alluvial fan over the erosional surface. This deposit now forms the basal conglomerate of the Paleocene Fort Union Formation. With the uplift of the Park Range and other Laramide structures, the ancestral Sand Wash Basin was defined, and all Tertiary formations younger than the Fort Union are finer grained and show lateral facies changes consistent with deposition in a slowly subsiding basin. During middle and late Tertiary, extensional tectonism resulted in numerous normal faults and volcanism in the region.

Structural Features of the Yampa Coal Field

The Yampa coal field occupies the southeastern corner of the Sand Wash Basin (Fig. 6). In the western and central parts of the coal field, the structural dip is toward the north, but farther to the east, the regional structure swings counterclockwise until in the northeastern part of the coal field the dip is toward the west. Another regional structure of significance in the area of the coal field is the northwest-trending Axial Basin anticline

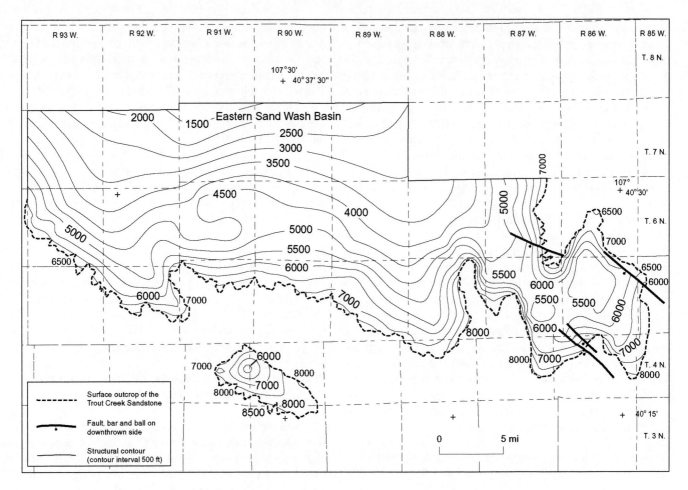

Figure 6. Structure contour map drawn on the top of the Trout Creek Sandstone Member of the Iles formation Yampa coal field. Surface outcrop of the Trout Creek modified after Tweto (1976).

(Fig. 7). This structure defines the western boundary of the coal field, and separates the Sand Wash Basin on the north from the Piceance Basin on the south. According to Stone (1986), this fold is a minor part of a much larger tectonic structure that extends from the Uinta Mountains in northeastern Utah to the Eagle Basin in north-central Colorado.

Folding in the Yampa coal field occurred after deposition of the Lance Formation but before deposition of the Fort Union Formation (Tweto, 1976). In general, folds within the coal field can be grouped into a western group and an eastern group (Fig. 7). On the east side of the coal field, southeast of Hayden, the fold axes trend north-northwest and north-north-east, and plunge in a southerly direction. These folds are asymmetrical with their axial planes inclined in a westerly direction. Starting from the west, the more significant folds are the Sage Creek anticline, the Fish Creek anticline, the Twentymile Park syncline, and the Tow Creek anticline. On the western side of the coal field, south and west of Craig, fold axes trend and plunge toward the northwest. Most of the folds are also asymmetrical with their axial planes inclined toward the northeast. Starting from the west, the more significant

folds are the Round Bottom syncline, the Williams Fork anticline, the Big Bottom syncline, and the Breeze Mountain-Buck Peak anticline.

In the southernmost part of the coal field, folds trend and plunge toward the northwest with their axial planes inclined toward the northeast following the structural pattern in the western part of the coal field. The dominant fold in this area is the Hart syncline. This structure is bounded on the north by the Beaver Creek anticline and on the south by the Seely anticline. The axes of these folds trend northwest, but the axis of the Hart syncline and the Seely anticline eventually veer westward.

Faulting in northwestern Colorado commenced during the middle or late Tertiary, and in the Sand Wash Basin units as young as the Miocene Browns Park Formation have been displaced (Tweto, 1976). Undoubtedly, faulting continued into the Quaternary and the region still experiences rare, mild earthquakes. Most of the faults in the coal field are high angle normal faults that trend northwest (Fig. 7). Displacements are down to the northeast or southwest, and horst and grabben structures are common. Overall, faulting has not disrupted the gross structure of the coal field to any significant degree.

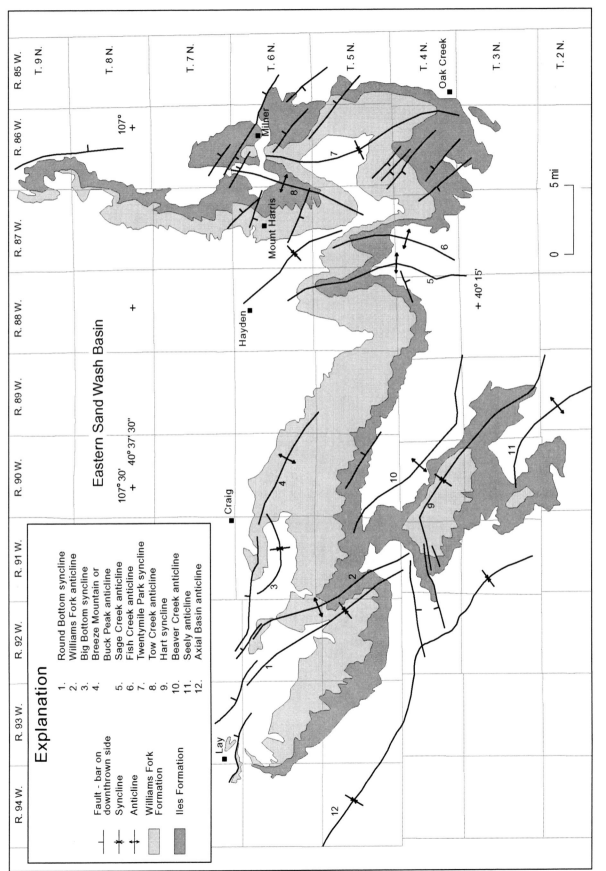

Figure 7. Generalized geology of the Yampa coal field showing the Iles and Williams Fork Formations, major faults, and folds. Modified after Tweto (1976).

COAL QUALITY

Most Cretaceous coal in the Yampa coal field is noncoking, high-volatile C bituminous, but some subbituminous A, B, and C is reported (Speltz, 1976). Rare anthracite is also found adjacent to Tertiary igneous intrusions. Coal analyses (as-received basis) from the lower coal group have a moisture content ranging from 6.3 to 12.2 percent, an ash yield ranging from 4.3 to 11.3 percent, a sulfur content ranging from 0.3 to 0.9 percent, and a caloric value ranging from 11,090 to 12,560 Btu/lb. Coals analyzed (as-received basis) from the middle coal group have a moisture content ranging from 7.7 to 11.8 percent, an ash yield ranging from 3.4 to 11.5 percent, a sulfur content ranging from 0.3 to 0.6 percent, and a caloric value ranging from 10,740 to 12,260 Btu/lb (Speltz, 1976).

The 3 major coal beds (Wolf Creek, Wadge, and Lennox) in the middle coal group were sample from the eastern part of the Yampa coal field and analyzed. The Wolf Creek bed has an average moisture content of 8.1 percent, an ash yield of 10.0 percent, a sulfur content of 0.72 percent, and a caloric value of 10,860 Btu/lb. The Wadge bed has an average moisture content of 9.3 percent, an ash yield of 7.8 percent, a sulfur content of 0.6 percent, and a caloric value of 11,180 Btu/lb. The Lennox bed has an average moisture content of 8.9 percent, an ash yield of 6.7 percent, a sulfur content of 2.6 percent, and a caloric value of 11,420 Btu/lb.

Little information is available on the quality of coal in the middle coal group in the western part of the coal field, but as-received data provided by the Eagle No. 9 Mine (Zook and Tremain, 1997) report the Eagle's F bed as subbituminous B, with ash yields ranging between 5.0 and 10.4 percent and a sulfur content ranging between 0.5 to 0.6 percent. The caloric value is reported to range between 10,380 and 11,570 Btu/lb.

The upper coal group is presently being mined in the western part of the coal field. The coal beds currently being extracted from the Trapper Mine are a mixture of subbituminous A, B, and C, and high-volatile bituminous C (Trapper Mine, oral commun., 1999). Coal beds mined during September 1998 have an ash yield of 7.05 percent and a sulfur content of 0.40 percent, with a caloric value of 9,930 Btu/lb (reported on an as-received basis).

A comparison of the mean content of trace elements of environmental concern (1990 Clean Air Act Amendment) is shown in Table 1. The Lennox, Wadge, Wolf Creek coal beds (eastern part of the coal field) of the middle coal group and the Q bed (Trapper Mine) of the upper coal group are compared to mean Cretaceous coal values for the Colorado Plateau (Affolter, in press).

COAL RESOURCES

The Yampa coal field contains an estimated remaining net coal resource of about 76 billion short tons in beds equal to or greater than 1.2 ft (Johnson and others, in press). Coal classified as identified makes up about 46 percent with the remaining 54 percent classified as hypothetical (Wood and others, 1983). Most of the hypothetical coal lies deep in the eastern Sand Wash Basin (Fig. 7).

Although a significant amount of coal remains in the Yampa coal field, this figure must be regarded with caution because it does not reflect geologic, land use, and environmental restric-

Table 1.—Comparison of the mean content of elements of environmental concern (1990 Clean Air Act Amendment) for the Lennox, Wadge, Wolf Creek, and Q beds with Cretaceous coal from the Colorado Plateau (Arizona, Colorado, New Mexico, and Utah). All elements are in parts per million (ppm) on a whole coal basis.

Element	Cretaceous Lennox bed mean[1] (n=61)	Cretaceous Wadge bed mean[1] (n=135)	Cretaceous Wolf Creek bed mean[1] (n=42)	Cretaceous Q bed mean[1] (n=12)	Cretaceous Colorado Plateau mean[2] (n=1265)
Antimony	0.2	0.3	0.3	0.3	0.5
Arsenic	3.9	1.0	1.3	0.6	1.6
Beryllium	1.5	1.2	1.3	0.49	1.2
Cadmium	0.04	0.04	0.08	0.01	0.1
Chromium	4.4	3.1	6.4	3.7	4.5
Cobalt	0.94	1.2	1.5	1.1	1.5
Lead	4.0	5.6	7.8	3.1	6.5
Manganese	14	9.5	60	17	22
Mercury	0.05	0.02	0.04	0.01	0.06
Nickel	3.5	2.9	4.2	4.3	3.7
Selenium	1.0	0.9	1.2	0.4	1.2
Uranium	0.45	0.92	1.4	0.67	1.3

[1] This report.
[2] Affolter, in press.

Figure 8. The Victor-American Fuel Company's Pinnacle Mine located about 1 mile southwest of Oak Creek, eastern Yampa coal field. The mine produced coal from the Pinnacle bed, lower coal group, Iles Formation (Bass and others, 1955). Photograph by Marius R. Campbell, 1921.

tions that might limit coal availability. For example, about 48 percent of the coal is under more than 3,000 ft of overburden and thus is unavailable for underground mining. In addition, coal beds that dip more than 12 degrees are unavailable for underground mining at any depth, and such resources in the vicinity of the Sage Creek, Fish Creek, and Tow Creek anticlines are considerable. But even more significant, about 90 percent of the coal is under more than 500 ft of overburden, and is thus unavailable for surface mining. In addition, thin coal beds would be spoiled during surface mining and the coal in beds greater than 14 ft would be left behind in underground mines. To conclude, it is apparent that only a small percentage of the 76 billion tons of remaining net coal resource could be recovered.

Over the next decade most of the coal produced in the Yampa coal field will come from areas presently under lease and from coal mines already in existence. The middle coal group will continue to dominate production in the eastern part of the field while the upper coal group will continue to dominate in the western part of the field. Coal mining in northwestern Colorado will remain an important social and economic factor in northwest Colorado well into the 21st century.

ITINERARY

The following is a brief description of the stops to be visited on the trip. Weather permitting, the field trip will visit all stops, but some may be modified because of road conditions and mining operations. Figure 1 is a location map of the Yampa coal field with the trip stops.

Day 1. Denver to Steamboat Springs

The trip begins at the Colorado Convention Center. Driving west. Drive west on I-70, through the Eisenhower Tunnel, to Silverthorne and then turn north on State Highway (SH) 9 to Kremmling. Proceed west on U.S. 40 to SH 134 (about 7 miles), then west over Gore Pass to Toponas. Turn right on SH 131 and drive north through Yampa and Phippsburg to Oak Creek. Turn west on Routt County Road (RC) 27 just north of Oak Creek.

Stops 1 and 2. Oak Creek and Edna Mine. Oak Creek is located on the southeastern margin of the Yampa coal field and is the eastern limit of outcrops of the Mesaverde Group. Fenneman and Gale (1906) conducted the first detailed study of the Oak Creek area. Development of coal in the Oak Creek area followed the arrival of the railroad in 1906. Several large mines were developed north and west of Oak Creek and continued into the 1940's. The Moffat Coal Company (Argo or Oak Hills Mine) and the Victor-American Fuel Company (Pinnacle Mine, Fig. 8) mined the Pinnacle bed (No. 2 bed of Bass and others, 1955, Fig. 9). Most of the coal mined in the Oak Creek district was from the lower coal group of the Iles Formation.

The Edna Mine is located about 3 miles west of Oak Creek and east of Trout Creek. The mine produced coal from the Wadge bed (Fig. 9). The Edna Mine is now abandoned and the mine area was reclaimed. About 1.5 miles south along Trout Creek, the Apex Mine produced coal from the No. 2 bed until it closed in the middle 1980's (Fig. 9). Fenneman and Gale (1906) named the Trout Creek Sandstone Member of the Iles for exposure on Trout Creek. The Trout Creek forms the prominent white sandstone just east of Trout Creek.

Within the Oak Creek and Edna Mine area, the Upper Cretaceous rocks are dominated by marine facies including offshore marine, lower and upper shoreface, estuarine, and mire. Gaffke (1979) studied a portion of the section located along the road to the old Oak Creek dump. Figure 10 shows a composite section with depositional environments interpreted for rocks exposed in the southern part of the Edna Mine. The section was constructed from USGS coal exploration drill-hole data and data collected by the authors. The middle coal group is about 300 ft thick in the vicinity of the Edna Mine. Estuarine deposits associated with a bay or lagoon dominate the section. Estuarine deposits contain brackish-water pelecypods and brachiopods. Coals in this section are overlain by splay deposits, or in the case of the Wadge coal, overlain by brackish water sediments. The splays have sharp contacts with the underlying coal suggesting that 1) the deposits are distal splays occurring in a low-energy environment, and 2) the peat is resistant to erosion. The rocks underlying the coals have sharp contacts and are not rooted suggesting that the initial peat formation may have been in the form of floating mats.

About 2 miles west of Stop 2, at Middle Creek, is an excellent exposure of rippled sandstone and siltstone below the Lennox coal bed. The ripples are symmetrical indicating oscillatory flow caused by small waves. The ripple marks are flat-topped, indicating that they were eroded by small waves that are best explained by tidal action. Associated with the wave ripples are *Arenicolites* (a marine to brackish-water trace fossil) and sea urchin burrows. The rippled deposit, therefore, indicates an intertidal (lagoonal) setting.

After examining the intertidal deposit, time permitting, walk west along the road (RC 27) and observe the facies change from coastal plain to offshore marine.

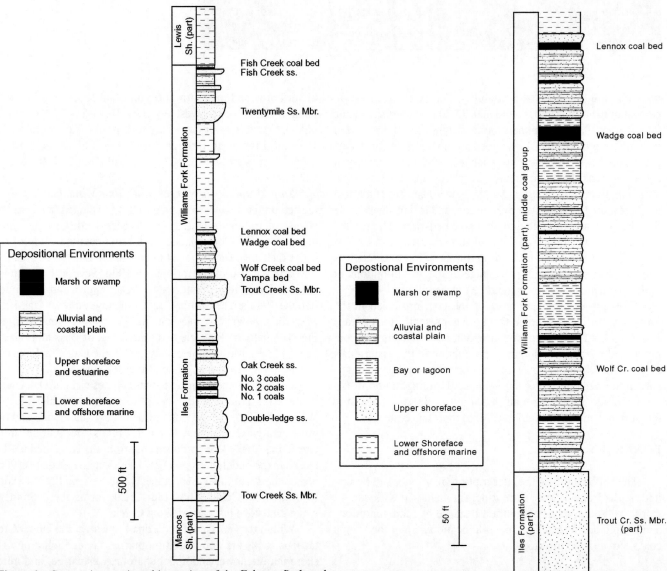

Figure 9. Composite stratigraphic section of the Eckman Park and Edna Mine area, eastern Yampa coal field, showing interpreted depositional environments and major coal beds. Compiled from USGS coal-exploration drill holes and field studies. Coal bed names from Bass and others (1955).

Figure 10. Stratigraphic section of the middle coal group in the southern part of the Edna Mine showing interpreted depositional environments and coal beds. Compiled from USGS coal-exploration drill holes and field data. Coal bed names from Bass and others (1955).

Stop 3. Eckman Park and Foidel Creek Mine. Eckman Park is about 7 miles west of Oak Creek on RC 27 and is developed on gently northward dipping strata of the middle coal group of the Williams Fork Formation. Eckman Park is the site of the old Energy 1 (strip) and the present day Foidel Creek (longwall) mines. The Energy 1 Mine produced coal from the Lennox, Wadge, and Wolf Creek coal beds and the Foidel Creek Mine produces coal from the Wadge. Using longwall mining methods the Foidel Creek Mine cut three one-mile long panels setting a monthly coal production world record in 1996 of just over one million tons (Wynn and Coates, 1998; Fiscor, 1998).

Within Eckman Park the Trout Creek Sandstone is conformably overlain by a 330 ft thick section of the middle coal group. The Twentymile Sandstone Member of the Williams Fork Formation forms massive white sandstone cliffs on the north and west sides of Eckman Park. The Twentymile overlies a 600 ft thick sequence of marine sediments containing Late Campanian *Baculites reesidei* (Izett and others, 1971). Underlying the marine shale section is about 330 ft of the middle coal group sediments. The overlying coal-bearing rocks (upper coal group) are about 190 feet thick and contain the Fish Creek coal bed (Fig. 8). The Fish Creek coal bed ranges from 1 to 6 ft thick and was mined by Energy Fuels Corporation in the 1970's. The Lewis Shale conformably overlies the Williams Fork. The Twentymile Sandstone, upper coal group, and Fish Creek sandstone are well exposed in the rail cut in Foidel Creek Canyon (Fig. 1).

Continue west on RC 27 to the junction with RC 33 and turn left passing the old Foidel Creek School. Driving south on RC 27 the Twentymile Sandstone forms massive cliffs to the right and the Foidel Creek Mine facilities are on the left. Normal faults can be seen offsetting the top of the Twentymile. Pinnacle Peak to the south is capped by the Twentymile Sandstone and bounded on the north and south by faults. RC 27 turns west passing through the Twentymile Sandstone. Drive west to RC 37 and turn left following Fish Creek southwest to Fish Creek Canyon.

Stop 4. Fish Creek Canyon. Fish Creek Canyon contains one of the best stratigraphic sections (Fig. 11) of the uppermost Mancos Shale and the Mesaverde Group in the eastern part of the coal field. On the west end of Fish Creek Canyon there are excellent exposures of the marine sandstone units in the upper Mancos Shale. The canyon section is dipping about 40° to the east on the western end and increases to 55° at the eastern end. The Iles Formation within the canyon is dominated by 4 progradational shoreface sandstone units: the Tow Creek Sandstone Member, the double-ledge sandstone (Bass and others, 1955), the Oak Creek sandstone (Kucera, 1959), and the Trout Creek Sandstone Member. The Tow Creek Sandstone Member of the Iles Formation is the basal unit of the Iles Formation and is about 70 ft thick. About 340 ft above the Tow Creek is a massive ledge-forming sandstone unit called the double-ledge sandstone, which is 250 ft thick. Kucera (1959) applied the name Oak Creek Sandstone to a 100 ft thick cliff-forming sandstone that

crops out along Oak Creek near Haybro, about 3 miles north of Oak Creek. The Oak Creek Sandstone can be mapped through Trout and Middle Creeks to the Fish Creek Canyon. In the canyon the unit is about 85 ft thick. The strike valley above the Oak Creek Sandstone is formed in the marine sediments that underlie the Trout Creek Sandstone. The Trout Creek is well exposed in the section. Directly above the Trout Creek is carbonaceous shale that is interpreted as a salt marsh and is overlain by a washover fan consisting of rippled sandstone and siltstone. The Yampa bed (Brownfield and Johnson, 1986), an altered volcanic ash, is exposed above the washover fan. The Yampa was found within 5 feet of the Trout Creek Sandstone northeast of Eckman Park but was not preserved in the Edna Mine section. The Wolf Creek coal bed is burned in the canyon but the Wadge bed (about 11.8 ft thick) was mined for a short period while the Lennox bed was not found. The Twentymile Sandstone forms a massive white cliff and dip slope on the eastern end of the canyon and is about 185 ft thick. This shoreface sandstone and the thick occurrences of the Trout Creek are much thicker than their modern counterparts. Siepman (1985) believes that these very thick sandstone units represent vertical stacking of shoreface sand units. The overlying sediments of the Williams Fork Formation represent lagoon and tidal inlet.

After examining the Fish Creek Canyon section return to the Foidel School and drive north on RC 33 to Steamboat Springs. Between the Foidel Creek Canyon and Steamboat Springs the road will cross both limbs of the Twentymile Park syncline (Fig. 7). Just north of the bridge crossing the Yampa River in Steamboat Springs is a faulted outcrop of the Lower Cretaceous Dakota Sandstone and some associated hot springs.

Day 2. Steamboat Springs to Craig

Continue on west U.S. 40 from Steamboat Springs. The trip will be traveling through the Lower and Upper Cretaceous Mancos Shale and Upper Cretaceous Mesaverde Group. About 13 miles west of Steamboat Springs is the town of Milner (Fig. 1). Milner is located in the northern extension of the Twentymile Park syncline (Fig. 7). Several old mines were developed in the Wadge bed along the flanks of the syncline. Just north of Milner is the site of McGregor where one of the first coal fired electric plants was built in northwest Colorado. U.S. 40 crosses the asymmetrical Tow Creek anticline about 2.5 miles west of Milner. Oil was discovered in 1924 and production is from the fractured Niobrara. Most of the wells are now abandoned north of the highway but a few are still maintained to the south.

Stop 5. Mount Harris. Mount Harris was an important coal mining town from about 1915 to the early 1950's (Fig. 12). Two large mines were developed in the Wadge bed south of the Yampa River: the Colorado and Utah Coal Company's Harris Mine and the Victor-American Fuel Company's Wadge Mine (Campbell, 1923). These mines were closed after mining accidents killed several miners.

The Mount Harris section has been studied several times

M. E. Brownfield et al.

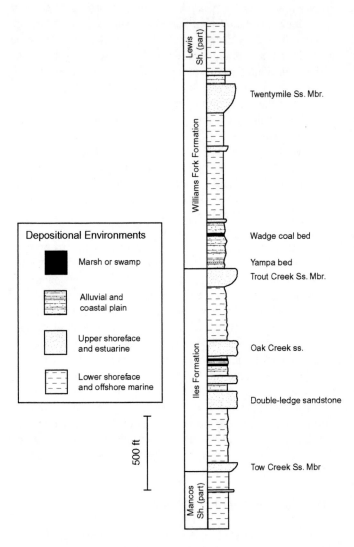

Figure 11. Stratigraphic section of the Fish Creek Canyon strata showing interpreted depositional environments, and location of Yampa bed (Brownfield and Johnson, 1986). Modified after Bass and others (1955).

(Masters, 1966, 1967 and Siepman, 1985). The Williams Fork Formation section exposed at Mount Harris can be divided into two parts: part 1 is a partial section of the middle coal group that includes the Lennox and Wadge beds, and part 2 includes the Twentymile Sandstone with the underlying transitional offshore and the overlying upper coal group. The Lennox and Wadge coal beds crop out in a roadcut just east of the rest stop. Associated with the coals are a series of crevasse splay, lagoon or bay, tidal (channel and delta), washover fan, and fluvial channel deposits. An interesting siltstone dike extends from the top to the bottom of the Lennox coal (north side of road) and allows an estimate of compaction of the Lennox swamp by comparing the length of the folded dike and and the thickness of the coal bed.

A large covered interval separates the middle coal group and Twentymile/upper coal group section. The covered interval consists of offshore marine shale and siltstone with minor sandstone about 550 ft thick. The Twentymile sandstone is about 160 ft thick and overlain by 230 ft of upper coal group rocks. The depositional environments related to the Twentymile include lower shoreface, middle shoreface, upper shoreface, tidal channel, foreshore, and backshore. The upper coal group consists of fluvial channel, crevasse splay, washover fan, lagoon, marsh, and swamp deposits. The Mount Harris section was deposited along a wave-dominated, interdeltaic, progradational shoreline and coastal plain setting.

After examining the Mount Harris section drive west to Craig and turn south on SH 13. Drive south crossing the Yampa River and turn left following the Trapper Mine access road to the mine headquarters.

Stop 6. Trapper Mine. The Trapper Mine is located about 7 miles south of Craig (Fig. 1). The mine supplies coal to the Craig power plant (Tri State Generation and Transmission, Inc.) 3 miles north of the mine (Fig 13). This three-unit power plant has a net capacity of 1,264 megawatts (the largest capacity in Colorado), and features state-of-the-art environmental controls (electrostatic precipitators and desulfurization scrubbers). The first part of the mine visit will be an oral presentation of mining and reclamation activities by Karl C. Kiehler, senior environmental engineer of Trapper Mining Inc. (P.O. Box 187, Craig, CO 81626). Karl will also discuss environmental issues controlling coal mining now and in the 21st century. The second part of the mine visit will be a tour of the mine property observing mine reclamation and coal deposits in the mine. *Remember once within the mine property, follow all mine safety regulations!*

Currently the mine is extracting the R and Q beds from the upper coal group of the Williams Fork Formation. The upper coal group has thickened to about 800 ft when compared to the section at Mount Harris (230 ft). The R bed is about 200 ft above the Twentymile Sandstone and has an average thickness of 4 ft. The distance between the R and the base of the Q ranges from 5 to 30 ft and generally is about 25 ft. According to Trapper Mine personnel the Q bed is in reality a coal zone consisting of 4 beds, the Q3, Q2, Q1, and Q0, in ascending order. The Q3 bed has an average thickness of 1.7 ft and is too thin to mine. The Q2 bed is the thickest bed in the Trapper Mine reaching a maximum thickness of 15.7 ft and averaging 8.2 ft thick. The Q1 bed ranges in thickness from 1 to 13 ft and is thickest in the eastern part of the mine. The Q0 bed is a thin, discontinuous, pod like bed that is not mined.

The Q2 bed was sampled and chemically analyzed by USGS personnel. A comparison of the mean content of trace elements of environmental concern (1990 Clean Air Act Amendment) for the Q2 is shown in Table 1. According to Trapper Mine personnel, the apparent rank of coals currently being mined are a mixture of subbituminous A, B, and C, and high-volatile bituminous C. Mean values for all of the beds mined

Figure 12. Tipple of the Colorado and Utah Coal Company's Harris Mine and the town of Mount Harris from the south side of the Yampa River. Photograph by L. C. McClure (Campbell, 1923). Twentymile Sandstone Member of the Williams Fork Formation forms the massive sandstone cliff in the background.

Figure 13. Looking north from the Trapper Mine at the Craig power plant located near the center of the photo. Cedar Mountain, to the left of the plant, consists of Miocene Browns Park Formation capped by basalt.

during September 1998 are ash yield 7.05 percent and sulfur content 0.4 percent, with a caloric value of 9,931 Btu/lb on an as received basis.

The upper coal group was deposited along wave-dominated, progradational shoreface, and coastal plain setting. The Twentymile Sandstone and upper coal group at the Trapper Mine area consist of shoreface, foreshore, lagoon, washover fan, splay, fluvial channel, marsh, and freshwater swamp deposits.

Return to Denver and the Convention Center.

REFERENCES CITED

Affolter, R.H., in press, Quality characterization of the United States western Cretaceous coal from the Colorado Plateau resource assessment area: U.S. Geological Survey Professional Paper

Bass, N.W., Eby, J.B., and Campbell, M.R., 1955, Geology and mineral fuels of parts of Routt and Moffat Counties, Colorado: U.S. Geological Survey Bulletin 1027-D, p. 143-250.

Boreck, D.L., and Murray, D.K., 1979, Colorado coal reserve depletion data and coal mine summaries: Colorado Geological Survey Open-File Report 79-1, 65 p.

Boyles, M.J., Kauffman, E.G., Kiteley, L.W., and Scott, A.J., 1981, Depositional systems Upper Cretaceous Mancos Shale and Mesaverde Group, northwestern Colorado: Society of Economic Paleontologists and Mineralogists, Rocky Mountain Section, Field Trip Guidebook, 146 p.

Brownfield, M.E., and Anderson, K., 1979, Geologic map and coal sections of the Lay SE quadrangle, Moffat County, Colorado: U.S. Geological Survey Open-File Report 79-1680, scale 1:24,000.

Brownfield, M.E., and Anderson, K., 1988, Geologic map and coal sections of the Lay SE quadrangle, Moffat County, Colorado: U.S. Geological Survey Coal Investigation Map C-117, scale: 1:24,000.

Brownfield, M.E., and Prost, G.L., 1979, Geologic map and coal sections of the Lay quadrangle, Moffat County, Colorado: U.S. Geological Survey Open-File Report 79-1679, scale: 1:24,000.

Brownfield, M.E., and Johnson, E.A., 1986, A regionally extensive altered airfall ash for use in correlation of lithofacies in the Upper Cretaceous Williams Fork Formation, northwestern Piceance Creek and southern Sand Wash Basin, Colorado, in, Stone, D. S., ed., New interpretations of northwest Colorado geology: Rocky Mountain Association of Geologist, p. 165-169.

Brownfield, M.E., Roberts, L.N.R., Johnson, E.A., and Mercier, T.J., 1998, Assessment of the distribution and resources of coal in the Deserado coal area, Lower White River coal field, northwest Colorado: U.S. Geological Survey Open-File Report 98-352, 28p.

Brownfield, M.E., Roberts, L.N.R., Johnson, E.A., and Mercier, T.J., in press, Assessment of the distribution and resources of coal in the Deserado coal area, Lower White River coal field, northwest Colorado: U.S. Geological Survey Professional Paper.

Brownfield, M.E., Roberts, L.N.R., Johnson, E.A., and Mercier, T.J., in press, Assessment of the Distribution and Resources of Coal in the Fairfield Coal Group, Danforth Hills Coal Field, Northwest Colorado: U.S. Geological Survey Professional Paper

Campbell, M.R., 1906, Character and use of Yampa coals, in, Fenneman, N.M., and Gale, H.S., The Yampa coal field, Routt County, Colorado: U.S. Geological Survey Bulletin 297, p. 82-91.

_____1912, Miscellaneous analyses of coal samples from various fields of the United States: U.S. Geological Survey Bulletin 471-J, p. 629-655.

_____1923, The Twentymile Park district of the Yampa coal field, Routt County, Colorado: U.S. Geological Survey Bulletin 748, 82 p.

Chisholm, F.F., 1887, The Elk Head anthracite coal field of Routt County, Colorado: Proceedings of the Colorado Scientific Society, v. 2, p. 147-149.

Collins, B.A., 1976, Coal deposits of the Carbondale, Grand Hogback, and southern Danforth Hills coal field, eastern Piceance Basin, Colorado: Quarterly of the Colorado School of Mines, v. 71, no. 1, 138 p.

Crawford, R.D., Willson, K.W., and Perini, V.C., 1920, Some anticlines of Routt County, Colorado: Colorado Geological Survey, Bulletin 23, 59 p.

Eakins, W., and Coates, M.M., 1998, Focus: Colorado coal: Colorado Geological Survey, Rock Talk, v. 1, no. 3, p. 1-4.

Eby, J.B., 1924, Coal in Elkhead District of Yampa coal field, northwestern Colorado: U.S. Geological Survey Press Notice 16653.

_____1925, Contact metamorphism of some Colorado coals: American Institute of Mining and Metallurgical Engineers Transactions, v. 71, p. 250.

Fenneman, N.M., and Gale, H.S., 1906a, The Yampa coal field, Routt County, Colorado: U.S. Geological Survey Bulletin 285-F, p. 226-239.

_____1906b, The Yampa coal field, Routt County, Colorado: U.S. Geological Survey Bulletin 297, 96 p.

_____1906c, The Yampa coal field, Routt County, Colorado: Mining reporter, v. 54, p. 251-252.

Fiscor, S., 1998, U.S. longwalls thrive: Coal Age, v. 103, no. 2, p. 22-27.

Gaffke, T.M., 1979, Depositional environments of a coal-bearing section in the Upper Cretaceous Mesaverde Group, Routt County, Colorado: U.S. Geological Survey Open-File Report 79-1669, 15 p.

Gale, H.S., 1909, Coal fields of northwestern Colorado and northeastern Utah: U.S. Geological Survey Bulletin 341-C, p. 283-315.

_____1910, Coal fields of northwestern Colorado and northeastern Utah: U.S. Geological Survey Bulletin 415, 265 p.

Hail, W.J., Jr., 1974, Geologic map of the Rough Gulch quadrangle, Rio Blanco and Moffat Counties, Colorado: U.S. Geological Survey Geologic Quadrangle Map GQ-1195, scale 1:24,000.

Hancock, E.T., 1925, Geology and coal resources of the Axial and Monument Butte quadrangles, Moffat County, Colorado: U.S. Geological Survey Bulletin 757, 134 p.

Hancock, E.T., and Eby, J.B., 1930, Geology and coal resources of the Meeker quadrangle, Moffat and Rio Blanco Counties, Colorado: U.S. Geological Survey Bulletin 812-C, p. 191-242.

Haun, J.D., and Weimer, R.J., 1960, Cretaceous Stratigraphy of Colorado, in, Weimer, R.J., and Haun, J.D., eds., Guide to the geology of Colorado: Geological Society of America, Rocky Mountain Association of Geologists, and Colorado Scientific Society Guidebook, p. 58-65.

Hayden, F.V., 1877, Explorations made in Colorado under the direction of Professor Ferdinand V. Hayden in 1876: American Naturalist, vol. XI, no. 2, p. 73-86.

Hildebrand, R.T., Garrigues, R.S., Meyer, R.F., and Reheis, M.C., 1981, Geology and chemical analyses of coal and coal-associated rock samples, Williams Fork Formation (Upper Cretaceous), northwestern Colorado: U.S. Geological Survey Open-File Report 81-1348, 94 p.

Holmes, W.H., 1877, Report (on the San Juan District, Colorado): U.S. Geological and Geographical Survey of the Territories, 9th Annual Report for 1875, p. 237-276.

Horn, G.H., and Richardson, E.E., 1956, Geologic and structure map of the Williams Fork Mountains coal field, Moffat County, Colorado: U.S. Geological Survey unnumbered map, scale 1:24,000.

Izett, G.A., Cobban, W.A., and Gill, J.R., 1971, The Pierre Shale near Kremmling, Colorado, and its correlation to the east and west: U.S. Geological Survey Professional Paper 684-A, 19 p.

Johnson, E.A., 1987, Geologic map and coal sections of the Round Bottom quadrangle, Moffat County, Colorado: U.S. Geological Survey Coal Investigations Map C-108, scale 1:24,000.

Johnson, E.A., and Brownfield, M.E., 1984, Selected references on the geology of the Yampa coal field and Sand Wash Basin, Moffat, Routt, and Rio Blanco Counties, Colorado: U.S. Geological Survey Open-File Report 84-769, 42 p.

Johnson, E.A., Roberts, L.N.R., and Brownfield, M.E., in press, Geology and resource assessment of the middle and upper coal groups in the Yampa coal field, northwestern Colorado: U.S. Geological Survey Professional Paper.

Kerr, B.G., 1958, Geology of the Pagoda area, northwestern Colorado: Golden, Colorado, Colorado School of Mines Masters thesis, 124 p.

Kitely, L.W., 1983, Paleogeography and eustatic-tectonic model of Late Campanian (Cretaceous) sedimentation, southwestern Wyoming and northwestern Colorado, in, Reynolds, M. W., and Dolly, E. D., eds., Mesozoic paleogeography of the west-central United States: Rocky Mountain Paleogeography Symposium 2, Society of Economic Paleontologists and Mineralogist, Rocky Mountain Section, p. 273-302.

Konishi, K, 1959, Upper Cretaceous surface stratigraphy, Axial Basin and Williams Fork area, Moffat and Routt Counties, Colorado, in, Haun, J.D., and Weimer, R.J., eds, Symposium on Cretaceous rocks of Colorado and adjacent areas, 11th annual field conference: Rocky Mountain Association of Geologists, p. 67-73.

Kucera, R.E., 1959, Cretaceous stratigraphy of the Yampa district, northwestern Colorado, in, Haun, J.D., and Weimer, R.J., eds., Symposium on Cretaceous rocks of Colorado and adjacent areas, 11th annual field conference: Rocky Mountain Association of Geologists, p. 37-45.

———1962, Geology of the Yampa District, northwestern Colorado: Boulder, Colorado, University of Colorado Ph.D. dissertation, 675 p.

Lakes, A, 1903, Coal and asphalt deposits along the Moffat railroad: Mines and Minerals, v. 24, p. 134-136.

———1904, The Yampa coal field; a description of the anthracite, bituminous, and lignite field transversed by the Moffat road in Routt County: Mines and Minerals, v. 24, p. 249-251.

———1905a, The Yampa coal field of Routt County: Mining Reporter, v. 51, p. 404-405.

Landis, E.R., 1959, Coal reserves of Colorado: U.S. Geological Survey Bulletin 1072-C, p. 131-232.

Lauman, G.W., 1965, Geology of Iles Mountain area, Moffat County, northwestern Colorado: Golden, Colorado, Colorado School of Mines Masters thesis, 129 p.

Massoth, T.W., 1982a, Depositional environments of a surface coal mine in northwest Colorado, in, Gurgel, K.D., ed., Proceedings, 5th symposium on the geology of Rocky Mountain coal 1982: Utah Geological and mineral Survey Bulletin 118, p. 115-120.

———1982b, Depositional environments of some Upper Cretaceous coal-bearing strata at Trapper Mine, Craig, Colorado: University of Utah Masters thesis, 124 p.

Masters, C.D., 1959, Correlation of the post-Mancos Upper Cretaceous sediments of the Sand Wash and Piceance Basins, in, Haun, J.D., and Weimer, R.J., eds., Symposium on Cretaceous rocks of Colorado and adjacent areas, 11th annual field conference: Rocky Mountain Association of Geologists, p. 78-80.

———1966, Sedimentology of the Mesaverde Group and the upper part of the Mancos Formation, northwest Colorado: Yale University, Ph.D. dissertation, 88 p.

———1967, Use of sedimentary structures in determination of depositional environments, Mesaverde Formation, Williams Fork Mountains, Colorado: American Association of Petroleum Geologists Bulletin, v. 51, no. 10, p. 2033-2043.

Resource Data International, 1998, COALdat data base: 1320 Pearl Street, Suite 300, Boulder, Colorado 80302.

Roberts, L.N., and Kirschbaum, M.A., 1995, Paleogeography of the late Cretaceous of the Western Interior of Middle North America - coal distribution and sediment accumulation: U.S. Geological Survey Professional Paper 1561, 115 p.

Roehler, H.W., 1987, Surface -subsurface correlations of the Mesaverde Group and associated Upper Cretaceous formations, Rock Springs, Wyoming, to Mount Harris, Colorado: U.S. Geological Survey Miscellaneous Field Studies map MF-1937.

Roehler, H.W., and Hansen, D.E., 1989, Surface and subsurface correlations showing depositional environments of the Upper Cretaceous Mesaverde Group and associated formations, Cow Creek in southwestern Wyoming to Mount Harris in northwest Colorado: U.S. Geological Survey Miscellaneous Field Studies map MF-2077.

Ryer, T.A., 1977, Geology and coal resources of the Foidel Creek EMRIA site and surrounding area, Routt County, Colorado: U.S. Geological Survey Open-File Report, 77-303, 31 p.

Sears, J.D., 1925, Geology and oil and gas prospects of part of Moffat County, Colorado, and southern Sweetwater County, Wyoming: U.S. Geological Survey Bulletin 751, p. 269-319.

Siepman, B.R., 1985, Stratigraphy and petroleum potential of trout Creek and Twentymile Sandstones (Upper Cretaceous), Sand Wash Basin, Colorado: Colorado School of Mines Quarterly, v. 80, no. 2, 59 p.

Speltz, C.N., 1976, Strippable coal resources of Colorado - location, tonnage, and characteristics of coal and overburden: U.S. Bureau of Mines Information Circular 8713, 70 p.

Stone, D.S., 1986, Seismic and borehole evidence for important pre-Laramide faulting along the Axial arch in northwest Colorado, in, Stone, D. S., ed., New interpretations of northwest Colorado geology: Rocky Mountain Association of Geologist, p. 19-36.

Tremain, C.M., Hornbaker, A.L., Holt, R.D., Murray, D.K., and Ladwig, L.R., 1996, 1996 summary of coal resources in Colorado: Colorado Geological Survey Special Publication 41, 19 p.

Tweto, O.,1976, Geologic map of the Craig 1o X 2o quadrangle, northwestern Colorado: U. S. Geological Survey Miscellaneous Investigations Series I-972, scale 1:250,000.

——— 1979, Geologic map of Colorado: U. S. Geological Survey, scale 1:500,000.

Wood, G.H., Kehn, T.M., Carter, M.D., and Culbertson, W.C., 1983, Coal resource classification system of the U.S. Geological Survey: U. S. Geological Survey Circular 891, 65 p.

Zook, J.M., and Tremain, C.M., 1997, Directory and statistics of Colorado coal mines with distribution and electric generation map, 1995-96: Colorado Geological Survey Resource Series 32, 55 p.

Zapp, A.D. and Cobban, W.A., 1960, Some Late Cretaceous strand lines in northwestern Colorado and northeastern Utah: U.S. Geological Survey Professional Paper 450-D, p. 52-55.

Geological Society of America
Field Guide 1
1999

Field guide to the continental Cretaceous-Tertiary boundary in the Raton basin, Colorado and New Mexico

C. L. Pillmore and D. J. Nichols
U.S. Geological Survey, MS 913, Federal Center, Denver, Colorado 80225, United States
R. F. Fleming
Denver Museum of Natural History, 2001 Colorado Boulevard, Denver, Colorado 80205, United States

This guide consists of three general sections: an introduction that includes discussions of Raton basin stratigraphy and the Cretaceous-Tertiary (K-T) boundary; descriptions of the geology along the route from Denver, Colorado, to Raton, New Mexico; and descriptions of several K-T sites in the Raton basin. Much of the information is from previous articles and field guides by the authors together with R. M. Flores and from road logs co-authored with Glenn R. Scott, both of the U.S. Geological Survey.

INTRODUCTION

Geologists have recognized for nearly a century that a remarkable mass-extinction event occurred at the end of the Cretaceous Period, about 65 million years ago. Thousands of species of plants and animals that lived during the Cretaceous no longer existed during the Tertiary Period immediately following. This disappearance has led to much speculation and controversy among scientists as to possible causes. Among the causes suggested are changes in climate or sea level, neither of which has been universally accepted. In 1980, researchers from the University of California at Berkeley advanced the startling hypothesis that a large asteroid about 10 km in diameter struck the Earth about 65 million years ago, causing a worldwide biospheric catastrophe (Alvarez and others, 1980).

The Berkeley team found anomalously high concentrations of iridium (Ir) and other noble elements in a claystone layer that marks the K-T boundary in certain marine rocks in Italy, Denmark, and New Zealand. Ir is extremely rare in the Earth's crust relative to its abundance in certain types of meteorites. Because of this crustal Ir deficiency, the team proposed that the source of the anomalously high Ir concentrations in the boundary layer was extraterrestrially derived and that the layer was formed from fallout of ejecta after the asteroid impact occurred. Since this original proposal, more than 100 sites scattered around the globe have been reported where abundance anomalies of Ir occur in rock layers at the major extinction horizon of marine invertebrate and continental plant and animal fossils that defines the K-T boundary.

Significantly, anomalously high concentrations of Ir were first discovered at the K-T boundary in continental rocks in 1981. A team consisting of chemists from Los Alamos National Laboratory aided by geologists from the U.S. Geological Survey (USGS) reported an Ir abundance anomaly precisely at the palynological K-T boundary in nonmarine rocks in the Raton basin, New Mexico (Orth and others, 1981). The discovery was made in core samples from a hole drilled specifically for palynological and chemical analyses at the York Canyon coal mine about 35 miles west of Raton. Prior to this, all reports concerning the hypothesized asteroid impact at the K-T boundary were from marine rocks, and some geologists had suggested that marine processes concentrated the anomalous Ir; therefore, the Ir need not be extraterrestrial The discovery of an Ir abundance anomaly at the K-T boundary in nonmarine rocks in the Raton basin removed this objection and provided important supporting evidence for the Berkeley team's hypothesis. In later studies, shock-metamorphosed quartz grains were found coincident with the Ir anomaly (Bohor and others, 1984; Pillmore and others, 1984; Izett and Pillmore, 1985a, b); these unusual grains are further evidence of a major impact event. Shock-metamorphosed mineral grains are found only at known meteorite impact sites or nuclear bomb craters, never in volcanic rocks. Thorough and quite readable accounts of the K/T boundary impact and extinction theory and the evidence in support of it can be found in two recently published, popular books: Alvarez (1997) and Powell (1998); these books are highly recommended.

Since the original Raton basin discovery, many more K-T boundary sites have been found in the eastcentral and southern parts of the basin (Fig. 1), and other continental K-T sites have

Pillmore, C. L., Nichols, D. J., and Fleming, R. F., 1999, Field guide to the Continental Cretaceous-Tertiary boundary in the Raton basin, Colorado and New Mexico, *in* Lageson, D. R., Lester, A. P., and Trudgill, B. D., eds., Colorado and Adjacent Areas: Boulder, Colorado, Geological Society of America Field Guide 1.

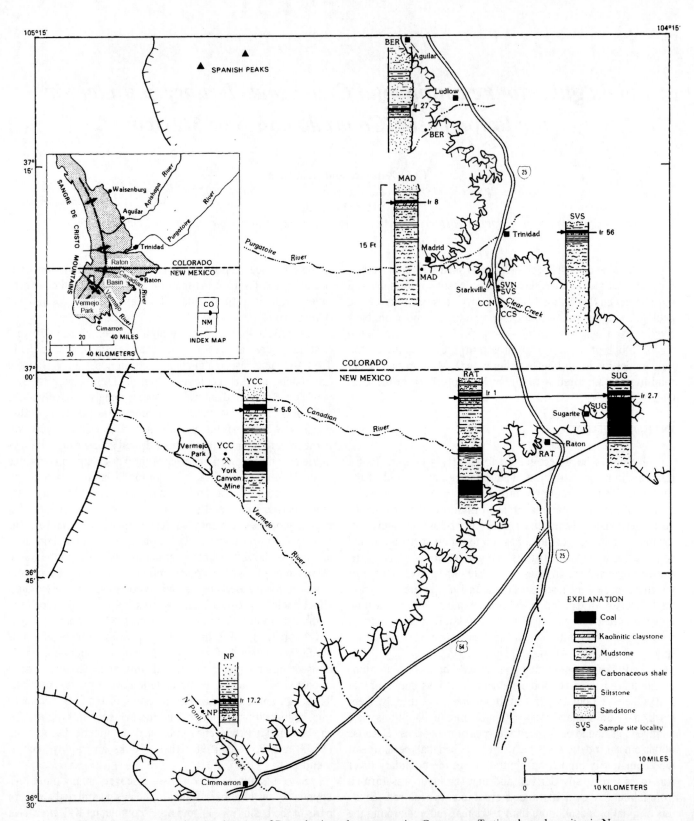

Figure 1. Index map showing location of Raton basin and representative Cretaceous-Tertiary boundary sites in New Mexico and Colorado. Columnar sections at or near top of lower zone of Raton Formation show lithology of the boundary interval at selected sites: YCC, York Canyon; RAT, Raton; SUG, Sugarite; SVS, Starkville South; BER, Berwind Canyon; and MAD, Madrid. Arrows indicate the K-T boundary. Measured Ir anomalies shown in ng/g (10^{-9} g/g). Hachured line shows top of Trinidad Sandstone. Modified from Pillmore and Flores (1987).

been located in Wyoming, Montana, North Dakota, Alberta, and Saskatchewan. On this trip we will visit several of the K-T boundary sites in the Raton basin to observe similarities and differences in the character of the boundary sequence and the enclosing interval of rocks.

RATON BASIN

The Raton basin (Fig. 1) is a large, asymmetric syncline (2,500 sq. mi. area) that extends from Huerfano Park, Colorado, to Cimarron, New Mexico. The Cretaceous and Tertiary rocks dip steeply and form hogbacks along the western margin of the basin on the east flank of the Sangre de Cristo Mountains. These rocks dip more gently inward along the other margins of the Raton basin and are highly dissected.

Stratigraphy. The sedimentary rocks of the Raton basin are shown in Figure 2. The marine Pierre Shale (Campanian to Maastrichtian) and overlying marginal-marine Trinidad Sandstone (Maastrichtian) underlie the nonmarine Upper Cretaceous and Tertiary rocks in the basin. Nonmarine sedimentary rocks in the Raton basin, from oldest to youngest, include the coal-bearing Vermejo Formation (Maastrichtian) and Raton Formation (Maastrichtian and Paleocene), and the non-coal-bearing Poison Canyon Formation (also Maastrichtian and Paleocene), which overlies and intertongues with the Raton Formation.

Depositional and tectonic history. The formations present in the Raton basin are the Pierre Shale and Trinidad Sandstone, the Vermejo and Raton Formations, and the Poison Canyon Formation as portrayed in the stratigraphic section (Fig. 2). Their depositional and tectonic history is shown in Figure 3. The Cretaceous epeiric sea covered the area during most of Late Cretaceous time. The sea was filled by a thick sequence of calcareous, deep marine or basinal shales, overlain by the Pierre Shale, 1,800 to 1,900 feet of shallow marine or shelf shale and siltstone. These beds range in age from Cenomanian to Maastrichtian as determined from ammonite fossils studied by G.R. Scott and W.A. Cobban (G.R. Scott, written commun.). Most of the Pierre consists of gray, noncalcareous shale that coarsens upward into silty shale and siltstone, reflecting the influx of silt as the paleoshoreline of the sea retreated to the east across Colorado and New Mexico in late Campanian and early Maastrichtian time. The upper part of the Pierre grades into the overlying Trinidad Sandstone

The Trinidad Sandstone is a tabular body, in most places about 80-100 ft thick, composed mostly of fine to very fine-grained sandstone that contains *Ophiomorpha, Diplocraterion,* and other trace fossils. It forms persistent, conspicuous, light-colored cliffs at the east and south edges of the basin. The

AGE	FORMATION NAME	GENERAL DESCRIPTION	SYMBOL	APPROXIMATE THICKNESS IN FEET
PALEOCENE	POISON CANYON FORMATION	SANDSTONE--Coarse to conglomeratic; beds 15-50 ft thick; interbeds of yellow-weathering, clayey sandstone. Thickens to west at expense of underlying Raton	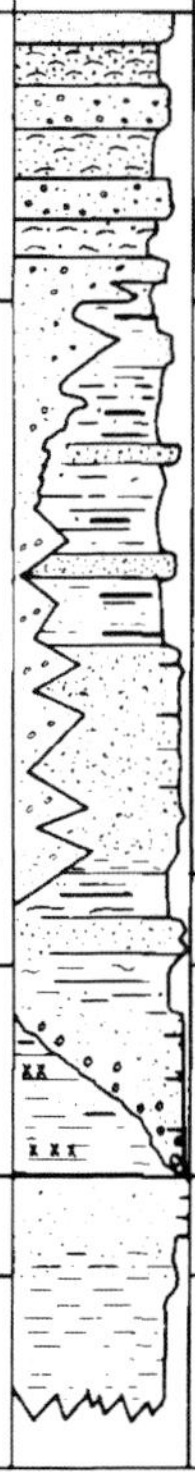	500+
PALEOCENE / UPPER CRETACEOUS	RATON FORMATION	Formation intertongues with Poison Canyon Formation to the west UPPER COAL ZONE--Very fine grained sandstone, siltstone, and mudstone with carbonaceous shale and thick coal beds BARREN SERIES--Mostly very fine to fine grained sandstone with minor mudstone, siltstone, carbonaceous shale, and thin coal beds LOWER COAL ZONE--Same as upper coal zone; coal beds mostly thin and discontinuous. Conglomeratic sandstone at base; locally absent		0?–2100 K T BOUNDARY
UPPER CRETACEOUS	VERMEJO FORMATION	SANDSTONE--Fine- to medium-grained; also mudstone, carbonaceous shale, and extensive, thick coal beds. Local sills		0–380
UPPER CRETACEOUS	TRINIDAD SANDSTONE	SANDSTONE--Fine- to medium-grained; contains casts of *Ophiomorpha*		0–300
UPPER CRETACEOUS	PIERRE SHALE	SHALE--Silty in upper 300 ft. Grades up to fine-grained sandstone. Contains limestone concretions		1800–1900

Figure 2. Columnar section of rocks in the Raton basin. Thickness of zones in the Raton Formation are: lower coal zone, 100-300 ft; barren series, 180-700 ft; and upper coal zone, 590-1100 ft.

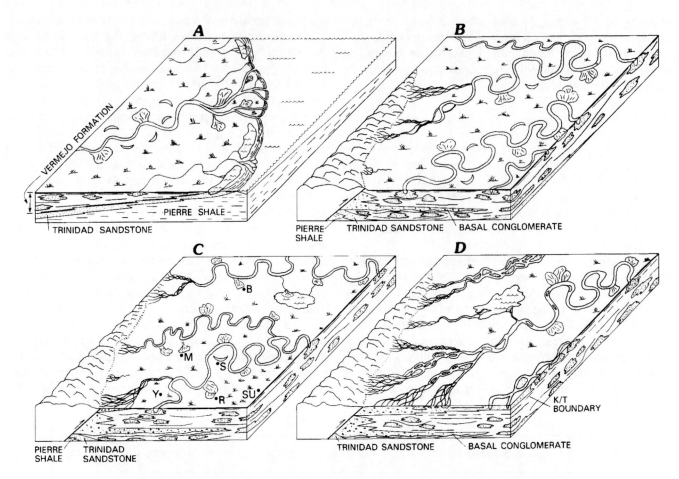

Figure 3. Diagrammatic block diagrams depicting paleoenvironments of Late Cretaceous and Paleocene rocks of the Raton basin: A. The offshore environments of Pierre Shale; the contemporaneous
delta front and barrier environments of the Trinidad Sandstone; and the fluvio-deltaic plain of the
Vermejo Formation. Diagram shows oxbow lakes, crevasse splays, and coal-forming swamps related
to meandering streams of the lower delta plain. B. The floodplain and fine-grained meander-belt
environments on the alluvial plain that developed during the Late Cretaceous after deposition of the
basal conglomerate of the lower coal zone of the Raton Formation. Coal-forming swamps and oxbow
lakes characterized the floodplain. C. The depositional environments on the alluvial plain of the
lower coal zone of the Raton Formation at the end of the Cretaceous Period. The K-T boundary fallout material was deposited on a surface such as that shown here. The letters show representative
depositional sites: B, Berwind site (floodplain sequence, dominated by crevasse splays from nearby
stream channel); M, Madrid site (channel-floodplain-crevasse splay sequence); S, Starkville sites
(floodplain sequence developed on abandoned channel sequence); Y, York Canyon (floodplain-
crevasse splay sequence); R, Raton site (floodplain, marginal to major backswamp); SU, Sugarite
site (dominated by an extensive swamp). D. The braided stream and coarse meander-belt environments of the Paleocene barren series of the Raton Formation.

Trinidad was deposited in contemporaneous delta-front and
interdeltaic barrier-bar environments as the sea continued to
regress eastward. The delta-front deposits include distributary
mouthbar and distributary channel sandstone beds (Flores and
Tur, 1982). The barrier-bar deposits consist of middle
shoreface, river-estuarine-inlet, and beach sandstone beds
(Leighton, 1980) that are overlain by the fluvial-deltaic and
back-barrier deposits of the Vermejo Formation.

The Vermejo Formation consists of interbedded sandstone,
siltstone, shale, carbonaceous shale, and coal that together form

steep, generally debris-covered slopes above the cliffs of the
Trinidad Sandstone. The Vermejo varies in thickness from 370
to 390 along the western border of the Raton basin to 0 ft in the
eastern part of the basin, south and east of Raton, New Mexico.
It contains coal beds as thick as 10-13 ft near the top and bottom of the formation (Pillmore, 1976). The sedimentary rocks
and coal beds of the Vermejo were deposited in contemporaneous fluvio-deltaic and back-barrier coastal plains fronted by
barrier-bar and delta-front sandstone of the Trinidad (Flores and
Tur, 1982). Lower alluvial plains dissected by meandering

streams separated by flood basins characterized the upper part of the Vermejo. Here, sand-filled stream channels and fine-grained sequences of silt and mud were laid down in floodplains associated with crevasse-splay and minor crevasse-channel sandstone. Coal beds in the lower part of the formation formed in poorly drained, back-barrier, coastal swamps and in swamps adjacent to distributary channels of delta plains. These alluvial deposits grade upward into the more landward deposits of the Raton Formation.

The Raton Formation contains the K-T boundary interval. The formation consists of sandstone, siltstone, mudstone, coal, carbonaceous shale, and conglomerate. It ranges in thickness from more than 2,100 ft in the west-central part of the basin to 1,100 ft in the eastern part. A basal pebble-conglomerate bed commonly rests unconformably directly on the Vermejo Formation, but in the vicinity of Raton and Trinidad, the conglomerate is commonly absent and no unconformity is evident. This scour-based sandstone forms a persistent cliff throughout much of the Raton basin, especially in the western and southern parts. Lee (1917) originally divided the Raton into a basal conglomerate, a lower coal zone, a barren series, and an upper coal zone. A similar subdivision that better fits our purposes, especially in the area around Raton and Trinidad where the basal conglomerate is lacking, includes the basal conglomerate in the lower coal zone. The three subdivisions are mostly consistent and identifiable throughout the Raton basin and thicken from east to west.

The Raton Formation was deposited on an upper alluvial plain (Flores, 1984) characterized by various modes of aggradation and erosion. Deposition on the alluvial plain was interrupted by several styles of fluvial sedimentation. Following the deposition of fluvio-delta plain and back-barrier sediments of the Vermejo Formation, rapid uplift of the source area to the west (the present San Luis valley, termed the San Luis highland, Tweto, 1987) caused widespread erosion of the highland and concurrent deposition of the basal conglomerate of the Raton. The basal conglomerate, which consists of medium- to coarse-grained channel sandstone with lenses and stringers of conglomerate, was probably deposited in basin-margin braided streams merging basinward into meandering streams. The uplift of the highland during Laramide time may have moved along thrust faults that bordered the western margin of the basin (Flores and Pillmore, 1987; Woodward and Snyder, 1976). After deposition of the basal conglomerate, more stable tectonic conditions returned and sand was deposited in meandering streams at the same time that interbedded thin coal, carbonaceous shale, mudstone, and sandstone accumulated in floodplains and backswamps. Crevasse splays periodically interrupted and infilled low-lying floodplains during river floods. Though some thick coal beds were deposited locally in the lower coal zone, most coal swamps that formed in the flood basins were well-drained, small, and shallow, which limited peat accumulation and resulted in thin lenticular coal beds mostly less than 8-12 in thick. At the close of the Cretaceous, the extensive alluvial plain was an ideal environment for deposition and preservation of the impact ejecta and fallout resulting from the K-T boundary asteroid impact.

Shortly after the close of the Cretaceous, tectonic conditions in the Raton basin changed as uplift of the source area was reinitiated in the west (Fig. 3c). It is probable that episodic upthrusting along the fault belt to the west created extensive erosion and sediment input into the basin. Sediment load increased and a fluvial system characterized by braided streams merging basinward into meandering streams and well-drained flood basins once again characterized the depositional basin. Streams aggraded broad belts across the alluvial plain, resulting in sheetlike to vertically stacked channel deposits. These deposits are locally interbedded with and laterally grade into floodplain mudstone and siltstone deposits. Associated carbonaceous shale and thin, lenticular coal beds formed in backswamps. Along the eastern and southern margins of the basin, cliffs of the barren series commonly stand high above slopes of the lower coal zone and the Vermejo Formation. The upper coal zone of the Raton supports the hills and ridges above the barren zone in the interior part of the basin.

The barren series was succeeded by the upper coal zone, which was deposited on a low-gradient alluvial plain, resulting from a decrease in tectonic movement in the source area. This tectonic pause, perhaps combined with basin subsidence, led to aggradation of the alluvial plain by a meandering fluvial system. The system was accompanied by floodplains that developed poorly drained backswamps in which coal beds as thick as 12 ft accumulated. The floodplains were locally filled by overbank and crevasse-splay detritus during episodes of floods. These deposits extended into the backswamps and caused splits of coal beds where the peat swamps reestablished after detrital influx. Deposits of this setting grade upward into deposits of more landward environments of the Poison Canyon Formation.

The Poison Canyon Formation overlies the Raton Formation throughout most of the Raton basin, but to the west and southwest, it intertongues with the Raton. The Poison Canyon consists of thick to massive, lenticular, ledge-forming beds of coarse-grained to conglomeratic arkosic sandstone intercalated with beds of nonresistant, yellow-weathering, sandy, micaceous mudstone and siltstone. The contact with the underlying Raton generally is indefinite and gradational. The Poison Canyon was deposited on a high-gradient alluvial plain (Strum, 1984), characterized by non-coal-bearing floodplain and braided- to meandering-stream alluvial-fan deposits. The high sediment input into the alluvial plain indicated by these high-bedload streams and fans probably reflects intensive erosion of a rapidly rising source area to the west. These tectonic and depositional conditions probably correspond to a pulse of upthrusting that exposed core rocks, as indicated by the abundance of potassium feldspar grains in the channel sandstone. The Poison Canyon deposits coarsen to the west, forming the alluvial fans of the piedmont environment marginal to the rising ancestral San Luis highland.

CRETACEOUS-TERTIARY BOUNDARY

The Cretaceous-Tertiary (K-T) boundary has been placed at different stratigraphic horizons by different workers. Modern workers believe the boundary is now more accurately placed than ever, based on three lines of independent and corroborating evidence: fossil pollen; trace element chemistry; and mineralogy. Lee (1917) originally placed the K-T boundary at the unconformity at the base of the Raton Formation. Later, Brown (1962) identified Cretaceous plant fossils in the lower part of the Raton, at a site about 3.5 mi north of Trinidad, Colo., and indicated that the boundary should be placed at least 50 ft above the base of the formation. In 1967, the late R.H. Tschudy of the USGS bracketed the K-T boundary on the basis of palynology between two coal beds about 270 ft above the base of the Raton Formation in a core hole drilled at York Canyon, New Mexico (R.H. Tschudy, written comm., 1967, and Pillmore, 1969). The position of the boundary was precisely established by Tschudy four years later, as reported in Orth and others (1981). In the Raton basin, this extinction horizon coincides with the top of a 0.5- to 1.0-in-thick claystone bed termed the boundary claystone (Orth and others, 1981; Pillmore and others, 1984; Tschudy and others, 1984). This horizon also coincides with an anomalous concentration of Ir and shock-metamorphosed minerals (Orth and others, 1981; Izett and Pillmore, 1985a, 1985b) and a sudden change in the relative proportion of fern spores to angiosperm pollen (Tschudy and others, 1984). This unique claystone bed has been found at more than 25 sites throughout an area of about 1000 mi^2 in the east-central and southern parts of the Raton basin

Palynology. In the northern part of the Rocky Mountain region, the palynological K-T boundary was originally defined on the basis of the disappearance of fossil pollen of *Proteacidites* spp. and most species of *Aquilapollenites* (Leffingwell, 1970; Tschudy, 1970). This horizon occurs within a few feet above the disappearance of dinosaurs. In the Raton basin, the K-T boundary is defined solely on the basis of palynology due to the absence of dinosaur bones. *Aquilapollenites* pollen occurs only rarely in the southern part of the Rocky Mountain region, and Tschudy (Tschudy, 1973; Orth and others, 1981; Tschudy and Tschudy, 1986) used the extinction of the species *"Tilia" wodehousei, Trisectoris* sp., and *Trichopeltinites* sp. in addition to the extinction of *Proteacidites* spp. to locate the K-T boundary in the Raton basin. These taxa are herein called the *Proteacidites* assemblage.

Tschudy and others (1984) concluded that the K-T boundary event, the hypothetical asteroid impact, caused massive destruction of vegetation, disrupted the terrestrial ecosystem, and resulted in the extinction of the plants that produced the *Proteacidites* assemblage. Plants that survived exhibit three basic patterns of survival. The first pattern is shown by pollen of *Kurtzipites* spp., which is common in the latest Cretaceous, survives the K-T boundary event, and persists into the Paleocene until it disappears about the middle Paleocene (Tschudy

and Tschudy, 1986). *Psilastephanocolpites* sp. exemplifies the second pattern. This species is rare in the Cretaceous but becomes more abundant in the Paleocene, perhaps because the plants that produced this particular fossil pollen were better adapted to new ecological conditions. The last pattern is characterized by species little affected by the K-T boundary event and includes such fossil pollen as *Ulmipollenites* sp. and *Pandaniidites radicus*. The patterns of extinction and survival within the Raton basin indicate that different plant species responded in different ways to environmental stress caused by the K-T boundary event.

Detailed palynological sampling across the K-T boundary revealed the presence of an anomalously abundant fern spores just above the extinction level, which appears to be unique to the K-T boundary event. This abundance of fern spores (termed the "fern spike") occurs in K-T boundary sections from the Raton basin to south-central Saskatchewan (Tschudy and others, 1984; Nichols and others, 1986; Tschudy and Tschudy, 1986). In Cretaceous assemblages of the Raton basin, fern spores usually constitute 15-30 percent of the palynomorph assemblages. Just above the K-T boundary claystone, the fern spore percentage increases dramatically to as much as 99 percent. Usually the percentage returns to the 15-30 percent level within 3-5 in above the boundary.

The fern-spore spike is an unusual palynological assemblage in comparison with typical Upper Cretaceous and Paleocene palynological assemblages found in nonmarine rocks (Fleming and Nichols, 1990). Comparison of these assemblages from just above the K-T boundary at many localities with typical assemblages reveals that the fern-spore spike is characterized by: (1) relatively abundant spores, ranging from 70 percent to 100 percent of the assemblages (in contrast with 10 percent to 40 percent for typical Upper Cretaceous and Paleocene assemblages in the same sections); (2) dominance of only one of a few species at each locality; (3) restriction of the anomaly to a layer 0-6 in above the K-T boundary (usually only an inch or two above the boundary); (4) independence of lithology (the anomaly occurs in coal, carbonaceous shale, and mudstone); and (5) isochroneity (based on palynological and geochemical evidence) and wide distribution (from northern New Mexico to south-central Saskatchewan, a distance of more than 800 mi). Within the Raton basin, comparison of three K-T boundary localities reveals the pattern of relative abundance of fern spores and the independence of lithology characteristic of this unique assemblage (see Fig. 4).

Tschudy and others (1984) pointed out the importance of this phenomenon with respect to the destruction of terrestrial vegetation. They attributed the "fern spike" to early colonization of the devastated landscape by ferns. The temporary dominance of ferns at the K-T boundary is due to the "early arrival of wind-dispersed spores, the removal of competitors, and the known tolerance of ferns to soils deficient in mineral nutrients" (Tschudy and others, 1984, p. 1031). In general, palynological observations of patterns of extinction and survival suggest that

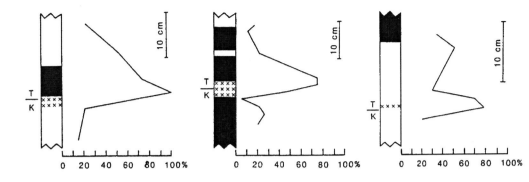

Figure 4. Fern-spore relative abundances from three K-T boundary localities in the Raton basin. Left: Starkville North section, Colorado. Middle: Sugarite section, New Mexico. Right: Raton Pass section, New Mexico. (Black = coal; white = mudstones and shales; xxxx = K-T boundary claystone; T = Tertiary; K = Cretaceous).

the terrestrial ecosystem was stressed by a significant, though geologically brief, event (Tschudy and others, 1984; Tschudy and Tschudy, 1986).

Paleobotany. Early in the K-T boundary controversy, Hickey (1981) asserted that the megafloral record was inconsistent with the hypothesis that a catastrophe caused terrestrial extinctions. However, in their joint paper, Johnson and Hickey (1990) reversed this position by presenting new evidence that the megafloral change is about 80 percent and that it coincides with a peak in palynofloral extinctions and the occurrence of Ir and shock-metamorphosed mineral grains. They state that the results of their analysis of the terrestrial plant record are "compatible with the hypothesis of a biotic crisis caused by extraterrestrial impact at the end of the Cretaceous" (Johnson and Hickey, 1990, p. 433).

Wolfe and Upchurch (1986) analyzed fossil leaves and dispersed fragments of leaf cuticles from K-T boundary sequences in the Raton Formation. Their results suggest a brief low-temperature excursion (mean temperature near 0°C) that caused a masskill and ecological disruption of terrestrial vegetation at the K-T boundary. Leaf size and shape indicate that a major increase in precipitation occurred across the boundary. Their conclusions are consistent with the bolide impact hypothesis.

The K-T boundary claystone bed. The boundary claystone bed resembles a tonstein (tonsteins are kaolinitic claystone partings thought to result from alteration of volcanic ash beds in acidic coal swamps), but, unlike typical tonsteins, it usually weathers to a lighter, pinkish, color and exhibits a fine-grained to amorphous texture and a distinctive hackly to conchoidal fracture. The claystone is mostly gray and grayish pink to grayish yellow and commonly contains tiny specks and thin, contorted lenses or layers of organic matter, especially near the margins. Small spheroidal structures can be seen on fracture surfaces of some specimens and in thin section. X-ray diffractograms show that, like many tonsteins, the boundary claystone is nearly pure, well-crystallized kaolinite with lesser amounts of randomly stratified illite-smectite clay and some quartz and feldspar (Pollastro and others, 1983; Pollastro and Pillmore, 1987). However, the boundary

claystone is texturally and chemically different from typical Raton basin tonsteins. As seen in ultrathin section, it is fine grained to amorphous but may exhibit an imbricate fabric and relics of small bubbles in a fine crystalline matrix of kaolinite. Microspherules (40-120 microns in diameter) consisting of calcium, aluminum, strontium, cerium, rare earth elements, and phosphorus (similar to goyazite, a hydrous strontium alumino-phosphate in composition; microprobe analysis by Ralph Christian, USGS, 1984) have been observed in samples of the boundary claystone from the Raton site. These phosphatic spherules are rarely seen at other sites but they form discrete layers in the boundary claystone at the Dogie Creek and Teapot Dome localities in the Powder River basin in Wyoming (Izett, 1990). Smit (1984) has referred to similarly shaped grains in the boundary claystone from the Raton basin and other areas as microtektite-like structures, implying an impact origin. The microtektite grains from Raton basin sites are mostly spheroidal to subspheroidal and resemble dull to shiny resinous little balls under the microscope. Some are hollow. Under the scanning electron microscope, they have an uneven surface texture and commonly are pitted by irregularly shaped cavities. On the basis of textures and shapes observed in goyazite spherules in the Powder River Basin, that are identical to those seen in microtektites, the spherules are thought to result from the alteration of glassy ejecta material (microtektites) blown out of the crater during the K-T impact event.

High concentrations of Ir and shock-metamorphosed mineral grains, both compelling evidence of impact origin (Bohor and others, 1984; Izett, 1990), are present in a discrete layer at the top of the boundary claystone. This layer was called the flaky shale layer by Pillmore (Pillmore and others, 1984), the K-T boundary impact bed by Izett and Bohor (1986), and the fireball layer by Hildebrand and Boynton (1988). The shocked grains consist mainly of quartz, with rare microcline and plagioclase. The shock-metamorphosed quartz grains contain as many as nine intersecting sets of closely spaced planar features per grain (Izett, 1990). Figure 5 is a photograph by Izett of one of the shocked grains of quartz from the Starkville South site, showing two sets of planar lamellae.

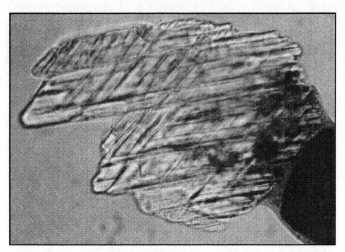

Figure 5. Photomicrograph by G.A. Izett of a 0.28-mm diameter shock-metamorphosed quartz grain from the Madrid site. The grain is mounted in index oil on the needle (dark part of photograph) of a spindle stage. The two sets of planar lamellae that are prominent in the photograph are strong evidence of impact origin (Izett, 1990).

Table 1. Elemental abundances in the thin kaolinitic K-T boundary claystone bed compared to elemental abundances in kaolinitic tonstein beds found in coal beds above and below the K-T boundary in the Raton basin. From Gilmore and others (1984).

Element†	K-T boundary bed		Beds above/below K-T boundary	
	Range	Average	Range	Average
Al_2O_3 (%)	24–36	32.9	23–36	31.7
K (%)	0.2–1.1	0.52	0.2–2.5	0.81
Sc	21–26	23.3	3–12	6.0
TiO_2 (%)	1.38–2.67	2.00	0.40–0.95	0.757
V	110–187	137	10–67	27
Cr	46–102	67.3	0.9–5.0	3.3
Co	1.2–53	9.8	0.7–4	2.7
As	1–95	36	0.2–46	4.1
Se	2–19	8	<0.1–6	4.8
Sb	0.3–11.5	6.3	0.1–0.8	0.38
La	16–80	43	9–88	28
Yb	0.7–2.2	1.5	0.8–4.1	2.1
Hf	3.3–6.4	4.5	3.2–10.6	7.6
Ir (ng per g)	0.90–2.7	1.7	0.005–0.020	0.010
Th	5–21	7.8	5–34	9.2
La/Yb	15–57	28.7	5–67	13.3
TiO_2/Al_2O_3	0.056–0.074	0.060	0.016–0.027	0.024
Cr/Al_2O_3‡	1.64–2.83	2.04	0.036–0.147	0.103

* Boundary data from six sampling sites, non-boundary data from 21 sites.

† Concentrations in µg per g unless noted otherwise.

‡ Cr/Al_2O_3 relative (p.p.m./%)

Geochemistry. Both the ejecta layer of the boundary claystone and tonsteins found in coal beds are high-alumina clays that characteristically contain about 32 percent Al_2O_3 (Gilmore and others, 1984). Ir abundance anomalies as high as 56 ng/g (56 x 10^{-9} g/g, about 8000 x background) have been measured in samples from the fireball layer at the top of the boundary claystone collected from the study area (Pillmore and others, 1984). In comparison, background Ir concentrations of only 0.004-0.040 ng/g are observed in tonsteins and other beds of coal and shale not associated with the boundary (Gilmore and others, 1984). In addition, titanium, scandium, vanadium, chromium, and antimony in the boundary claystone are enhanced by factors of about four or more over their concentrations in all other Raton basin tonsteins that were analyzed by Gilmore and others (1984). Table 1 shows a comparison of these and other elements in the boundary claystone with those in tonstein beds in the Raton basin (Gilmore and others, 1984). The boundary claystone has been found at scattered localities from Cimarron, New Mexico, to Red Deer Valley, Alberta, remarkably consistent in chemical composition and always in direct contact with the high-Ir, shocked-quartz-bearing fireball or K-T boundary impact layer (Izett, 1990). The Ir abundance anomalies in the boundary claystone bed, its wide geographic extent, and its unusual textural and chemical character indicate that the boundary claystone bed was derived from a different source than were the tonstein beds. It is widely accepted now that the boundary claystone consists primarily of altered glass ejecta material from the Chicxulub impact site on the northern tip of the Yucatan Peninsula in Mexico and that the fireball layer contains material from the vaporized bolide together with some vaporized target material.

DESCRIPTION OF THE ROUTE FROM DENVER TO RATON

Denver to Castle Rock

South of Denver on I-25, the route is underlain by the Denver Formation (Maastrichtian and Paleocene); about 5 mi south of the I-225 bypass around the city, the Denver Formation intertongues with rocks of the Dawson Arkose (Paleocene and Eocene), which is overlain by the Castle Rock Conglomerate (Oligocene). At Castle Rock, the small butte northeast of town is capped by the conglomerate and is the type section.

The structure of the Denver Basin is largely concealed by the relatively flat-lying Denver and Dawson formations. The basin is asymmetric, with the deepest part located only a few miles east of the mountain front. Its axis lies beneath Denver and extends south-southeastward to Castle Rock and through points about 15 mi east of Colorado Springs and Pueblo. The Front Range, which consists mainly of Precambrian igneous and metamorphic rocks, was uplifted and thrust eastward over the west flank of the basin during the Laramide orogeny. Between Denver and Castle Rock, the thrusts are located a short distance east of the mountain front, where they are poorly exposed due to the soft nature of the Denver Formation, or concealed by the Dawson, which is post-Laramide.

The town of Kiowa, Colorado, lies about 20 miles to the

east of Castle Rock and is situated approximately above the deepest part of the Denver basin. In March of this year (1999), a 2,200-foot well was core-drilled at the Elbert County Fairgrounds in Kiowa. This 2.5-inch-diameter core was obtained as part of the Denver Basin Project, which is a Denver Museum of Natural History project funded in part by the National Science Foundation. The goals of the project are to reconstruct the sedimentary and tectonic history of the Denver basin during the Late Cretaceous and early Tertiary. The hydrogeology of the Denver basin is also a primary focus of the project and detailed studies of the aquifers cored in the Kiowa well will provide valuable information about water resources in the Denver basin.

The multidisciplinary project includes hydrogeology, paleontology (palynology and paleobotany), stratigraphy, sedimentology, paleomagnetism, fission-track dating, and other disciplines. Results from these various fields will be integrated to provide a basis for interpreting the history of the Denver basin and implications of that history for topics ranging from groundwater use to the extinction of the dinosaurs.

In this project the K-T boundary in the Denver basin and extinction patterns across it will be examined. The study will integrate paleobotanical and palynological data within the general sedimentological and stratigraphic framework in hopes of documenting the evolution of terrestrial vegetation from the Late Cretaceous to the early Tertiary in the Denver basin.

The K-T boundary claystone has been documented in numerous sites to the south and north of the Denver basin. However, attempts to locate the boundary claystone in the Denver basin have been unsuccessful. Preliminary palynological results place the K-T boundary approximately 880 feet below the town of Kiowa. No K-T boundary claystone similar to the K-T boundary layer in the Raton basin was located in the core. However, the horizon of the palynological boundary will be projected to surface exposures east of Kiowa and suitable sites will be examined in detail to locate the K-T boundary.

Castle Rock to Colorado Springs

The Castle Rock itself and many of the mesas south of the town of Castle Rock are capped by the Castle Rock Conglomerate. In this area, except at Castle Rock, the conglomerate is locally underlain by the silicic Wall Mountain Tuff. The Wall Mountain is an ashflow tuff that occurs on the Rampart Range to the west and is also present in the Thirtynine Mile volcanic field. The tuff was a widespread ash flow that erupted 34.8 ± 1.1 Ma (J.D. Obradovich, USGS, written commun., 1969) from its presumed source near Mt. Aetna in the Sawatch Range, about 100 mi. to the west.

The route continues in the Dawson Arkose nearly to Colorado Springs. In the vicinity of the Air Force Academy, outstanding examples of pediments of three ages can be seen west of the highway, capped from highest to lowest by the Rocky Flats, Verdos, and Slocum Alluviums. Precambrian rocks of the Rampart Range, which lies west of the pediments, are uplifted

about 10,000 ft against Dawson Arkose by the Rampart Range reverse fault. This fault, the Ute Pass fault to the south, and other faults along the Front Range may be low angle thrusts, on the basis of geophysics and the outcrop pattern (A.F. Jacob, consulting geologist, written commun., 1985). The large scars visible on the mountain front of the Rampart Range north of Colorado Springs are quarries that were developed for concrete aggregate in limestone and dolomite of the Williams Canyon Formation (Devonian) and the Manitou Limestone (Ordovician).

Just past the south entrance to the Air Force Academy, I-25 crosses Monument Creek and passes down section through outcrops of the coalbearing Laramie Formation and the Fox Hills Sandstone (mostly covered), which is lithologically equivalent to the Vermejo and Trinidad Sandstone Formations of the Raton basin, (Fig. 2) and into the Pierre Shale, all of Late Cretaceous age. Much of the area immediately east of Monument Creek is the site of abandoned coal mines in the Laramie. Continuing south through Colorado Springs, good views are afforded of Pikes Peak, which is underlain by the Precambrian Pikes Peak Granite (about 1,000 Ma), and Cheyenne Mountain, underlain by granodiorite of the Routt Plutonic Suite (about 1,700 Ma; Tweto, 1987). Cheyenne Mountain houses the large underground installation of NORAD (North American Air Defense Command). The Ute Pass fault, a large reverse and thrust fault that dips to the west as low as 30 degrees and trends along the eastern base of the mountain, places Pikes Peak Granite against Pierre Shale. To the west of Colorado Springs, the Fountain Formation is spectacularly displayed in erosional forms at the Garden of the Gods. These arkosic sandstone and conglomerate units are about 4,000 ft thick and were deposited as alluvial fans on the east flank of the northwest-trending ancestral Rockies during Pennsylvanian and Early Permian time.

Colorado Springs to Pueblo

Proceeding south from Colorado Springs, the route continues on the Pierre Shale to Pueblo. The immense training grounds of Fort Carson Army Base lie to the west of the Interstate for many miles south of Colorado Springs. About midway to Pueblo, the sharp eye will observe the Tepee Buttes in the Pierre Shale outcrops on the horizon to the east across Fountain Creek. These conical-shaped buttes are capped by large resistant concretions in the Pierre Shale.

The Canyon City Embayment, which is about 20 mi west of the Interstate, is a topographic and structural feature that reflects the en-echelon arrangement of mountain ranges. The Front Range terminates south of Cheyenne Mountain and the Wet Mountains form the eastern front of the Rocky Mountains farther south. The Canyon City Embayment occupies the intervening space.

The Florence oil field, which was discovered in 1862, was the first oil field found west of the Mississippi River. Oil from this field was produced from fractured Pierre Shale (Upper Cre-

Table 2. Generalized stratigraphic section of Cretaceous and Tertiary rocks seen from Colorado Springs to Raton.

AGE		FORMATION	GENERAL DESCRIPTION	APPROXIMATE THICKNESS (ft)	(m)
TERTIARY	PALEOCENE	POISON CANYON FORMATION	Sandstone, coarse to conglomeratic, beds 5 ft (1.5 m) to more than 50 ft (15 m) thick, interbeds of soft yellow-weathering clayey sandstone; thickens to west at expense of underlying rocks.	500+	(150+)
		RATON FORMATION	Sandstone, very fine grained to fine grained, with interbeds of clay-stone, siltstone, and coal; commercial coal beds in upper part. Lower few feet conglomeratic; intertongues with Poison Canyon to the west. Generally sharp erosional contact with underlying Vermejo Formation.	0-2,000	(0-610)
CRETACEOUS	LATE CRETACEOUS	VERMEJO FORMATION	Sandstone, very fine grained to medium grained, interbedded with mudstone, carbonaceous shale, and coal; extensive thick coals top and bottom.	0- 380	(0-115)
		TRINIDAD SANDSTONE	Sandstone, very fine grained to medium grained; contains casts of Ophiomorpha sp.	0- 130	(0- 40)
		PIERRE SHALE	Black shale, limestone concretions, silty in upper part; grades up to sandstone.	2,500+	(760+)
		NIOBRARA FORMATION	Limestone and calcareous shale; consists of the Smoky Hill and Fort Hays Limestone Members.	500+	(150+)
		CARLILE FORMATION	Black shale, gray calcareous shale, and calcarenite; consists of the upper black shale unit, and Juana Lopez, Blue Hill Shale, and Fairport Members.	250	(76)
		GREENHORN FORMATION	Limestone and calcareous shale. Consists of the Bridge Creek Limestone Member and the Hartland and Lincoln Members.	130	(39)
		GRANEROS SHALE	Black shale and shaly limestone.	110	(33)
	EARLY CRETACEOUS	DAKOTA SANDSTONE	Quartzitic sandstone.	145	(44)

taceous) and was in great demand as a lubricant for wagon wheels and sold for more than $5 per gallon.

Pueblo to Walsenburg

In Pueblo we pass from the Pierre Shale into the underlying Cretaceous Niobrara Formation (table 2) at about the Arkansas River. South of the river the old Colorado Fuel and Iron (CF&I) steel-making plant is on the left and we can see the new Comanche power generating plant to the southeast. This power plant was constructed in large part to provide power for electric steelmaking furnaces, which totally cut off the demand for Raton basin coking coal that had been produced for many years at CF&I's Allen and Maxwell mines (more recently called the New Eagle and Aztec mines, now closed), west of Trinidad. Coal for the steam plant is brought by unit train from the Belle Ayr mine near Gillette, Wyoming.

South of Pueblo we will drive across gently rolling, dissected terrain underlain by marine Upper Cretaceous rocks (table 2) of the Niobrara, Carlile, and Greenhorn Formations, past ridges and mesas supported mainly by the Fort Hays Lime-

stone Member of the Niobrara and the Bridge Creek Limestone Member of the Greenhorn. Benches formed by these limestone units can be seen in the breached anticline a few miles south of Pueblo and across the drainage divide at the abandoned rest stop on the left that is mostly on the pinon- and juniper-covered Bridge Creek Member beneath mesa crests formed by the Fort Hays Member.

At the Colorado City Exit from I-25, we cross the structural divide, the Apishapa arch, that separates the Denver and Raton basins. Just south of the excellent road-cut exposure of the Fort Hays Limestone at the exit, the route descends the dip slope of the Fort Hays Limestone, which crops out at the bottom of the hill along Greenhorn Creek, east of the bridge. After crossing the creek, we go up section through the Smoky Hill Member of the Niobrara Formation until we cross the Apishapa fault near the top of the hill. The sandstone beds exposed in roadcuts of the highway to the left are on the south side of the fault and consist of the Dakota Sandstone, the oldest Cretaceous unit in this vicinity; the large ridge at the crest of the hill is supported by Dakota Sandstone. The Lower Dakota Sandstone (table 2) represents the transgression of the Cretaceous epeiric sea, which extended across the North American continent from the Arctic Ocean to the Gulf of Mexico in the Early Cretaceous and existed during most of the Late Cretaceous (Cenomanian to Maastrichtian). The Dakota shows several hundred feet of reactivated Laramide movement along the fault, which originated during the late Paleozoic (Ogden Tweto, USGS, oral commun, 1980). The Apishapa fault trends about east across Las Animas County and forms the northern boundary of the late Paleozoic Apishapa arch. The fault is a major fracture zone and water wells in the zone provide a large part of the water supply for Colorado City.

South from the top of the Dakota ridge we leave the Denver Basin and proceed across the valley back up section through the progressively younger, south-dipping marine shale, limestone, and limy mud of the Graneros, Greenhorn, and Carlile to the next ridge, capped once again by Fort Hays Limestone (table 2). At the top of this ridge we look south into the Raton basin, a large arcuate structural basin that extends south into New Mexico (Fig. 1). The Fort Hays and Smoky Hill Members of the Niobrara Formation form the dip slope for several miles to the south from the ridge crest. Farther along, the Pierre Shale overlies the Smoky Hill and underlies the route into Walsenburg and on south to Trinidad.

The Wet Mountains, which dominate the skyline to the west of the Interstate highway, are composed mainly of Precambrian crystalline rocks. The summit of Greenhorn Mountain, which is the prominent peak at the southern end of the range (elevation 12,334 ft), is unusual in that a remnant of Oligocene volcanic rocks uncomformably overlies the Precambrian. Several prominent pediment surfaces are visible to the west along the route.

East of the highway and visible for several miles is Huerfano Butte, a conspicuous conical peak that rises about 100 feet above the surrounding plain. According to Penn (1994), the butte is a biotite olivine alkali-gabbro plug bisected by two east-trending dikes, rather than a volcanic edifice of some sort as indicated in popular literature. The age of the intrusives is about 25.2 Ma, similar in age to the East and West Spanish Peaks intrusives.

Walsenburg to Trinidad

Leaving Walsenburg, we continue on Pierre Shale for about the next 37 mi, past the town of Aguilar to Trinidad. Aguilar is located adjacent to coal-bearing Cretaceous and Tertiary rocks of the Raton basin. These rocks (Fig. 2), which define the eastern limits of the Walsenburg and Trinidad coal fields, are marked by abrupt cliffs formed by the Trinidad Sandstone (Upper Cretaceous). Overlying the Trinidad are the coalbearing Vermejo Formation (Upper Cretaceous) and the Raton Formation (Upper Cretaceous and Paleocene). The Raton is overlain by the Poison Canyon Formation (Paleocene). The Dawson Arkose of the Denver Basin is roughly equivalent to the Poison Canyon Formation of the Raton basin, and the Denver Forma-

tion is approximately equivalent to the Raton Formation. The scenery along this part of the route is dominated by the Spanish Peaks to the west. The Spanish Peaks, in part held up by intrusives, are well known for many dikes that radiate from them for several miles. The Spanish Peaks plutons have been dated at 21.7 ± 1.0 Ma (Stormer, 1972, East Peak Granite) and 22.9 ± 2.0 Ma (Smith, 1975, West Spanish Peak stock). Other dates ranging from 19.8 ± 1.6 Ma to 39.5 Ma have been reported from intrusive rocks in the Spanish Peaks area; the 39.5 Ma date appears questionable and most dates fall into the 20-25 Ma range (Marvin and others, 1974, p. 3233). Ages obtained by Penn (1994) indicate continued intrusive activity in the Spanish Peaks region from 26.6 Ma to 21.3 Ma with the middle period of activity (24.6-22.8 Ma) involving West and East Spanish Peaks and the radial dikes. Most of the sills and dikes in the Raton and Vermejo Park areas were intruded at about the same time (J.D. Obradovich, C.W. Naeser, and H.H. Mehnert, USGS, written commun., 1976-1983).

West of the interstate and along the route to Trinidad, the Pierre Shale is exposed beneath cliffs formed by the Trinidad Sandstone. Just past the turnoff to Ludlow, the view to the south is dominated by what appears (to some) to be the silhouette of an early 1920's Franklin automobile, outlining the hood and trunk, which is formed by the basalt flows on Fisher's Peak, the northern segment of Raton Mesa. Though not dated as yet, the flows that form the crest are considered to be about 3.5 m.y., the same as Bartlett Mesa north of Raton, New Mexico (Stormer, 1972; Stroud, 1998).

Trinidad

The site of Trinidad was an ancient Indian ceremonial ground. The area was later visited and periodically inhabited by Onate and other Spanish explorer-colonizers, priests, and soldiers; French and American trappers and traders; and Kearny and other subsequent U.S. military expeditions to New Mexico. The first permanent structure, a sheepherder's cabin on the south bank of the Purgatoire River (pronounced "pergatwar"), was erected in 1859, and the area soon attracted numerous farmers and traders. The early years of Trinidad were turbulent; conflicts between American and Mexican settlers and between the settlers and the Ute Indians were common in the 1860's.

Today the population of Trinidad is about 9,000. Coal mining, which began in the late 1800's, was the most important industry in the area until the 1950's when coal-burning steam locomotives were superceded by diesel locomotives. Coal continued to be produced at the New Eagle and Maxwell mines several miles west of Trinidad to fire the coke ovens in Pueblo until the late 1970's and later as fuel for electric power generation, but mining activity ceased in 1994.

In the Colorado part of the Raton Basin, more than 200 wells have been drilled in the past several years to develop the methane potential in coal beds of the Vermejo and Raton Formations. The center of activity is about 20 miles west and northwest of Trinidad. The main operators at present (1999) are Evergreen Resources, GeoMet, Inc., and Chandler and Associates. Most wells initially flow less than 100 MCFGPD and a few hundred barrels of water per day, but with all of the producing wells in the area the gas stream in the Colorado Interstate Gas pipeline is about 70 MM cu. ft./ day. Gas is also being produced from the abandoned workings of some of the old coal mines in the area. Since the installation of a pipeline into the area in 1994, drilling activity has greatly increased and contin-

ues to increase with recent activity (1999) southwest of Trinidad and in the New Mexico portion of the basin.

DESCRIPTION OF STOPS AND ROUTES TO STOPS

Trinidad to the Long Canyon K-T site

Note mileage and turn off I-25 to Colorado Highway 12 and proceed west out of Trinidad. Trinidad Sandstone caps the

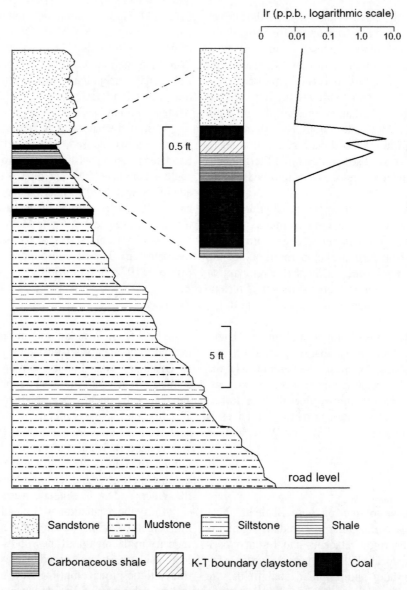

Figure 6. Diagram showing the lithology of the K/T boundary sequence at the Long Canyon K/T boundary site in the Watchable Wildlife area of Trinidad Lake State Park. The inset shows the detail of the boundary interval and the variation in iridium concentration. (Iridium analyses courtesy of F. Asaro, Lawrence Berkley Laboratory, written comm., 1999). Palynologic analyses of samples collected across the boundary at the Madrid site (Fig. 1) show that more than 80 percent of the assemblage in the coal bed directly above the boundary claystone bed consists of fern spores (R. H. Tschudy, USGS, written comm., 1982).

high ridge to the north of town and crops out along the highway. The entrance to the road across Trinidad Lake dam is near the top of the formation, and sandstone, mudstone, and carbonaceous shale and coal beds of the Vermejo Formation are well exposed in high roadcuts on the right. At about 5.7 miles is the approximate position of the base of the Raton Formation. Beds and stringers of pebble conglomerate generally occur at the base of the Raton, but they are commonly absent in this area and the base is represented by beds of sugary-textured, quartzose sandstone.

At the town of Cokedale is the large Asarco coal mine waste pile on the right, and rows of brick coke ovens on the left that produced coke for the Asarco smelters for 40 years until the mine closed in 1946.

About 1.5 miles past Cokedale watch for the sign to the Watchable Wildlife Area and turn left across the new bridge over the Purgatoire River, cross the railroad tracks, and bear left down the gravel road. Upstream along the railroad the K-T boundary sequence crops out in railroad cuts for several hundred yards beneath a tabular splay sandstone bed that is easily visible from the highway across the valley.

Where the road turns south up a small valley (about a 1/4 mi), the K-T boundary claystone is exposed directly beneath a prominent sandstone bed in a steep roadcut known as the Madrid East site. The boundary claystone (the thin, 1-in-thick, white claystone bed at the top of the cut) lies beneath a 2-in-thick coal bed that is directly overlain by the sandstone. The claystone bed is underlain by a brownish-orange-weathering, boney coal bed about 12 to 16 in thick. The claystone bed at this site and in this area exhibits all of the characteristics of the boundary claystone in the Raton basin: anomalously high concentrations of Ir, shocked quartz, the fern-spore spike, and the palynological extinction horizon, all indications of asteroid impact.

Proceed on the road about 2 mi to Long Canyon and turn left at the road junction about ½ mi to the parking area for the Watchable Wildlife Area.

Stop 1: Long Canyon K-T boundary site

We will walk about 300 to 400 yards up the trail to the south end of the Long Canyon K-T boundary site, an extended exposure of the K-T boundary sequence similar to the Madrid Road K-T boundary site. On the left of the trail are the floodplain/back-swamp boney coal and carbonaceous shale beds and interbedded siltstone and mudstone beds that contain the K-T boundary. This sequence is overlain by the prominent splays-sandstone ledge at the top of the exposure. Continue along the trail for a hundred yards or so and observe one of the best exposures of the K-T boundary sequence in the world. The section shown in Figure 6 was measured behind the park bench about midway along the exposure.

The rocks in the slope are typical of the mudstone and siltstone and thin coal beds found in the lower coal zone of the Raton Formation (Fig. 7). Near the top of the slope, immedi-

ately beneath the thick sandstone ledge, is the K-T boundary claystone bed overlain by a 1- to 2-in-thick coal bed and underlain by 7 to 8 in of brownish-orange-weathering carbonaceous shale and boney coal typical of the Madrid area. This coal sequence suggests that the ejecta cloud from the impact deposited a thin bed of glassy debris in a large coal-forming swamp mire where it eventually altered to kaolinitic clay. The thin coal bed indicates that plant material that accumulated following the impact was inundated by a splay deposit of sandstone, as determined by its nearly flat base, which is overlain by a splay-channel sandstone.

Return to the parking lot and retrace the route to the entrance of the Trinidad Lake dam. Turn right and proceed across the dam through Starkville across I-25 and turn right along the service road to Stops 2a and b, the Starkville K-T boundary sites.

Proceed south on the service road past waste piles of several abandoned coal mines in the Vermejo Formation about 2 miles to the Starkville K-T boundary sites. The prominent channel sandstone in the roadcut at about 1 mi rests directly on the upper Starkville coal bed, which is approximately 65 ft above the Trinidad Sandstone.

Stop 2a: Starkville North K-T boundary site

A short distance on down the road is the Starkville North K-T boundary site. This first recognized Colorado location of the K-T boundary is now obscured by a landslide. The K-T sequence was exposed in a carbonaceous shale bed 20 in thick, 111 ft above pavement level at top of the high cut on the east side of I25. The rocks exposed in the roadcut consists of interbedded channel sandstone and floodplain/back-swamp silt-

Figure 7. Photograph of the Long Canyon K/T boundary site showing mudstone and siltstone beds in the lower part of the sequence overlain by the carbonaceous coaly sequence of the K/T boundary interval and the white K/T boundary claystone directly beneath 7-8-ft-thick splay/channel, ledge-forming sandstone. Arrow points to the K/T boundary claystone layer.

Figure 8. Photograph of the Starkville North K/T boundary site showing the sedimentary units. The arrow at the top of the exposure shows the position of the K/T boundary sequence, but this upper part of the exposure is now covered by a landslide.

stone, mudstone, carbonaceous shale, and coal deposits (Fig. 8).

    The contact between the Raton and Vermejo Formations is estimated to lie a few feet beneath the level of the service road. At the level of the palynological extinctions (Figure 9), 6 ng/g (ppt) Ir was measured in the K-T boundary fireball layer, a dark, kaolinitic, carbonaceous layer 0.2 in thick that contains shock-metamorphosed mineral grains (Pillmore and others, 1984; Pillmore and Flores, 1987). This layer, which consists mainly of kaolinite, contains sparse pollen yet marks the change in the fern spore-angiosperm ratio from 21 percent fern spores to more than 99 percent fern spores (Tschudy and others, 1984; Pillmore and Flores, 1987). The ratio recovers to 22 percent fern spores within 4 in above the shale bed. The fireball layer overlies the ejecta layer, a distinctive, light-colored layer of kaolinitic claystone 0.8 in thick that forms the lower part of the boundary claystone. The boundary claystone is overlain by a thin blocky coal bed about 2 in thick that appears to persist along the outcrop for a few miles. The boundary interval lies

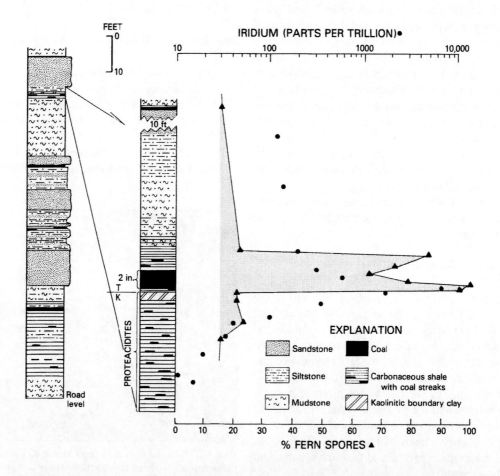

Figure 9. Diagram showing the lithology of the K-T boundary interval at the Starkville North K-T boundary site, 3 mi south of Trinidad, Colorado. The large black dots show the variation in Ir concentration, the solid line and triangles show the fern-spore percentage, and the inset shows the detail of the boundary interval. From Tschudy and others (1984).

about 200 ft above the Trinidad Sandstone in a sequence of fine-grained deposits of the lower coal zone of the Raton Formation.

Stop 2b: Starkville South K-T boundary site

Just a ¼ mile farther south is the Starkville South K-T boundary site (Fig. 10). This site has probably been the most heavily sampled site in the Raton basin. In 1984, a team from the Smithsonian Institution collected a 2½-ton sample of the boundary interval and shipped it back to Washington, D.C. to archive this significant rock sequence. The Starkville South section shown in Figure 1 represents several sites in roadcuts along I-25. At the Starkville South site, the strongest Ir anomaly ever measured in continental rocks (56 ng/g) was measured in the flaky shale at the top of the ejecta layer of the boundary claystone (Pillmore and others, 1984). The Ir is concentrated at the top of the K-T boundary claystone in this thin bed now known as the fireball layer (Fig. 11).

Palynology of the Starkville South K-T boundary site

The K-T boundary at Starkville South is placed at the level of the disappearance of the Cretaceous palynomorphs of the *Proteacidites* assemblage. These fossils are present in the Cretaceous rocks below but disappear at the top of a kaolinitic claystone, which here is located about 6 ft above the top of the point bar sandstone (Fig. 10). Cretaceous palynomorph assemblages in mudstone below the boundary (Fig. 11) are diverse and include many species of fossil pollen and spores that also occur in the lower Paleocene (e.g., *Pandaniidites radicus, Kurtzipites* spp., and *Ulmipollenites* sp.), but more importantly they include the *Proteacidites* assemblage, which does not occur in the Paleocene.

The relative abundance of fern spores just below the kaolinitic claystone is about 30 percent and comprises several species. Just above the kaolinitic claystone is a 2-cm-thick interval that to date has not yielded any palynomorphs and has been called the barren interval. Palynomorph assemblages recovered from the coal just above the barren interval are dominated by fern spores—this is the "fern-spore spike" interval (see section on the Cretaceous-Tertiary Boundary for a discussion of the significance of the fern-spore spike). Fern spores constitute 80 percent of the assemblage from the lower half and 77 percent of the assemblage from the upper half of the coal; a single species of fern spore dominates the assemblages. Angiosperm (flowering plant) pollen in the fern spore spike interval is greatly reduced in relative abundance and diversity. Above the coal, angiosperm pollen gradually increases in relative abundance and diversity until it again dominates.

A 3-ft-thick, olive-gray mudstone just above the K-T boundary coal bed contains fossil planktonic green algae (*Pediastrum* and *Scenedesmus*), which also occur in modern freshwater lakes and ponds. The presence of these algae indicates that the mudstone was deposited in a floodbasin lake or pond (Fleming, 1986, 1987). This interpretation is corroborated by the presence of *Pandaniidites radicus* in the brown mudstone immediately overlying the olivegray mudstone. The plants that produced *P. radicus* are attributed to modern *Pandanus* (screwpine), which grows in coastal and marshy areas of the tropics and subtropics. At Starkville South, the presence of *Pandaniidites radicus* indicates deposition in a marshy area, probably on the margin of the floodbasin lake that contained planktonic green algae. The palynological and sedimentological evidence suggests that the K-T boundary layers and associated sediments were deposited in a floodbasin environment that included peat-

Figure 10. Photograph of the Starkville South site, looking east, showing the position of the K/T boundary claystone interval about 6 ft above a 23-ft-thick point-bar sandstone. Arrow points to the K/T boundary claystone layer.

Figure 11. Photograph of the K/T boundary sequence at the Starkville south site showing the light gray boundary claystone and overlying thin coal bed that contains the fern spore spike. Point of hammer head is at the level of the K/T boundary claystone layer. The fireball layer is the very thin bed at the top of the boundary claystone interval, just below the coal bed.

forming swamps (represented by the coal), marshes (represented by brown, carbonaceous mudstone containing *Pandaniidites radicus*), and floodbasin lakes or ponds (represented by olive-gray mudstone containing *Pediastrum* and *Scenedesmus*).

The depositional framework at Starkville South is somewhat different from that at Starkville North. At Starkville North, the boundary lies near the top of a sequence of mudstone beds 20 ft thick that rests on a channel sandstone 7 ft thick; at Starkville South, the mudstone sequence has thinned to 10 ft, and the channel sandstone has thickened to 23 ft.

Go back up the service road, cross the Interstate, and turn south.

Starkville sites to Raton, New Mexico

Leaving the Trinidad area, we will go south over Raton Pass on I-25, through rocks of the Trinidad Sandstone (delta-front and barrier-bar environments) and Vermejo and Raton Formations (deltaic-fluvial environments), to the contact of the Raton and Poison Canyon Formations (fluvial environments), which is near the top of the pass. Several coal beds, dikes, and sills can be observed in roadcuts along the route. As we go south from Trinidad, coal dumps along the way are from early mining in the area.

We pass gently dipping alluvial floodplain sequences of sandstone, mudstone, shale, and coal beds of the Vermejo and Raton Formations and, about six miles south of Trinidad, the Morley dome becomes evident, marked by subtle changes in dip that bring the Trinidad Sandstone and Pierre Shale to the surface. The Morley dome has about 450 ft of closure and was apparently formed by the intrusion of a Tertiary plug or laccolith, which crops out about 1 mi northeast of the abandoned town of Morley. The crest of the fold is between the town and the exposed plug, and the closing contours extend for some distance around the north and east sides of this igneous intrusion. Coal seams lacking evidence of natural coking or metamorphism have been mined to the contact with the igneous plug. A minor oil seep was recognized in one mine and in 1948 Stanolind drilled a dry hole on the crest of the structure to a total depth of 6,831 ft into "granite." The K-T boundary is exposed in several I-25 roadcuts between the Starkville exit and Morley. Palynological examination of samples from a sequence of thin coal beds beneath a south-dipping sequence of sandstone beds of the barren zone of the Raton Formation just south of the Morley coal waste pile indicates they are of Cretaceous age and the K-T boundary must lie above the coal beds, perhaps in the basal sandstone. *Proteacidites* spp. is abundant in these samples.

Morley

Morley began as a railroad town on the A.T.&S.F. line in 1879, and trains were commonly shortened here before beginning the difficult climb over Raton Pass. Inactive during the late 1800's, Morley became a coal-mining company town in 1906, when CF&I opened the Morley mine. The fine-grade coking coal mined here was utilized by the steel mills in Pueblo and by Santa Fe Railroad locomotives. Peak coal production reached 500,000 tons per year in the late 1920's, when the town had a population of over 1,000. The Morley mine was never mechanized; the use of blasting powder and machinery was prohibited because of significant amounts of methane gas. Instead, the coal was extracted by hand labor and an underground herd of donkeys was maintained to haul the coal out of the mine; in fact, the Morley was reportedly the last mine in the U.S. to use these beasts. Cutbacks in steel production at Pueblo temporarily halted coal production in the early 1950's, and the mine was closed permanently on May 4, 1956, when all workable deposits had been exhausted. From 1907 until 1956 the Morley mine produced 11 million short tons of coking coal.

Top of Raton Pass

The view from the parking area includes the Trinidad coal field, the Spanish Peaks, and the high peaks of the Sangre de Cristo Mountains. The Spanish Peaks form a landmark that is visible throughout hundreds of square miles. The area lying between Raton Pass and the Spanish Peaks contains several hundred million tons of coal resources and probably trillions of cu. ft of related coalbed methane. The sign at rest area reads:

Raton Pass, named by the Spanish for the rock rats found there, was crossed by the Mountain Branch (also called the Bent's Fort Branch) of the Santa Fe Trail. Originally part of an old Indian trail, the pass was used by Spanish expeditions at least as early as 1718 and probably much earlier. When the Cimarron Branch of the Santa Fe Trail was abandoned because of its dangerous desert stretches, pioneer traffic increased over Raton Pass. Kearny's Army of the West came this way in 1846 with some of the first vehicular supply wagons to cross the pass. "Uncle Dick" Wootton, frontier scout, built a road over the pass in 1865 and collected tolls, often at the point of a gun until the coming of the railroads in 1878. For two years, a controversy raged between the Denver and Rio Grande and the Santa Fe Railroads to determine which company had the right-of-way over Raton Pass. After several section crew fights and much legal maneuvering, the Rio Grande gave up its claim to Raton Pass and the Santa Fe agreed not to contest another disputed right-of-way through the Royal Gorge.

The contact between the Raton and Poison Canyon Formations is gradational through several feet of section. It is mapped approximately at the color change from grayish-yellow-weathering rocks to grayish-orange and grayish-red-and-brown-weathering rocks just below the north exit from the Rest Area. The top of the Trinidad Sandstone is about 1,250 ft below the top of the pass. The thickness of the Vermejo Formation is estimated to be about 70 ft in this vicinity, making the Raton about

1,180 ft thick. To the west, in the central part of the basin, thicknesses are greater than 2,000 ft for the Raton and about 300-400 ft for the Vermejo.

Proceeding on south from Raton Pass, we follow the route of the Mountain Branch of the Santa Fe Trail (Scott, 1986) and pass back down section through the three zones of the Raton Formation. Thick coal beds are exposed in roadcuts in the upper coal zone and stacked channel sandstone beds make up the barren zone toward the bottom of the pass. The lower coal zone of the Raton and the Vermejo Formation are present in roadcuts near the mouth of the valley. The Trinidad Sandstone crops out just before the Raton exit. The town of Raton is built on the Pierre Shale. Enter Raton.

Turn right at the first intersection past the Melody Lane Motel to Moulton Avenue, which will take us up Goat Hill on the old Raton Pass road to stop 3, the Raton Pass K-T boundary site (Fig. 1). Proceed up Moulton past the Shady Mountain Park trailer court through the Trinidad Sandstone. At the junction, keep right. The top of the Trinidad Sandstone is approximately at the next loop in the road. The upper and lower coal beds of the Raton coal zone are exposed in roadcuts above the Trinidad. The basal conglomerate of the Raton is not present and its contact with the underlying Vermejo Formation is indefinite. Continue on up section through the lower coal zone of the Raton Formation to the saddle. Just below the saddle the Sugarite coal zone (Upper Cretaceous) is exposed in the roadcut.

Stop 3: *The Raton Pass K-T boundary site*

The Raton site is at the top of a saddle about 1 mi west of Raton on Southwell Mountain Road (the old Raton Pass Road). The roadcut exposes the first outcrop discovery of the Ir anomaly (about 1 ng/g) in the Raton basin. The anomaly is coincident with the disappearance of the *Proteacidites* assemblage. It is at the top of a 1-in-thick bed of rusty-weathering kaolinitic claystone about 7 to 8 in below the base of a thin coal bed. The boundary here is about 150 ft above the Trinidad Sandstone.

At the Raton Pass site three coal beds lie in an 11ftthick sequence of mudstone, siltstone, and carbonaceous shale. These three coal beds can be quite definitely correlated with the Sugarite coal bed on the basis of the presence of the K-T boundary in both sections. The boundary claystone occurs in a detrital claystone sequence and does not directly underlie a coal bed, as is the usual case elsewhere in the basin—in fact, its position beneath the coal bed ranges from 1.5 to 8 in. The boundary is below the base of the uppermost coal bed, which is 6 in thick; a 16-in-thick zone of coal and carbonaceous shale lies about 6 ft beneath the boundary, and a 28-in-thick coal bed is at the base of the sequence. This coal zone appears to be correlative with the coal bed mined at Sugarite, about 3 mi across the high ridge to the northeast.

After examining the K-T boundary at the Raton site, we will return to Raton and proceed south on Main Street through town to the Holiday Inn Express for the night.

SECOND DAY: ROUTE TO SUGARITE, NEW MEXICO

This segment of the field trip depends on weather more than the others. It involves a hike along a good trail through some rugged terrain and, in the event of inclement weather, we will scrub the climb and substitute a different K-T site. To reach Sugarite (Fig. 1) we will go north on I-25 and turn east from Raton via New Mexico highway 72. Our route crosses outcrops in creek bottoms of the upper (*Exiteloceras jenneyi*) part (Campanian) of the Pierre Shale (Pillmore and Scott, 1994) and gravelly alluvium on the broad flats formed by the Barilla and Beshoar pediment surfaces (Scott and Pillmore, 1993; Pillmore and Scott, 1976). Along the route, to the north, large landslides cover much of the slopes below the 3.5 Ma basalt caps of Bartlett Mesa. To the east, landslides also cover the slopes beneath the 7.2 Ma basalt cap of Johnson Mesa (Stormer, 1972). As a side observation, about a mile from the Interstate, the view to your left reveals a broad valley containing a small hill that contains a diatreme (breccia-filled volcanic pipe formed by a gaseous explosion). A fission-track date by C. Naeser of the USGS on zircon grains in a quartzitic sandstone (Trinidad Sandstone or possibly Dakota Sandstone) block yielded 25 ± 1.4 m.y., younger than the sedimentary rocks in the area but older than the oldest basalt. This number dates the last time the rock was heated past the annealing temperature of zircon, 200 ± 25°C.

About 3.5 miles from the Interstate is the junction with New Mexico 526. Bear left and proceed left up Chicorica Canyon on highway 526. On the left, cliffs formed by sandstone beds of the barren zone of the Raton Formation and the Trinidad Sandstone are visible in the upper slope. Basaltic lava flows cap Horse Mesa on the right and Bartlett Mesa on the left.

A short distance up the creek, large coal waste piles from the Sugarite mines become evident. We will leave the vans at the visitor's center, follow the trail through the abandoned Sugarite town site, and continue up the east side of the valley to examine an exposure of the Sugarite coal bed that contains the K-T boundary. The coal is about 6 ft thick and was mined for several years from entries on both sides of the canyon.

Sugarite

Sugarite is an abandoned mining town. Coal from the Sugarite bed was mined in this valley from about 1902 until the mines were closed in 1941. Coal was produced from Wagon Mine No. 2 until the main Sugarite mine (No. 1) was opened in 1912. The coal is high-resin, noncoking, and was prized for domestic fuel (Lee, 1917). The Sugarite coal camp was established in 1908 by the Chicorica Coal Co., and began full operation the following year

The opening of the Sugarite mine was hailed as one of the

important developments in northern New Mexico, and the mine produced high-quality domestic coal for more than 30 years. When the camp first opened it consisted of scattered tents, but within a short time the construction of a full-fledged company town began. Managed by the St. Louis Rocky Mountain and Pacific Company, concrete and stone dwellings were built on slopes and terraces along the canyon sides, along with a mercantile store, schoolhouse, post office, and community center. The population of the camp fluctuated between 400 and 1,000 during its years of operation, and this mining community was considered one of the best coal camps in the area because of its beautiful setting along a running stream, its everyday amenities and community activities, and its first class equipment.

The Sugarite mines were located high on both sides of Chicorica Creek. Miners in the canyon relied on mules and burros to do the heavy work of pulling carts loaded with coal from the underground mines to the surface. From there, the coal was initially hauled by wagon to Raton, where it was used for steam generation and domestic purposes. A few years later, railroads served both the Sugarite and Yankee camps nearly every day, with coal runs to and from Raton and on to other markets.

Following closure of the mines, some of Sugarite's houses and buildings were moved to Raton, while others were simply torn down for salvage. Now all that remains of the camp are rows of rock foundations, the old post office, a mule barn, and a few residences and structures associated with the operation of the Lake Alice and Lake Maloya water systems. Further description of the mines and Sugarite Canyon State Park is given by Virginia McLemore (McLemore, 1990).

The area is mostly covered with landslide debris, soil, and vegetation. The Sugarite coal bed is exposed near the top of the lower coal zone of the Raton Formation in a landslide scar at about 7,300 ft elevation on the east wall of the canyon, about 450 ft above the valley floor. It has also been uncovered in a pit dug beside the trail. The Trinidad Sandstone forms ledges about mid-slope. The Sugarite coal bed is only about 90 ft above the top of the Trinidad and about 10 ft below the base of the cliff-forming sandstone of the barren series.

The Sugarite coal accumulated in a large, poorly drained backswamp on a broad, low-gradient, alluvial plain near the eastern edge of the Raton basin. The thickness of the coal indicates a great accumulation of peat and a long period of stable conditions in the area. The presence of the isochronous K-T boundary clay allows us to correlate with confidence the coal zone observed at the Raton site with the Sugarite coal bed at Sugarite. The swamp or perhaps several smaller, coeval swamps extended over an area of roughly 35 sq mi. The site differs from all other sites located to date in that beds of coal sandwich the K-T boundary claystone.

Stop 4: The Sugarite K-T boundary site

At the Sugarite site the coal bed is 72 in thick and contains two thin partings: a thin carbonaceous shale 15 in above the

base of the coal, and the kaolinite-rich K-T boundary claystone, 12 in thick, 6 in below the top (Fig. 12). As shown in the diagram (fig 13), the boundary claystone contains Ir concentrations of 2.7 ng/g and also marks the disappearance of the *Proteacidites* assemblage and change in fern spore/angiosperm pollen ratios (Pillmore and others, 1984; Gilmore and others, 1984; and Pillmore and Flores, 1987)

Palynology of the Sugarite K-T boundary site

The palynology of the K-T boundary at the Sugarite site was first studied by R. H. Tschudy of USGS on the basis of samples collected by C. L. Pillmore. Tschudy located the boundary within the coal using the palynological extinction datum (Pillmore and others, 1984). This was an unexpected result because, in prior investigations in the Raton basin, the boundary was found below a thin coal bed, never within coal. These results had important implications because they showed clearly that the palynological K-T boundary datum is independent of lithofacies.

The original collection consisted of 15 samples from within and above the Sugarite coal bed. The original samples, which were supplemented by 13 others collected later, became the basis for a more detailed study of the palynological and

Figure 12. Photograph of the upper part of the Sugarite coal bed showing the K/T boundary claystone. The knife lies at the top of the boundary claystone layer.

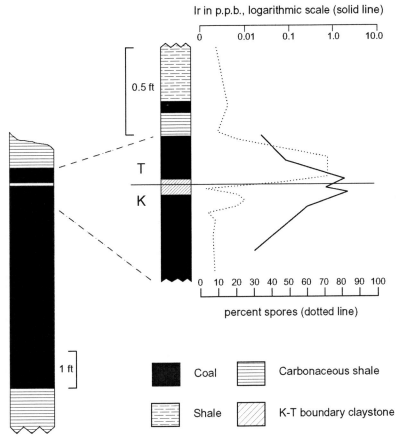

Figure 13. Diagram showing the position of the K/T boundary claystone bed in the Sugarite coal bed. Inset shows the variation in Ir values; the double peak in the Ir curve suggests migration of Ir from the boundary claystone into the adjacent coal.

paleobotanical changes across the boundary at this unique site (Nichols and others, 1985). The total interval eventually studied encompassed 1.6 m of section, including the entire thickness of the Sugarite coal bed and carbonaceous shale intervals above and below it (Fig. 13).

Both the original study and the detailed study of the Sugarite site utilized palynology to locate the boundary—specifically, the disappearance of the *Proteacidites* assemblage. At Sugarite this assemblage was found to include not only the previously reported species *Proteacidites retusus*, *"Tilia" wodehousei*, *Trisectoris* sp., and *Trichopeltinites* sp., but also *Aquilapollenites reticulatus*, *Libopollis jarzenii*, *Liliacidites complexus*, and *Tricolpites microreticulatus*. These species are well known from the uppermost Maastrichtian at K-T boundary sites elsewhere in the Western Interior of the United States and Canada (Nichols and Fleming, 1990). In all, 48 palynomorph taxa were identified, and about 20 percent disappeared at the K-T boundary at the Sugarite site.

The occurrence of a species of *Aquilapollenites* pollen at Sugarite is especially interesting. Several other species of the genus *Aquilapollenites* disappear at the K-T boundary at localities north of those in the Raton basin (Nichols, 1996), but in the Raton basin only *Aquilapollenites reticulatus* is present, and it is quite rare. In contrast, *Proteacidites retusus* and perhaps other species of the genus *Proteacidites* are quite common in

the Raton basin. At Sugarite, *Proteacidites* pollen is present in almost every sample below the boundary but in none above it. Two important patterns are revealed by these observations: (1) as already noted, *Proteacidites* pollen represents species that became extinct abruptly at the K-T boundary; and (2) pollen evidence shows that the vegetation varied in composition from south to north in latest Cretaceous time, yet all plant communities were similarly and simultaneously affected by the impact event.

In addition to the extinction of the pollen species named, the K-T boundary at Sugarite is marked by the fern-spore spike, the anomalous abundance (here up to 78 percent) of fern spores (mostly *Cyathidites* sp.) in samples just above the extinction horizon. As interpreted by Tschudy and others (1984), the fern-spore spike is evidence of the temporary replacement of a "normal" plant community by one dominated by an opportunistic species—one that can quickly reoccupy a devastated landscape (see also Fleming and Nichols, 1990).

The detailed study at Sugarite (Nichols and others, 1985) involved analysis of dispersed fragments of leaf cuticles as well as pollen and spores. Cuticle fragments, which represent leaf fossils not otherwise preserved in coal, are recovered by laboratory techniques similar to those used in preparing palynologic samples. The palynologic and dispersed-cuticle data suggest that in latest Cretaceous time an ecological transition—a hydrosere—was taking place in the mire that eventually formed the Sugarite coal. A mire community dominated by dicot angiosperms was changing to one dominated by monocots, probably palms. Sphagnum moss was also becoming more common in the mire. This gradual change in plant communities was abruptly terminated by the K-T extinction event. Ferns dominated the mire immediately after the K-T event. Samples from above the coal show that, although palms survived extinction, dicot angiosperms returned to dominance as dicots recolonized the area and the fern-dominated vegetation was replaced by the profoundly and permanently altered plant communities of earliest Tertiary time.

The Sugarite K-T boundary site remains unique in many ways. Further studies of the palynology and micro-paleobotany are continuing in conjunction with isotopic studies.

Return to Raton on highway 72.

REFERENCES CITED

Alvarez, W., 1997, *T. rex* and the crater of doom: Princeton, N.J., Princeton University Press, 185 p.

Alvarez, L.W., Alvarez, W., Asaro, F., and Michel, H.V., 1980, Extraterrestrial cause of the Cretaceous-Tertiary extinction: Science, v. 208, p. 1095-1108.

Bohor, B.F., Foord, E.E., Moires, P.J., and Triplehorn, D.M., 1984, Mineralogic evidence for an impact event at the Cretaceous-Tertiary boundary: Science, v. 224, no. 4651, p. 867-869.

Brown, R.W., 1962, Paleocene Flora of the Rocky Mountains and Great Plains: U.S. Geological Survey Professional Paper 375, 199 p.

Fleming, R.F., 1986, Fossil *Scenedesmus* (Chlorophyta) and its paleoecological significance: Nineteenth Annual Meeting, American Association of Stratigraphic Palynologists, Program and Abstracts, p. 10.

———, 1987, Paleoenvironmental significance of fossil Chlorococcalean algae from the Raton Formation, Colorado and New Mexico: Annual Meeting of the Rocky Mountain Section of the Geological Society of America.

Fleming, R.F., and Nichols, D.J., 1990, The fern-spore abundance anomaly at the Cretaceous-Tertiary boundary: A regional bioevent in western North America, in, Kauffman, E.G., and Walliser, O.H., (editors), Extinction events in Earth history: Lecture Notes in Earth Sciences, Springer-Verlag, p. 347-349.

Flores, R.M., 1984, Comparative analysis of coal accumulation in Cretaceous alluvial deposits, southern United States Rocky Mountain Basins, in Stott, D.F., and Glass, D., eds., The Mesozoic of Middle North America: Canadian Society of Petroleum Geologists Memoir 9, p. 373385.

Flores, R.M., and Pillmore, C.L., 1987, Tectonic control on alluvial paleoarchitecture of the Cretaceous and Tertiary Raton basin, Colorado and New Mexico, in Ethridge, F.G., Flores, R.M., and Harvey, M.D., eds., Recent developments in fluvial sedimentology: Society of Economic Paleontologists and Mineralogists Special Publication 39, p. 311-320.

Flores, R.M., and Tur, S.M., 1982, Characteristics of deltaic deposits in the Cretaceous Pierre Shale, Trinidad Sandstone, and Vermejo Formation, Raton Basin, Colorado: The Mountain Geologist, v. 19, p. 2540.

Gilmore, J.S., Knight, J.D., Orth, C.J., Pillmore, C.L., and Tschudy, R.H., 1984, Trace element patterns at a nonmarine Cretaceous-Tertiary boundary, Western Interior: Nature, v. 307, no. 5948, p. 224-228.

Hickey, L.J., 1981, Land plant evidence compatible with gradual, not catastrophic, change at the end of the Cretaceous, Nature, vol. 292, p. 529-531.

Hildebrand, A.R., and Boynton, W.V., 1988, Provenance of the K-T boundary layers, in Global catastrophes in Earth history - An interdisciplinary conference on impacts, volcanism, and mass mortality [abs.]: Lunar and Planetary Institute Contribution 673, p. 78-79.

Izett, G.A., 1990, The Cretaceous/Tertiary boundary interval, Raton Basin, Colorado and New Mexico, and its contents of shock-metamorphosed minerals; evidence relevant to the K-T boundary impact-extinction theory: Geological Society of America Special Paper, v. 249, 100 p.

Izett, G.A., and Bohor, B.F., 1986, Microstratigraphy of continental sedimentary rocks in the Cretaceous-Tertiary boundary interval in the Western Interior of North America: Geological Society of America Abstracts with Programs, v. 18, no. 6, p. 644.

Izett, G.A., and Pillmore, C.L., 1985a, Abrupt appearance of shocked quartz at the Cretaceous-Tertiary boundary, Raton basin, Colorado and New Mexico [abs.]: Geological Society of America Abstracts with Programs, vol. 17, no 7, p. 617.

———, 1985b, Shock-metamorphic minerals at the Cretaceous-Tertiary boundary, Raton basin, Colorado and New Mexico, provide evidence of asteroid impact in continental crust: EOS Transactions of the American Geophysical Union, v. 66, p. 1149-1150.

Johnson, K.R., and Hickey, J.J., 1990, Megafloral change across the Cretaceous/Tertiary boundary in the northern Great Plains and Rocky Mountains, U.S.A., in Sharpton, V.L. and Ward, P.D., eds., Global catastrophies in Earth history: an interdisciplinary conference on impacts, volcanism, and mass mortality: Geological Society of America Special Paper 247, p. 433-444.

Lee, W.T., 1917, Geology of the Raton Mesa and other regions in Colorado and New Mexico, p. 9221, in Lee, W.T., and Knowlton, F.H., Geology and paleontology of Raton Mesa and other regions in Colorado and New Mexico: U.S. Geological Survey Professional Paper 101, 450 p. [1918].

Leffingwell, H.A., 1970, Palynology of the Lance (Late Cretaceous) and Fort Union (Paleocene) Formations of the type Lance area, Wyoming, in Kosanke, R.M., and Cross, A.T., eds., Symposium on palynology of the Late Cretaceous and early Tertiary: Geological Society of America Special Paper 127, p. 1-64.

Leighton, V.L., 1980, Depositional environments and petrography of the Trinidad Sandstone and related formations, Raton area, New Mexico [M.S. thesis]: Fort Collins, Colorado State University, 105 p.

Marvin, R.F., Young, E.J., Mehnert, H.H., and Naeser, C.W., 1974, Summary of radiometric age determinations on Mesozoic and Cenozoic igneous rocks and uranium and base metal deposits in Colorado: Isochron West, 1974, no. 11, p. 3233 p.

McLemore, V.T., 1990, Sugarite Canyon: New Mexico Geology, v. 12, no. 2, p. 38-42.

Nichols, D.J., 1996, Vegetational history in Western Interior North America during the Cretaceous-Tertiary transition, Chapter 29D in Jansonius, Jan, and McGregor, D.C., eds., Palynology; Principles and Applications: Dallas, Texas, American Association of Stratigraphic Palynologists, v. 3, p. 1189-1195.

Nichols, D.J., and Fleming, R.F., 1990, Plant microfossil record of the terminal Cretaceous event in the western United States and Canada, in Sharpton, V.L., and Ward, P.D., eds., Global catastrophes in Earth history; an interdisciplinary conference on impacts, volcanism, and mass mortality: Geological Society of America Special Paper 247, p. 445-455.

Nichols, D.J., Fleming, R.F., Upchurch, G.R., Tschudy, R.H., and Pillmore, C.L., 1985, Paleobotanical changes across the Cretaceous-Tertiary boundary at Sugarite, New Mexico; new data and interpretations [abs.]: Society of Economic Paleontologists and Mineralogists, 1985 Annual Midyear Meeting Abstracts, v. 2, p. 68.

Nichols, D.J., Jarzen, D.M, Orth, C.J., and Oliver, P.Q., 1986, Palynological and Ir anomalies at Cretaceous-Tertiary boundary, south-central Saskatchewan: Science, v. 231, no. 4739, p. 714-717 p.

Orth, C.J., Gilmore, J.S., Knight, J.D., Pillmore, C.L., Tschudy, R.H., and Fassett, J.E., 1981, An Ir abundance anomaly at the palynological Cretaceous-Tertiary boundary in northern New Mexico: Science, v. 214, no. 4527, p. 1341-1343.

Penn, B.S., 1994, An Investigation of the temporal and geochemical characteristics, and the petrogenetic origins of the Spanish Peaks intrusive rocks of south-central Colorado [Ph.D. thesis]: Golden, Colorado School of Mines, 198 p.

Pillmore, C.L., 1969, Geologic map of the Casa Grande quadrangle, Colfax County, New Mexico, and Las Animas County, Colorado; U.S. Geological Survey Geologic Quadrangle Map GQ-823, scale 1:62,500.

Pillmore, C.L., 1976, Commercial coal beds of the Raton coal field, Colfax County, New Mexico, in Ewing, R.C., and Kues, B.S., eds., Guidebook of Vermejo Park, northeastern New Mexico: New Mexico Geological Society Guidebook, 27th Field Conference, p. 227-247.

Pillmore, C.L., and Flores, R.M., 1987, Stratigraphy and depositional environments of the Cretaceous-Tertiary boundary interval and associated rocks, Raton basin, New Mexico and Colorado, in Fassett, J.E., and Rigby, K.B., Jr., eds., Cretaceous-Tertiary boundary of the San Juan and Raton basins, northern New Mexico and southern Colorado: Geological Society of America Special Paper 209, p. 111-130.

Pillmore, C.L. and Scott, G.R., 1976, Pediments of the Vermejo Park area, in Ewing, R.C., and Kues, B.S., eds., Guidebook of Vermejo Park, northeastern New Mexico: New Mexico Geological Society Guidebook, 27th Field Conference, p. 111-120.

Pillmore, C.L., and Scott, G.R., 1994, Map showing the geology of the Clifton House 7½' quadrangle, Colfax County, New Mexico, and fossil zones in the Pierre Shale: U.S. Geological Survey Geologic Quadrangle Map GQ-1737, Scale 1:24,000.

Pillmore, C.L., Tschudy, R.H., Orth, C.J., Gilmore, J.S., and Knight, J.D., 1984, Geologic framework of nonmarine Cretaceous-Tertiary boundary sites, Raton basin, New Mexico and Colorado: Science, v. 223, no. 4641, p. 1180-1182.

Pollastro, R.M., and Pillmore, C.L., 1987, Mineralogy and petrology of the Cretaceous-Tertiary boundary clay bed and adjacent clayrich rocks, Raton basin, New Mexico and Colorado: Journal of Sedimentary Petrology, v. 57, no. 3, p. 456-466.

Pollastro, R.M., Pillmore, C.L., Tschudy, R.H., Orth, C.J., and Gilmore, J.S., 1983, Clay petrology of the conformable Cretaceous/Tertiary boundary interval, Raton Basin, New Mexico and Colorado [abs.]: Annual Clay Minerals Conference, 32nd, Buffalo, New York, 1983, Programs and Abstracts, p. 83 (2 pages).

Powell, J.L., 1998, Night comes to the Cretaceous: dinosaur extinction and the transformation of modern geology: New York, N.Y., W.H. Freeman and Company, 250 p.

Scott, G.R., 1986, Historic trail maps of the Raton and Springer 30' x 60' quadrangles, New Mexico and Colorado: U.S. Geological Survey Miscellaneous Investigations Series Map, I-1641, Scale 1;100,00.

Scott, G.R., and Pillmore, C.L., 1993, Geologic and structure-contour map of the Raton 30' x 60' quadrangle, Colfax and Union Counties, New Mexico, and Las Animas County, Colorado: U.S. Geological Survey Miscellaneous Investigations Series Map, I-2266, Scale 1:100,000.

Smit, Jan, 1984, Evidence for worldwide microtektite strewn field at the Cretaceous-Tertiary boundary [abs.]: Geological Society of America Abstracts with Programs, v. 16, no. 6, p. 659.

Smith, R.P., 1975, Structure and petrology of Spanish Peaks dikes, south-central Colorado [Ph.D. thesis]: Boulder, University of Colorado, 191 p.

Stormer, J.C., 1972, Ages and nature of volcanic activity on the southern high plains, New Mexico and Colorado: Geological Society of America Bulletin, v. 83, p. 2443-2448.

Stroud, J. R., 1998, Geochronology of the Raton-Clayton volcanic field, New Mexico, with implications for volcanic history and landscape evolution [M.S. thesis]: Socorro, New Mexico Institute of Mining and Technology, 51 p.

Strum, S.R., 1984, Depositional environments and lithofacies of the Raton Formation, eastern Raton Basin, New Mexico [M.S. thesis]: Raleigh, North Carolina State University, 81 p.

Tschudy, R.H., 1970, Palynology of the Cretaceous-Tertiary boundary in the northern Rocky Mountain and Mississippi Embayment regions, in Kosanke, R.M., and Cross, A.T., eds., Symposium on palynology of the Late Cretaceous and early Tertiary: Geological Society of America Special Paper 127, p. 65-111.

_____, 1973, The Gasbuggy Core—a palynological appraisal in Fassett, J.E., ed., Cretaceous and Tertiary rocks of the southern Colorado Plateau: Four Corners Geological Society Memoir, p. 131-143.

Tschudy, R.H., Pillmore, C.L., Orth, C.J., Gilmore, J.S., and Knight, J.D., 1984, Disruption of the terrestrial plant ecosystem at the Cretaceous-Tertiary boundary, Western Interior: Science, v. 225, no. 4666, p. 1030-1034.

Tschudy, R.H., and Tschudy, B.D., 1986, Extinction and survival of plant life following the Cretaceous/Tertiary boundary event, Western Interior, North America: Geology, v. 14, no. 8, p. 667-670.

Tweto, O.L., 1987, Rock units of the Precambrian basement in Colorado: U.S. Geological Survey Professional Paper 1321-A, 54 p.

Wolfe, J.A., and Upchurch, G.R., 1986, Vegetation, climatic and floral changes at the Cretaceous-Tertiary boundary: Nature, v. 324, no. 6093, p. 148-152.

Woodward, L.A., and Snyder, D.O., 1976, Structural framework of the southern Raton Basin, New Mexico: New Mexico Geological Society Guidebook, 27th Field Conference, Vermejo Park, p. 125-127.

Printed in U.S.A.

Geological Society of America
Field Guide 1
1999

Stratigraphy, sedimentology, and paleontology of the Cambrian-Ordovician of Colorado

Paul M. Myrow
Department of Geology, Colorado College, Colorado Springs, Colorado 80903, United States; pmyrow@coloradocollege.edu
John F. Taylor
Department of Geoscience, Indiana University of Pennsylvania, Indiana, Pennsylvania 15705, United States
James F. Miller
Department of Geography, Geology, and Planning, Southwest Missouri State University, Springfield, Missouri 65804, United States
Ray L. Ethington
Department of Geological Sciences, University of Missouri, Columbia, Missouri 65211, United States
Robert L. Ripperdan
Department of Geology, University of Puerto Rico, Mayaguez, Puerto Rico 00681
Christina M. Brachle
Department of Geology, Colorado College, Colorado Springs, Colorado 80903, United States

INTRODUCTION

The numerous extinctions that affected shallow marine faunas on the tropical shelves surrounding Laurentia in the Cambrian and Early Ordovician have been the focus of many detailed biostratigraphic, evolutionary, and paleoecologic studies. Data from carbonate platform and off-platform strata have been used to propose process-response models that invoke sea level change as a forcing mechanism for extinctions and/or radiations within the Cambrian and Early Ordovician. Some regressive features observed near horizons of faunal change within the Cambrian-Ordovician boundary interval on various continents have been used to propose a series of "eustatic events" (Nicholl et al., 1992). These include the "Lange Ranch Eustatic Event" and "Black Mountain Eustatic Event" of Miller (1984, 1992) and the *Acerocare* Regressive Event and *Peltocare* Regressive Event of Erdtmann (1986). There is much debate about the nature of these proposed events (Ludvigsen et al. 1986; Taylor et al. 1992; Landing 1993) based, at least in part, on the ambiguous nature of the sedimentological data and insufficient precision of correlation.

A rigorous test of the proposed linkage between extinctions and paleoceanographic events within the Cambrian-Ordovician boundary interval will require varied, high-resolution stratigraphic data from a complete onshore-offshore profile, including inner shelf, platform, and shelfbreak settings. Detailed data from inner shelf environments are particularly critical because facies and fauna in these settings are highly responsive to relative sea level changes and other environmental perturbations. However, the mixed carbonate and siliciclastic facies that dominate inner shelf successions in the Cambrian-Ordovician deposits of Laurentia are generally less fossiliferous than coeval carbonate platform facies, and in places contain numerous stratigraphic gaps. As a result, they have received less attention than outer shelf facies. Data on trilobite biofacies is sparse for faunas that occupied proximal shelf and cratonic environments. In addition, sedimentologic information in most previous biostratigraphic studies of inner shelf assemblages is either lacking or of insufficient precision to determine whether horizons and intervals of faunal change coincide with sequence boundaries or significant facies transitions. Chemostratigraphic techniques such as carbon isotope stratigraphy, when used in conjunction with high-resolution biostratigraphic information and detailed sedimentologic and sequence stratigraphic data, can significantly enhance temporal correlations between these deposits and the more heavily studied distal platform facies.

On this field trip we will examine a precise and integrated stratigraphic framework recently established for the Cambrian-Ordovician inner shelf deposits of Colorado. Within this framework we will observe a complex record of relative sea level changes (as recorded in the stratigraphic succession of lithofacies and sequence boundaries), paleoceanographic events (reflected in the isotope stratigraphy), and bioevents (extinction horizons and intervals of adaptive radiation).

Myrow, P. M., Taylor, J. F., Miller, J. F., Ethington, R. L., Ripperdan, R. L., and Brachle, C. M., 1999, Stratigraphy, sedimentology, and paleontology of the Cambrian-Ordovician of Colorado, *in* Lageson, D. R., Lester, A. P., and Trudgill, B. D., eds., Colorado and Adjacent Areas: Boulder, Colorado, Geological Society of America Field Guide 1.

BACKGROUND

The Cambrian-Ordovician rocks of Colorado were deposited in an inferred northwest-southeast oriented trough that may represent the only region of significant breaching along the length of the Transcontinental Arch. This trough, the "Colorado Sag" (Lochman-Balk 1956), was situated between the highlands areas of Siouxia to the north and the Sierra Grande to the south (Fig. 1). The Colorado Sag has been portrayed as consisting of two sub-basins, separated by a subdued arch (Fig. 1), with a northwestern sub-basin that presumably was connected to the paleo-Pacific margin, and a southeastern sub-basin that connected to the Midcontinental Sea (Gerhard 1972, 1974). The interface of these sub-zones has been described as a "basin high" (Gerhard 1972, 1974), but is more realistically considered by Allen (1992) to be the Homesteak Shear Zone (HSZ), a reactivated Precambrian northeast–southeast trending fault zone. Topography along the HSZ may have affected sedimentation at some times more than others. Work to date has confirmed that some prominent horizons hold great potential for correlation across the Colorado Sag, and to other regions in ancient North America. High resolution data are essential in these rocks owing to an abundance of stratigraphic breaks and condensed intervals, which is an expected consequence of deposition in shallow, nearshore environments in an epicratonic setting.

LITHOSTRATIGRAPHY

Lower Paleozoic rocks of Colorado unconformably overlie Proterozoic crystalline basement that include gneisses and granitic intrusives. The Upper Cambrian Sawatch Formation occurs at the base of this cover sequence (Fig. 2) and consists of white-weathering sandstone and pebbly sandstone which ranges from 5-160 meters in thickness across the state (Fig. 3). In the northwest sub-basin, the Sawatch is overlain by ~25–35 m of shale, dolostone, and flat-pebble conglomerate of the Dotsero Formation (Fig. 2). These lithologies typify units deposited in lagoonal areas of the Cambrian inner detrital belt (Lochman-Balk 1971; Sepkoski 1982). The top of the Dotsero is placed at the top of a regionally extensive 1-1.5-m-thick stromatolitic lithosome, the Clinetop Member (Bass and Northrop 1953; Campbell 1976). The lowermost part of the overlying carbonate unit, the Manitou Formation, is similar in character to the Dotsero Formation.

In the Mosquito Range of the southeast sub-basin, the Sawatch Formation is overlain by ~25 m of sandstone, dolomitic sandstone, and sandy dolostone of the Peerless Formation. Farther south, along the Front Range, extremely thin (< 5 m) nearshore and shoreline deposits of the Sawatch are overlain by nearshore, inner shelf glauconite-rich sandstone and dolostone with large tidal subaqueous dune deposits (Myrow 1998). Berg and Ross (1959) recovered the trilobites *Ptychaspis granulosa* (Owen), *Ellipsocephloides butleri* Resser, and *Idahoia wisconsensis* (Owen), which are diagnostic of the *Ptychaspis* Zone of

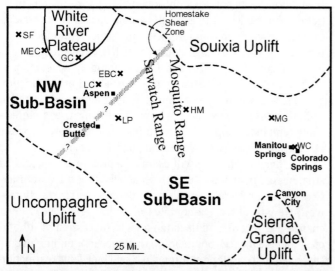

Figure 1. Cambrian-Ordovician paleogeography of Colorado. Location abbreviations: SF=South Fork, MEC=Main Elk Creek, GC=Glenwood Canyon, EBC=East Brush Creek, LC=Lime Creek, LP=Lambertson's Peak, HM=Horseshoe Mountain, MG=Missouri Gulch, WC=William's Canyon. Conodont biostratigraphy for these locations is given in Figure 5. Modified from Gerhard (1972).

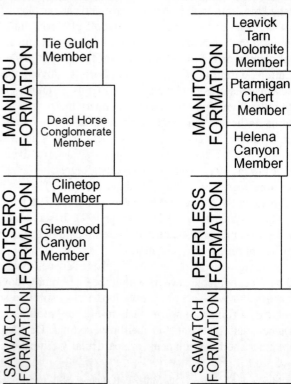

Figure 2. Lithostratigraphy of Cambrian and Ordovician deposits of Colorado. Modified from Gerhard (1972).

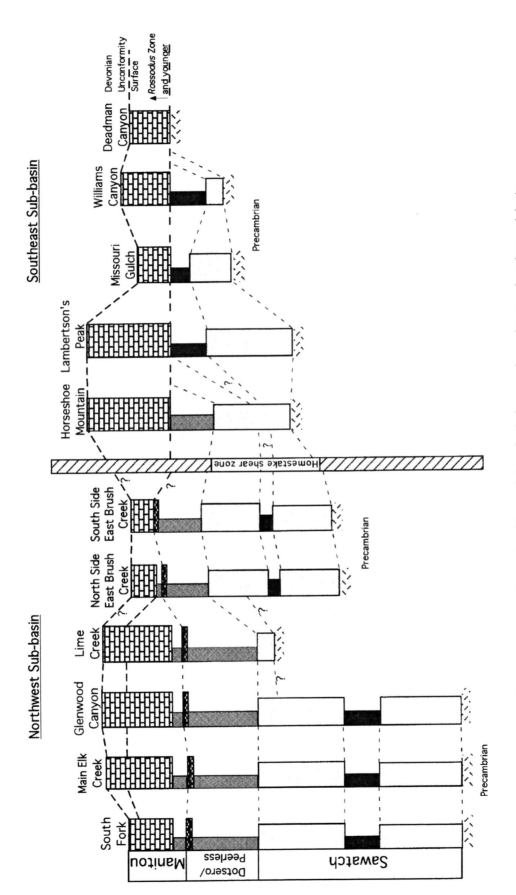

Figure 3. Schematic lithostratigraphic correlation of units between the northwest and southeast sub-basins.

the Franconian Stage of the Cambrian, from these younger strata and assigned them to the Peerless Formation. No fauna had ever been found in the type section (Horseshoe/Peerless Mountains) of the Peerless Formation until our recovery of conodonts of the *Eoconodontus* Zone from the upper Peerless (Myrow et al. 1995) and Lower Ordovician *Rossodus manitouensis* Zone conodonts from the base of the Manitou Formation. This indicates: (1) that the glauconitic unit along the Front Range is not the Peerless Formation, but the middle member of the Sawatch Formation (Stop 1; Fig. 4), and (2) a substantial Cambrian–Ordovician unconformity exists at Horseshoe Mountain. We now recognize that this is the same sub-Ordovician unconformity that Berg and Ross (1959) showed to progressively cut out underlying Cambrian strata along the Front Range. The Manitou Formation onlaps onto this erosion surface—from north to south (Berg and Ross 1959; Fig. 4)—and rests successively on the middle Sawatch (Stops 1 and 2), lower Sawatch, and Precambrian basement (Deadman Canyon, Colorado Springs) (Fig. 4).

The Manitou is of variable thickness and reaches up to 112 m in the Sawatch Range (Stevens 1961). There are differences in lithofacies within the Manitou between northwestern and southeastern sub-basins, and therefore separate member nomenclatures exist for each area (Gerhard 1972, 1974; Fig. 2). Everywhere in the southeast sub-basin, the entire Manitou is Lower Ordovician (*Rossodus manitouensis* Zone or younger). In the more complete western Colorado exposures, the Cambrian–Ordovician boundary occurs within the lower part of the Manitou as it is defined in that area with its base at the top of the Clinetop stromatolite bed. An interval of sandy dolomite and dolomitic sandstone occurs immediately above the Dotsero in the western part of the White River Plateau and near the Homestake Shear Zone in the northwestern part of the Sawatch Range. Recovery of lower Ibexian conodonts from this unit indicates that it is a proximal facies of the Manitou Formation.

PREVIOUS WORK

Most previous work on the Cambrian–Ordovician of Colorado (Anderson 1970; Bass and Northrop 1953; Brainerd et al. 1933; Campbell 1972; Gerhard 1974; Johnson 1934, 1944, 1945; Ross and Tweto 1980; Tweto 1949) are invaluable sources of basic information, but they contain very little process-oriented sedimentology or modern stratigraphic analyses. In addition, little paleontological work had been done since the trilobite study of Berg and Ross (1959). In this guidebook we present some of the results of our research on the Cambrian–Ordovician of Colorado over the past seven years. We examined nearly 40 measured sections, including nearly all those listed in Ross and Tweto (1980) and Bass and Northrop (1953). A few from each region were chosen for detailed description and high-resolution sampling for biostratigraphic and chemostratigraphic analyses (Figs. 1, 3, 5). Myrow provided the detailed lithologic descriptions and sedimentological interpretations. Taylor systematically sampled the sections for

macrofossils and is responsible for the biostratigraphic and paleogeographic interpretations based upon the trilobite data. A suite of macrofossils was also collected at Missouri Gulch (Stop 2) by James D. Loch of Central Missouri State University. Conodont samples were processed by Ethington and Miller. Carbon isotope study was undertaken by Colorado College students in collaboration with Ripperdan at the University of Puerto Rico-Mayaguez and Claudia Mora at the University of Tennessee.

STOP 1A: FRONT RANGE

Take Highway 24 West from Colorado Springs to Manitou Springs (Fig. 6). Just past mile marker 298 there is a stoplight for Manitou Springs/Cave of the Winds. Go left at the light and proceed downhill along the Fountain Creek to a stop sign. Take a SHARP right up a hill leading to Hwy 24 for 0.4 miles. CAREFULLY cross the road into a pulloff on the left.

Lower Sawatch Description

At this stop we will examine the sedimentology of the lower and middle members of the Sawatch Formation, which is summarized by Myrow (1998; Fig. 7). Along the Front Range these Upper Cambrian rocks rest nonconformably on 1.75-Ga metamorphic rocks of the Idaho Springs Formation and younger (1030 Ma) intrusive rocks of the Pikes Peak Granite. The lower member of the Sawatch Formation consists of 4.0 to 4.5 m of white-weathering, thin to medium bedded (5 to 20 cm thick), coarse, very coarse, and pebbly quartz arenite with < 5% feldspar (Lewis 1965). The nonconformity is remarkably flat at the outcrop scale, except where several of the large 0.5 to 3-m diameter corestones in the underlying granite project up to 40 cm above the nonconformity surface into the overlying Sawatch. Stratification in the Sawatch clearly abuts against the corestone, indicating that these are Precambrian-age weathering features.

The lower 2.3 m of the lower Sawatch is parallel laminated with widely-spaced, small-scale (<8 cm thick) trough cross-stratification, as well as thin pebble lags. Well-preserved polygonal cracks filled with pebbly sandstone occur at the 0.95-m mark. Above 2.3 m beds are dominantly bioturbated and separated by partially dispersed (bioturbated) pebble lags with sandstone intraclasts up to 10 cm in diameter. The upper 60 cm of the lower Sawatch is a distinctive white-weathering, bioturbated pebbly coarse sandstone bed with abundant burrows, including *Teichichnus* and lined burrows of cf. *Paleophycus*. The uppermost 10 cm of the bed consists of a hematite-coated lag with abundant quartz pebbles, large intraclasts of quartz sandstone, as well as pyritic steinkerns and pyrite-replaced and silicified cephalopod, gastropod, trilobite, and brachiopod shells. In thin section, small, mm-diameter burrows are seen projecting inwards radially around rounded (< 1 cm diameter) intraclasts of mudstone that must have been firm enough early on to withstand transport and rounding. Directly overlying the lag surface

MOSQUITO RANGE — FRONT RANGE CORRELATION

Traditional View

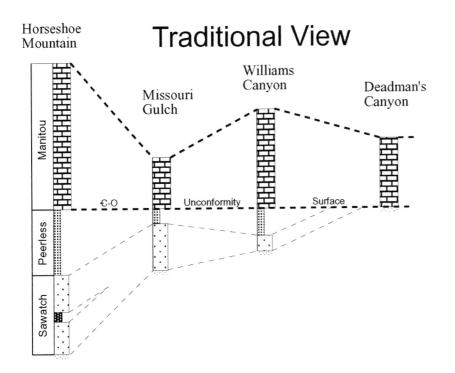

Proposed Revision

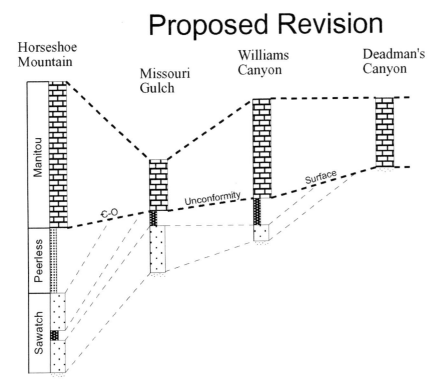

Figure 4. Comparison of correlation schemes for the Front Range and Mosquito Range. The "Traditional View" (Berg and Ross, 1959) correlates glauconitic and dolomitic deposits along the Front Range (Missouri Gulch, Williams Canyon, Deadman's Canyon) with the Peerless Formation whose type section is exposed at Horseshoe Mountain. Conodonts recovered from the Peerless at the type section indicate a much younger age, and hence the Front Range deposits are correlated with the glauconite-rich middle member of the Sawatch Formation, in the "Proposed Revision."

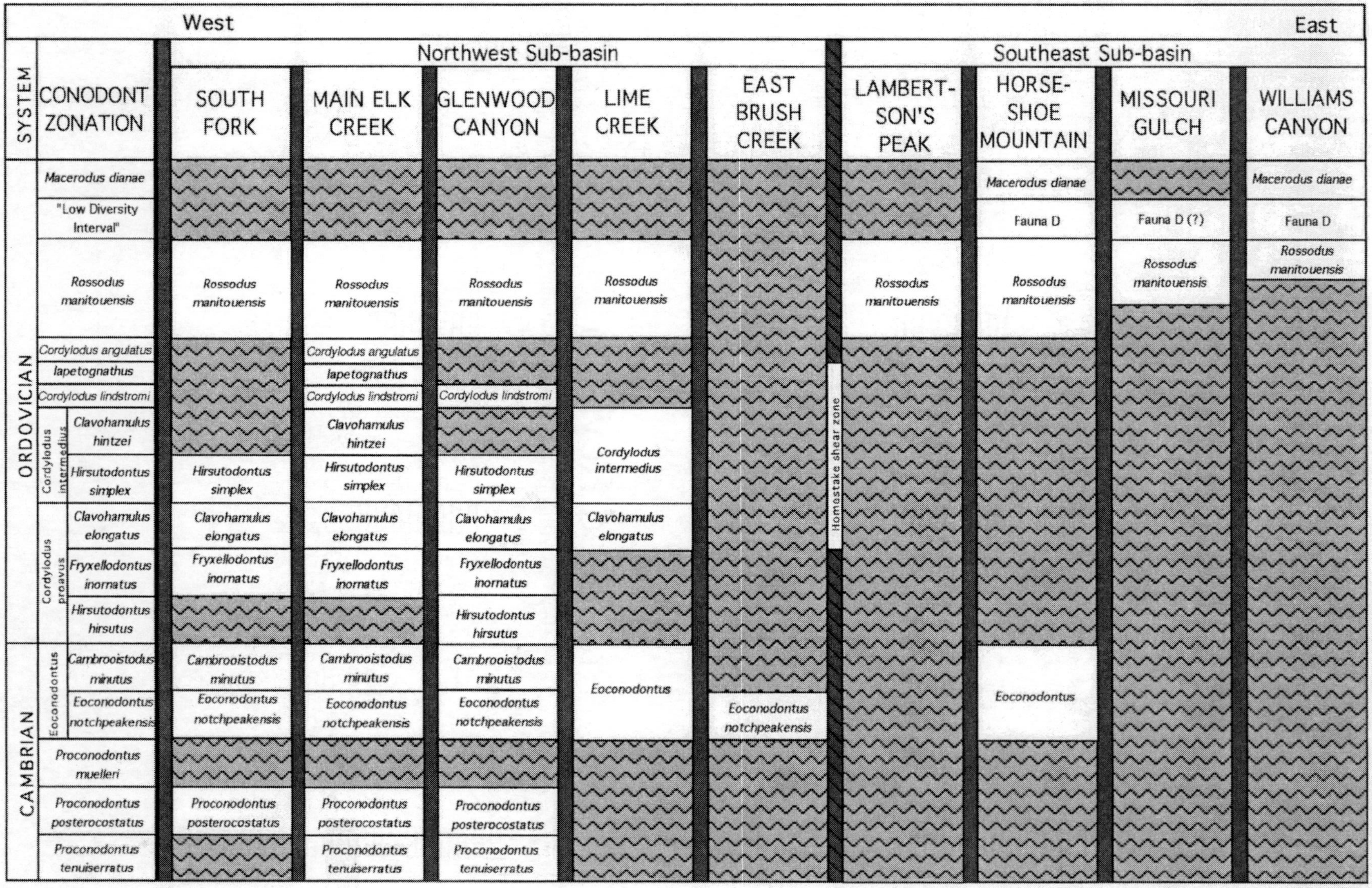

Figure 5. Conodont biostratigraphy for selected locations in western Colorado. Shaded areas denote intervals missing at unconformities. Note increase in number of zones/subzones present in more distal sections to northwest.

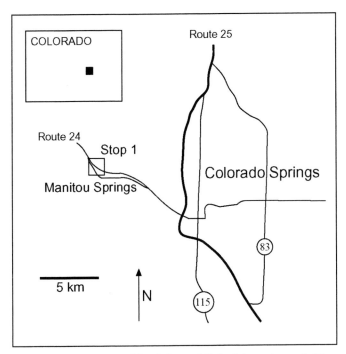

Figure 6. Location map of Cambrian and Ordovician outcrops in Manitou Springs, Colorado.

are the glauconite-rich, dune deposits of the middle Sawatch member.

The lower member contains no evidence of multi-cycle grains (e.g., with rounded overgrowths) (Lewis 1965), yet it is compositionally mature, even in poorly sorted pebbly sandstone beds. These observations suggest that the underlying basement was subjected to intense chemical weathering that produced a highly chemically mature regolith that was a primary sediment source for the Sawatch. Erosion of the regolith was uniform through a deeply weathered zone, except where resistant corestones projected up into the weathered profile. Paleoenvironments are difficult to interpret because of the paucity of diagnostic sedimentary structures. The strata up to 2.3 m do not contain structures diagnostic of braided stream deposition, such as lenticular beds or abundant dune-scale trough cross-stratified beds. The rarity of cross-strata is unusual because dunes are common in sediment of this grain size in most fluvial and marginal marine environments. Abundant horizontal stratification and the presence of desiccation cracks at 0.9 m suggest that these may be foreshore deposits.

The presence of bioturbation fabrics and burrows such as *Teichichnus* above 2.3 m indicate a marine paleoenvironment. The stratigraphic position of the first significant marine flooding surface appears to be at 2.3 m, at the sudden shift from laminated beds to massive bioturbated beds. This surface and most of the overlying pebble lags represent potential ravinement or marine erosion surfaces. Intervening sandstone beds represent redistributed sediment that was thin remnants of any number of possible nearshore and shoreline subenvironments. This is the

stratigraphic record one would expect from small changes in relative sea level in a cratonal setting with low sediment input, low accommodation space, and low equilibrium bottom slopes (Myrow 1998). As envisioned, the Sawatch records reworking, during transgression, of a landscape lacking significant fluvial systems and consisting instead of a low relief bedrock surface mantled with a thick regolith.

The uppermost bed of the lower member may contain an assemblage of trace fossils of the Glossifungites ichnofacies, which form above omission surfaces in firm but unlithified marine shoreline deposits (Frey and Seilacher 1980; Pemberton and Frey 1985). Although *Teichichnus* and cf. *Paleophycus* are not restricted to *Glossifungites* ichnofacies, the sandstone intraclasts and burrows in the small mudstone intraclasts are consistent with firmground conditions. Myrow (1998) interpreted the uppermost lag as a major ravinement surface formed by tidal currents (i.e., Allen and Posamentier 1993).

The facies transition to overlying glauconitic and dolomitic dune deposits of the middle member is very sharp. Such transitions are common across ravinement surfaces, particularly when zones of active erosion are wide (onshore–offshore) so that more distal marine sediment is deposited above the ravinement surface (Belknap and Kraft 1981, 1985). The Sawatch lag is interpreted to represent part of a condensed deposit in which slow sedimentation rates and unusual geochemical conditions produced pyritic, silicic, and hematitic replacement and coatings of grains.

Middle Sawatch Member

The middle Sawatch member consists of 13 m of dolostone, sandy dolostone, and dolomitic sandstone with scattered glauconite (Fig. 7). Its base is marked by the appearance of thick, coarse-grained, compound cross-bed sets that rest directly on the uppermost lag of the lower member. These consist primarily of dolomite-cemented detrital grains of quartz, glauconite and white-weathering inarticulate brachiopod shell fragments. Individual cross-bed sets range up to 3.5 m in thickness and consist of complete or near-complete formsets. Thick co-sets between 3 and 5 m in thickness consist of 2 to 3 stacked to shingled sets of cross-strata. The compound cross-bedded formsets contain nearly symmetrical cross-sectional shapes and low stoss (3-8°) and foreset dips (10-15°). Foreset dip directions are fairly consistent from NNE to NW (Fig. 8). Small-scale cross-bedding consists of cm- to dm-scale sets that record the migration of smaller secondary dunes across dunes. These show bimodal to polymodal paleocurrent orientations (Fig. 8).

The rest of the upper member consists of thin to medium bedded, complexly mixed, siliciclastic and carbonate strata that range from nearly pure micritic dolostone to quartz arenite. Sandstone beds are medium to coarse grained, < 6 cm to 50 cm in thickness, and have planar lamination and trough cross-bedding. Bioturbation is common in the middle part of the middle member and burrows include *Planolites, Chondrites, Cur-*

volithes, and possible *Thalassinoides horizontalis* (Myrow 1995). There is a decrease in bioturbation and an increase in the abundance of trough cross-stratified and intraclast-rich beds towards the top of the member.

A distinctive, 50–100-cm-thick, resistant-weathering, red, coarse dolostone unit marks the contact between the middle Sawatch and overlying Manitou Formation. This highly recrystallized coarse dolostone bed has a sharp erosional lower surface (~40 cm of relief) and a complex internal microstratigraphy that includes karstic cavities with collapse breccia and carbonate cement-filled vugs. Conodonts from this bed are from the *Rossodus manitouensis* Zone, similar to those from the base of the Manitou Formation at the Missouri Gulch section (Stop 2). The immediately overlying carbonate beds begins with conodonts of the low-diversity interval (= base of Fauna D of Ethington and Clark 1981). Two closely spaced sequence boundaries therefore bracket this bed. The basal surface is the Cambrian–Ordovician boundary unconformity, which along the Front Range of Colorado represents a considerable hiatus interpreted as a depositional sequence boundary within Palmer's (1981) Sauk III subsequence (Landing 1993).

The stratigraphic shift to the glauconitic tidal dune deposits of the lowermost middle member represents a significant paleoenvironmental change across a tidal (?) ravinement surface. Large tidal dunes in many locations globally rest directly on ravinement surfaces produced during Holocene transgression (e.g., Davis et al. 1993). Marine transgression has generally been implicated in the formation of dune complexes (e.g., Hine 1977) and in deposition of glauconite (Brasier 1980; Odin and Fullagar 1988). The dune deposits are considered to be condensed deposits, for they are rich in glauconite and formed in response to deepening above a ravinement surface. The initiation of bedform development is likely to have occurred as the result of amplification of tidal currents due to the interaction of rising relative sea level and the geometry of the transgressed landscape. The area around Manitou Springs must have been a large embayment in the Cambrian shoreline that had a funneling effect on tidal currents and possibly also locally created geochemical conditions (i.e., slightly reducing) that favored glauconite formation. Myrow (1998) calculated a minimum water depth for the tidal dune deposits of 21 m based on preserved bedform height, but considers 35 m to be a more reasonable estimate. Transgression was therefore potentially rapid, given that there is only 3.1 m of section between the surface with desiccation cracks and the base of the dune deposits.

The nearly symmetrical geometry and low dips of stoss and lee sides of the Peerless deposits resemble modern tidal dunes (e.g., Fenster et al. 1990) that form under less extreme conditions of tidal asymmetry. Deposits of such dunes (Allen's [1980] Class V and VI) are poorly documented or understood from the ancient. The uniform NNW dips of the large-scale foresets may represent the orientation of the dominant tidal current (flood or ebb) as tidal dunes are generally flow-transverse. Large-scale bedform migration resulted from net accumulation

over many tidal cycles. Because sediment transport increases nonlinearly with increasing shear stress (Middleton and Southard 1984), small tidal asymmetry could have produced locally consistent bedform migration directions.

The greater scatter in the paleocurrent data of small-scale cross-bedding (=superimposed dunes) is similar to modern tidal dunes (Houbolt 1982; Fenster et al. 1990; Davis et al. 1993). This is a reflection of the highly unsteady and nonuniform nature of the currents at the base of tidal flows. A large percentage of the smaller dunes migrated directly up and down the dune faces. This may indicate that the high threshold velocities required to move the coarse sediment of these dunes may have only been exceeded during peak flow conditions, during which time secondary flow may have been oriented nearly perpendicular to the dune crests.

Interpretations of depositional environments are difficult for the rest of the upper member because it lacks diagnostic sedimentary structures (e.g., desiccation cracks). The lack of wave- or storm-diagnostic features and the abundance of trough cross-stratification indicates that tidal currents may have continued to influence deposition. Coarse quartz-rich sand was likely mixed with locally derived grains of carbonate and glauconite during episodic flooding events. The upper member shoals to the Cambrian-Ordovician unconformity, as indicated by an upward increase in abundance of trough cross-stratification and flat-pebble conglomerate and much less evidence of bioturbation.

STOP 1B: FRONT RANGE

Retrace route for Stop 1A by going downhill 0.4 mi, taking a sharp left just past the gate to Manitou Springs at the first intersection. Proceed to hairpin turn and park in pulloff.

For those who wish to further examine the Upper Cambrian deposits, exposure of the same section as Stop 1A occurs at the sharp hairpin turn parking lot. Here the tidal dune deposits contain much more trough cross-bedding, most of which is highly oblique to the large-scale foresets. Several yards down hill from the parking lot along the creek are extensive bedding planes in strata from just above the dune deposits. These contain abundant eocrinoid columnals and bioturbated dolomitic and glauconitic deposits. The main outcrop of Stop 1B is further downhill on the east side of the road. Here there are exposures of the limestone and dolostone of the Lower Ordovician Manitou Formation.

At this stop, the lowermost Manitou contains well-devel-

Figure 7. Detailed measured section showing lithostratigraphy; stratigraphic changes in quartz and glauconite; and sedimentological, sequence stratigraphic, and paleoenvironmental interpretations. The left side of the grain size scale shows fine sand (FS) to cobble (Cob) for siliciclastic sediment. The right side shows mudstone, grainstone and flat pebble conglomerate in carbonate facies. Measured section from Stops 1A and 1B, Manitou Springs (southeast part of Sec. 31, T. 13 S., R. 67 W., Manitou Springs 7.5' quadrangle, El Paso County, Colorado).

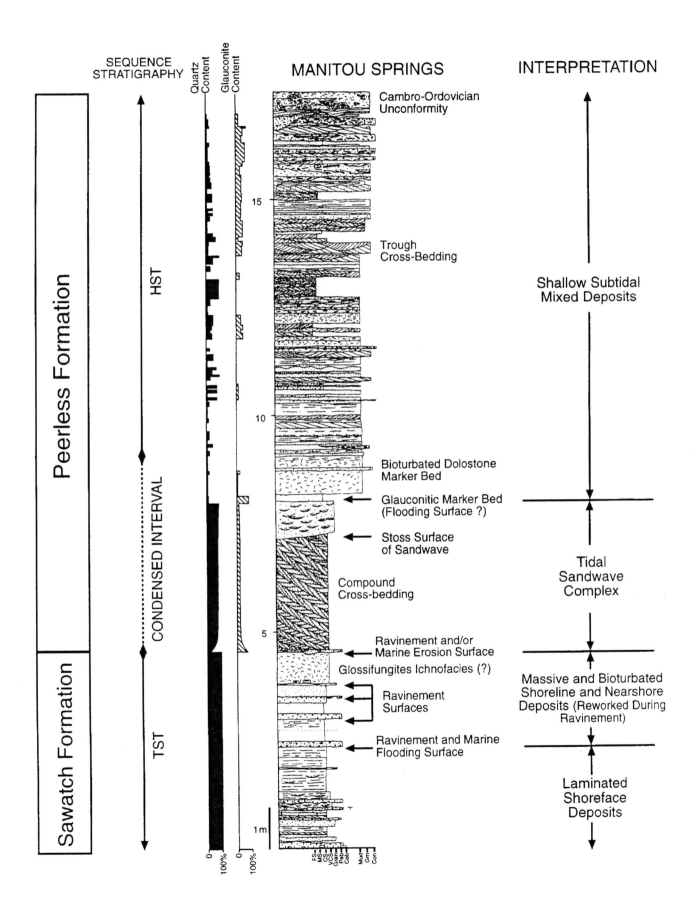

SEQUENCE
STRATIGRAPHY

Quartz
Content

Glauconite
Content

MANITOU SPRINGS

INTERPRETATION

Peerless Formation

Sawatch Formation

HST

CONDENSED INTERVAL

TST

15

10

5

1 m

Cambro-Ordovician
Unconformity

Trough
Cross-Bedding

Bioturbated Dolostone
Marker Bed

Glauconitic Marker Bed
(Flooding Surface ?)

Stoss Surface
of Sandwave

Compound
Cross-bedding

Ravinement and/or
Marine Erosion Surface

Glossifungites Ichnofacies (?)

Ravinement
Surfaces

Ravinement and Marine
Flooding Surface

Shallow Subtidal
Mixed Deposits

Tidal
Sandwave
Complex

Massive and Bioturbated
Shoreline and Nearshore
Deposits (Reworked During
Ravinement)

Laminated
Shoreface
Deposits

0   100%   0   100%

FS
MS
CS
VCS
Gran
Peb
Cob
Mud
Grm
Con

reworking formed surficial lags of large gastropods and trilobite spines on the upper surfaces of the wave ripples.

STOP 2: MISSOURI GULCH

Missouri Gulch: Take Highway 24 West from Manitou Springs for about 10 mi to Woodland Park. Turn north onto Hwy 67 towards Deckers. Continue 11.5 miles to Farm Road (FR) 350 and turn right following FR 350 for nearly 1/4 mi. Bear right onto FR 348 (FR 350 turns north at this point) and stop at the first sharp U-curve. A trail leads off to the right along the creek. The outcrop is less than 1/4 mile from the road.

This stop will focus on more northern and somewhat older basal Manitou strata. The lower member of the Sawatch is well exposed on the trail to the Manitou cliffs, but the middle member is only exposed in patches on the hillside across from the cliffs. The Manitou Formation at this locality consists largely of nodular lime mudstone and thin to medium beds of fossiliferous grainstone (Fig. 10). Trilobites and white-weathering inarticulate brachiopods are abundant in many of the grainstone beds. Much of the trilobite material described by Berg and Ross (1959) was recovered from this section. More recently, James D. Loch systematically sampled this section to refine the trilobite range data. Additionally, a few complete trilobite specimens recently were recovered by collector Ron Meyer from the nodular lime mudstone beds at a nearby locality. The lowest beds of

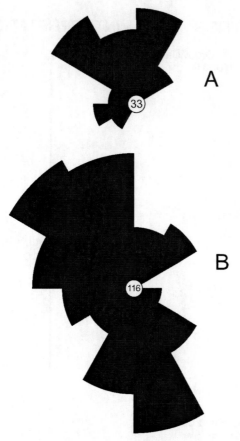

Figure 8. (A) Equal-area rose diagram showing dip directions of large-scale foresets of tidal dunes. (B) Rose diagram showing dip directions of superimposed small-scale cross-beds. North toward top.

oped, upward-shoaling, meter-scale peritidal cycles (Myrow 1995; Fig. 9). Cycles consist of, in ascending order, of (1) a basal wave-rippled grainstone; (2) thin, ribbon-bedded micrite; (3) very thin to thin, nodular-bedded micrite; (4) *Thalassinoides*-bioturbated micrite; and (5) a planar hardground surface. The shoaling nature of these cycles is reflected in the transition from coarse-grained grainstone to micrite, an upward thinning of bedding both within the ribbon-beds and between the ribbon and nodular beds, and evidence for subaerial exposure — microkarst and pseudomorphs of evaporite minerals — in the burrowed intervals below hardground surfaces. Myrow (1995) described a new species of the trace fossil *Thalassinoides* from these strata. *Thalassinoides* is most abundant in the nodular-bedded and bioturbated sub-hardground micritic lithofacies in the upper parts of these cycles (Fig. 9) that reflect an upward transition from soft sediment to firmground to hardground conditions.

The planar hardground surfaces contain truncated marcasite nodules that formed during early diagenesis and were planed off with the rest of the hardground during subsequent relative sea level rise. The overlying grainstone apparently abraded these hardground deposits as the grains were reworked by waves. This

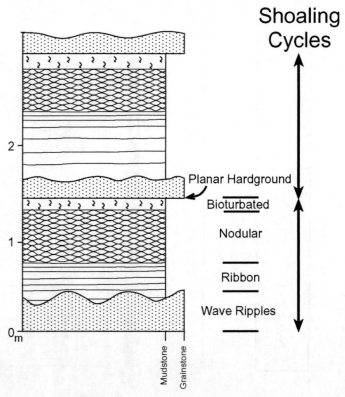

Figure 9. Generalized shoaling cycles in lower Manitou Formation. See text for details.

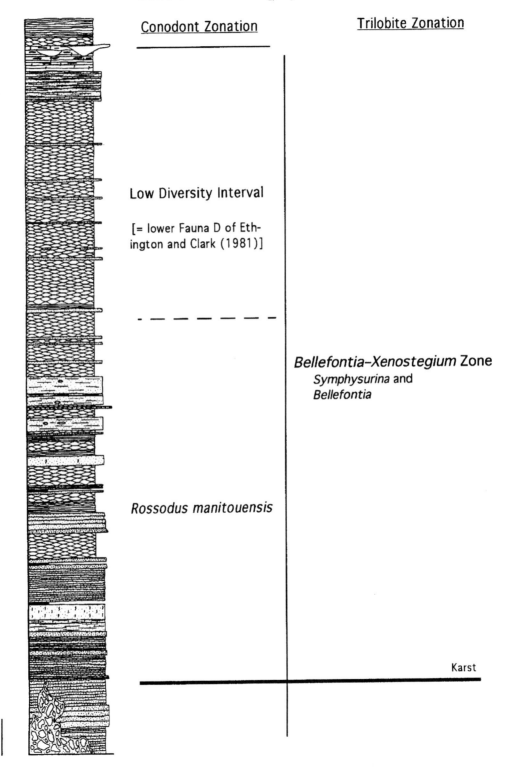

Figure 10. Generalized stratigraphic column of Missouri Gulch with conodont and trilobite zonations.

the Manitou at Missouri Gulch yield species of *Symphysurina* and *Bellefontia*, confirming the presence of the *Bellefontia-Xenostegium* Zone, which is missing farther south in the Front Range because of delayed onlap of the basal Manitou.

Conodont collections from the Manitou at this outcrop, the type locality for the conodont *Rossodus manitouensis*, are impressive in size and quality. The Color Alteration Index (CAI) is very low. Conodonts from the lower part of the section are from the Rossodus manitouensis Zone and those from the upper part of the section yield a less diverse assemblage characteristic of the low-diversity interval (lower Fauna D of Ethington and Clark 1981) (Figs. 5, 10).

A karstic breccia horizon has been recently discovered at Missouri Gulch within the basal few meters of the Manitou Formation. Conodont samples below and above the karst both yield *Rossodus manitouensis* Zone conodonts, so the duration of the hiatus associated with the paleokarst is not well resolved. A deeply channeled unconformity, present at the top of the Manitou Formation, is overlain by the Missisippian William's Canyon Member of the Leadville Limestone.

STOP 3: GLENWOOD CANYON

Heading west on I-70 past Vail and Eagle, take the Dotsero Exit 133. Go right (north) off the exit ramp to the first intersection and turn left (west) towards the Glenwood Canyon trailhead parking lot. Follow the service road parallel to the highway for nearly 3 miles to reach the parking lot. From here, take the paved trail by foot under the highway and along the Colorado River until it cuts back under the highway (about 0.9 miles). Emerging from the underpass turn right immediately onto a faint trail that leads to outcrops above the level of the highway.

This is the first of two sections to be visited (see also Stop 4) in the White River Plateau region to examine the Dotsero and Manitou formations (Fig. 11). The Dotsero Formation is divided into two members, a 25-35 m thick lower Glenwood Canyon Member and a 0.5-1.5 m thick stromatolitic biostrome, the Clinetop Member (Fig. 2). The Glenwood Canyon Member consists of a complexly interbedded very thin- to thin-bedded shale, very thin- to medium-bedded grainstone, and thin- to thick-bedded flat-pebble conglomerate. The shaly intervals range from a few cm to 60 cm in thickness and consist of alternating mm-scale thin laminae to very thin (< 3 cm) beds of shale and fine grainstone. The grainstone intervals are generally less than 25 cm in thickness, although some beds reach 90 cm. Thin, 1-3 cm thick, grainstone beds in some cases form amalgamated units up to 1.5 m in thickness. The grainstone beds are dominantly parallel-laminated although many beds contain well-developed wave-ripple and small- to medium-scale hummocky cross-stratification. The bases of grainstone beds are covered with sole markings, including groove and prod casts.

Carbonate flat-pebble beds are more abundant in the upper half of the Glenwood Canyon Member, locally making up close to 50% of the strata. These beds are tabular and generally range from 7-30 cm in thickness, although a few beds are > 1 m thick. The tabular clasts, which consist almost entirely of parallel-laminated limestone and dolostone, are usually less than a centimeter thick and measure a few centimeters in their long and intermediate axes. The matrix of the flat-pebble beds is fine to coarse grainstone. The flat pebbles are dominantly oriented with their long axes sub-parallel to horizontal.

The basal part of the overlying Manitou Formation is lithologically identical to the Glenwood Canyon Member of the Dotsero and varies in thickness from a few meters at Glenwood Canyon to nearly 10 m at Main Elk Creek (Stop 4). This facies is typical of inner shelf lagoon deposits of the inner detrital belt (Lochman-Balk 1971) of the paleo-Pacific Ocean during the Late Cambrian to Early Ordovician. The stratification in the grainstone beds reflects deposition by wave and storm processes. Flat-pebble beds are also generally considered to result from storm processes. The clasts are similar in lithology, texture, thickness, and sedimentary structures to the interbedded thin grainstone beds, which indicates that these beds were broken up by storms and redeposited.

Slump structures are abundant in the upper 15 m of the Dotsero Formation and the lower 2 m of the Manitou Formation (Fig. 11). These features include enigmatic, isolated, coherent blocks and contorted folds that appear to have originated as locally derived sea-floor slumps and slides. They are particularly abundant in the stratigraphic interval (+/- 2 m) above and below the Clinetop bed. Some large slide blocks rest directly on the upper surface of the stromatolitic biostrome. Well-preserved thrust-faulted beds show classic fault-bend folds, and some in cases these folds detached and came to rest well beyond the thrust ramp as isolated bodies. Sparse preliminary orientation data indicate bipolar orientations for thrust directions, which might argue against downslope, gravity-driven failures. Failure under shear stresses produced by wave oscillations would be consistent with this data, particularly given the abundant evidence for storm currents in this facies.

The Clinetop Member is a widespread marker bed that previously was believed to occur only in the White River Plateau area. Our discovery of the bed in the northwestern part of the Sawatch Range to the south (East Brush Creek and Lime Creek, Figs. 1, 3) extends its range to approximately 3500 km[2]. Several extremely widespread stromatolite beds occur in the Upper Cambrian–Lower Ordovician of the Great Basin as well, and these appear to mark a worldwide resurgence of stromatolites at this time (Shapiro 1998). This resurgence was likely due to relatively high sea levels and an equatorial position of Laurentia. The Clinetop bed contains several irregular hardground surfaces marked in part by truncation of stromatolitic lamination. In nearly all Clinetop localities, the upper surface is a flat hardground that is directly overlain by a wave-rippled grainstone bed that is locally glauconitic.

Red silt-filled fractures interpreted as possible paleokarst features occur within the Clinetop bed at Glenwood Canyon.

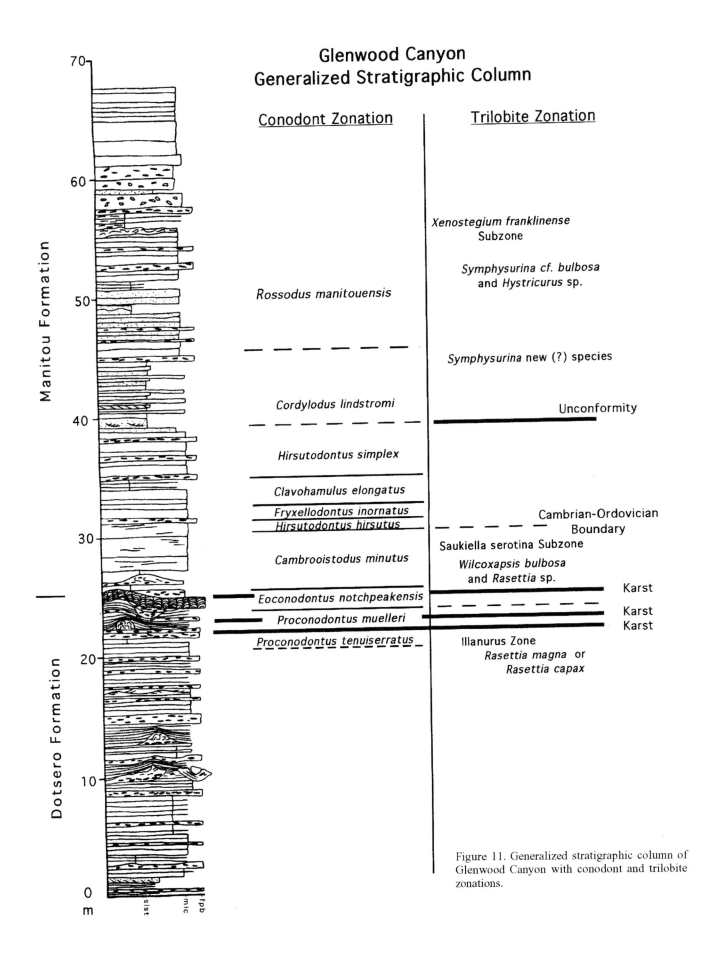

Glenwood Canyon
Generalized Stratigraphic Column

Figure 11. Generalized stratigraphic column of Glenwood Canyon with conodont and trilobite zonations.

Such paleokarst features also occur at several horizons within two to three meters of section directly below the Clinetop at this location and in a number of beds both below and above the Clinetop at Main Elk Creek (Stop 4). The Clinetop appears to be more strongly karsted at a remote location on top of the White River Plateau, where the top of the bed is also cut by a series of grainstone-filled channels. Conodonts recovered from up to a meter above the bed are from the same conodont subzone as the bed itself (*Eoconodontus notchpeakensis* Subzone), so any break at the top of the bed may be minor. A larger break may be represented at the irregular base of the bed because the highest conodont sample below the bed (~1 m below) is assigned to the *Proconodontus posterocostatus* Zone. Trilobite range data also suggest the presence of a stratigraphic break or condensed interval within or just below the Clinetop Member (see Stop 4). Deposition of the Clinetop Member is considered to be a three-stage process, as illustrated in Figure 12.

The abundance and spacing of thin karst horizons within the Dotsero and lower Manitou formations reflect generally low accommodation space and the occurrence of high frequency, low amplitude relative sea level changes. These fluctuations did not impart a cyclical stratigraphic facies pattern, in part because this was not a depositional system characterized by in situ carbonate production (i.e., a system driven to keep up with sea level). Instead, the carbonate beds shows clear evidence of traction transport (with selective size sorting) from adjacent environments during storms, which alternated with ambient deposition of mud from suspension.

The bulk of the overlying Manitou Formation consists of parallel laminated and hummocky cross-stratified dolomitic grainstone and a few thin shale and siltstone beds. These strata are the record of a long-standing, high-energy, storm-dominated, epicratonic carbonate setting. Some intervals particularly rich in quartz sandstone represent potential sequence boundaries.

STOP 4: MAIN ELK CREEK

Take Exit 105 off of I-70 west of Glenwood Springs. Go right (north) to the first intersection. Turn left onto the service road leading west to New Castle. Drive 1.1 miles along this road, which turns into Main Street. Before the end of town at 7th Street, turn right onto 245 RD (Buford Road) and follow for 3 miles to FR 243. Turn right at the sign indicating "National Forest Access, Main Elk Creek and Clinetop Road." FR 243 will follow Main Elk Creek for 5.8 miles (last 0.9 mi on dirt road). Just before small bridge and steep winding hill, look for Warren Stalts' Elk Creek Ranch on the left. Park here and walk dirt road to his house and ask for access to the outcrop on the west side of Main Elk Creek to the west of the rancher's property.

Sections in the northwest sub-basin are significantly more complete through the Cambrian-Ordovician boundary interval (Fig. 5) than those in the southeast sub-basin, and Main Elk Creek (MEC) (Fig. 13) is the most complete section of those studied to date. All conodont zones are represented from Procondontus tenuiserratus to *Rossodus manitouensis* zones, except for the *Proconodontus muelleri* Zone and the *Hirsutodontus hirsutus* Subzone of the *Cordylodus proavus* Zone. The section is highly condensed relative to those in the Great Basin and elsewhere because of low subsidence rates in this cratonal setting. Nonetheless, it is one of the most complete known from the inner detrital belt in North America and was deposited far enough into the inner shelf lagoon to record a protracted history of relative sea level changes, carbon isotope fluctuations, and conodont evolution.

A detailed $\delta^{13}C$ profile from the Main Elk Creek section (Figure 14) extends the depositional interpretations made from biostratigraphic and lithostratigraphic information. Carbon isotope data from outer platform sections show a strong cyclicity in $\delta^{13}C$ values during the Cambrian-Ordovician boundary interval, followed by a sharp decline to more negative values near the base of the *Rossodus manitouensis* Zone (Ripperdan and Miller, 1995). The rapidity of $\delta^{13}C$ variation, coupled with relatively continuous deposition on the outer platform, facilitate high-resolution correlation of the more fragmentary $\delta^{13}C$ record at Main Elk Creek. Examples of this are found at the 1st and 3rd karst horizons at Main Elk Creek. Overlying strata contain $\delta^{13}C$ variations and conodont faunas that can be precisely correlated to the contemporaneous section at Lawson Cove, Wah Wah Mountains, Utah ($\delta^{13}C$ profile events Ibe$_6$ and Ibe$_7$, Figure 14). The range of $\delta^{13}C$ variation found within each section is approximately 2.0 permil, permitting temporal approximation of intermediate $\delta^{13}C$ values between known maxima and minima. Equivalents to the Lawson Cove Ibe$_4$ and Ibe$_5$ $\delta^{13}C$ profile events are virtually absent at Main Elk Creek, thus constraining the duration of non-deposition (exposure?). A similar analysis of profile events Ibe$_1$ and Ibe$_2$ suggest that the 1st karst horizon also represents a depositional hiatus of substantial duration.

The basal Manitou Formation at MEC is quite shaly for about 10 m above the Clinetop bed before it gives way to more typical amalgamated grainstone of the upper Manitou (Fig. 13). This is in striking contrast to the Glenwood Canyon section where the amalgamated grainstone facies begins less than a meter above the Clinetop. This may reflect the more distal (=more western) position of this outcrop relative to the Homestake Shear Zone.

Extraordinary thin karsted beds occur in the section, one of which is 1 m below the Clinetop. Several more are scattered up to 12 m above the Clinetop (Fig. 13). Karsting features are developed within flat pebble and thick grainstone beds. These features include red silt-filled and cement-filled veins and cavities whose edges show sharp truncation of flat-pebble clasts and grainstone laminae. Large mud-crack features occur locally in float, and in a few places on bedding planes. The karst surfaces correspond with interruptions in isotopic curves that verify the loss or partial preservation of some conodont subzones.

Intensive sampling for trilobites in the Main Elk Creek section has documented the presence of several zones and subzones not previously reported from Colorado, including the two highest zones of the Cambrian (*Illaenurus* and *Saukia* Zones) and the two lowest zones of the Ordovician (*Missisquoia* and *Symphysurina* Zones). Several trilobite taxa from the MEC and GC sections strongly suggest that these strata were deposited on the western side of the Transcontinental Arch. For example, *Wilcoxaspis bulbosa* and *Clelandia typicalis* are common elements of the *Saukia* and *Illaenurus* Zones (respectively) in western North America, but have never been reported from sec-

tions deposited on the eastern (Appalachian) side of Laurentia. *W. bulbosa* was recovered by A.R. Palmer from the thin grainstone bed atop the Clinetop Member at Glenwood Canyon. *C. typicalis* is the commonest species present in the *Illaenurus* Zone collections recovered from MEC and from two sections in the Sawatch Range to the south.

The Cambro-Ordovician boundary at MEC occurs at a cryptic unconformity whose hiatus includes the highest subzone of the *Saukia* Zone and lowest subzone of the *Missisquoia*

Sea Level

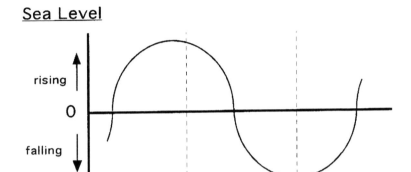

Figure 12. Three-stage model for deposition of the Clinetop Member stromatolite bed.

Stage 1

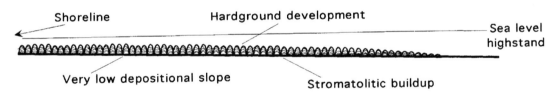

Stage 2

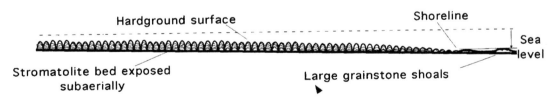

Stage 3

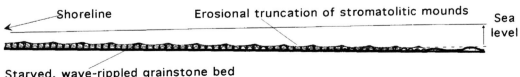

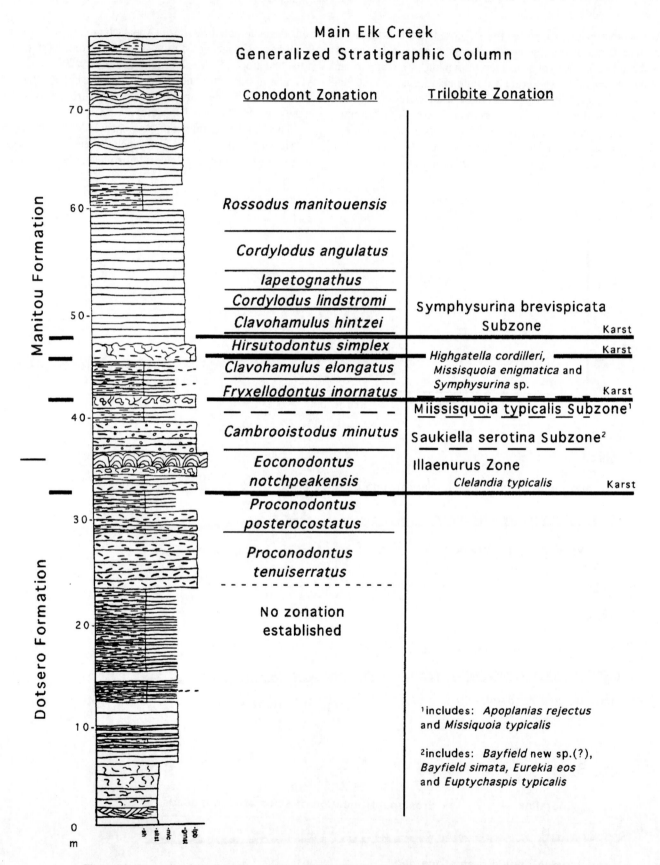

Figure 13. Generalized stratigraphic column of Main Elk Creek with conodont and trilobite zonations.

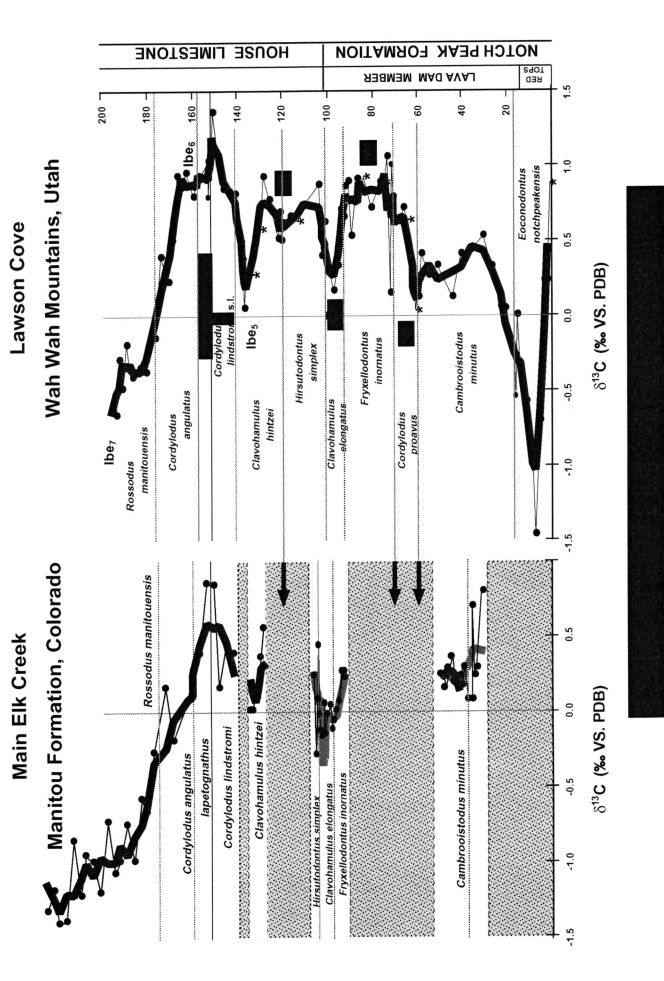

Lawson Cove
Wah Wah Mountains, Utah

HOUSE LIMESTONE

NOTCH PEAK FORMATION

LAVA DAM MEMBER

RED TOPS

200 180 160 140 120 100 80 60 40 20

Ibe₇

Ibe₆

Ibe₅

Rossodus manitouensis

Cordylodus angulatus

Cordylodus lindstromi s.l.

Clavohamulus hintzei

Hirsutodontus simplex

Clavohamulus elongatus

Fryxellodontus inornatus

Cordylodus proavus

Cambrooistodus minutus

Eoconodontus notchpeakensis

δ¹³C (‰ VS. PDB)

1.5   1.0   0.5   0.0   -0.5   -1.0   -1.5

Main Elk Creek
Manitou Formation, Colorado

Rossodus manitouensis

Cordylodus angulatus

Iapetognathus

Cordylodus lindstromi

Clavohamulus hintzei

Hirsutodontus simplex

Clavohamulus elongatus

Fryxellodontus inornatus

Cambrooistodus minutus

δ¹³C (‰ VS. PDB)

0.5   0.0   -0.5   -1.0   -1.5

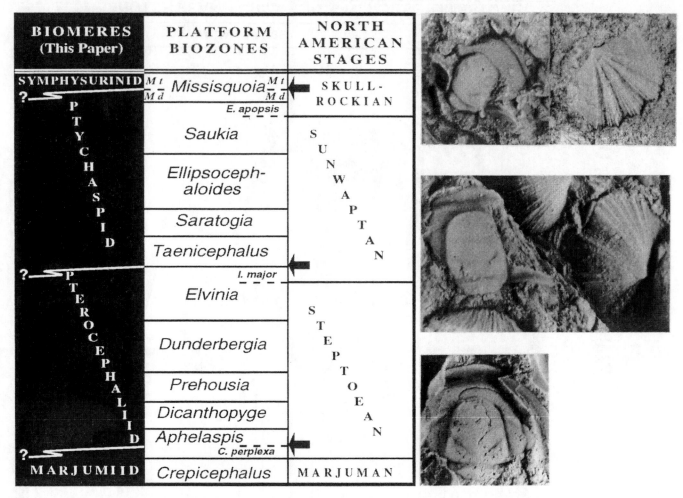

BIOMERES (This Paper)	PLATFORM BIOZONES	NORTH AMERICAN STAGES
SYMPHYSURINID ?	$\frac{Mt}{Md}$ *Missisquoia* $\frac{Mt}{Md}$	SKULL-ROCKIAN
	E. apopsis	
P T Y C H A S P I D	*Saukia*	S U N W A P T A N
	Ellipsoceph-aloides	
	Saratogia	
	Taenicephalus	
? P	*I. major*	
T E R O C E P H A L I I D	*Elvinia*	S T E P T O E A N
	Dunderbergia	
	Prehousia	
	Dicanthopyge	
?	*Aphelaspis*	
	C. perplexa	
MARJUMIID	*Crepicephalus*	MARJUMAN

Figure 15. Diagram contrasting positions of biomere and stadial boundaries in Laurentian platform successions. Black arrows in stage column indicate levels (bases of biomeres) dominated by olenimorphic trilobites illustrated to right of column. Cranidia illustrated (bottom to top) are *Aphelaspis, Parabolinoides,* and *Apoplanias.* Brachiopods to the right of *Parabolinoides* and Apoplanias are *Eoorthis* and *Apheoorthis,* respectively. Jagged lines and question marks used to represent biomere boundaries indicate potential diachroneity.

Zone. This relationship is similar to that reported by Miller (1984, 1992) in his description of his Lange Ranch Eustatic Event. Recovery of the trilobite *Apoplanias rejectus* from strata immediately above the unconformity assigns these beds to the *Missisquoia typicalis* Subzone and establishes a precise correlation with boundary sections throughout North America. The collections of *Apoplanias* from MEC are also worthy of mention because they contributed to a recent breakthrough in the study of biomeres, the stage-level biostratigraphic units originally defined by Palmer (1965) in Upper Cambrian platform successions in North America. Each biomere (a few million years in duration) records an adaptive radiation of Laurentian platform trilobite faunas, followed by a step-wise decline in species diversity through an extinction interval, culminating in a return to minimum diversity which marks the base of the next biomere.

Analysis of the MEC collections led to the realization (Taylor 1997; Taylor, in prep.) that the base of the *M. typicalis*

Subzone (marked by the appearance and dominance of *Apoplanias*) is a better horizon for designation as the top of the youngest Cambrian biomere (the *Ptychaspid* Biomere) than other zonal boundaries utilized for that purpose in previous studies (Longacre 1970; Stitt 1975; Palmer 1984; among others). Figure 15 shows the relationship of the biomere boundaries to the extinction horizons that currently serve as stadial boundaries for the Upper Cambrian of Laurentia. With this repositioning, the top of the Ptychaspid Biomere is marked by the appearance of an olenimorph-dominated fauna of minimum taxonomic and morphologic diversity similar to those at the bases of the *Aphelaspis* and *Taenicephalus* Zones, which define the bases of the underlying Marjumiid and Pterocephaliid Biomeres, respectively.

Interestingly, at MEC, the horizon of major turnover in conodont and trilobite faunas (the disconformity whose hiatus includes both the base of the Ibexian Series and top of the Ptychaspid Biomere) lies a short distance below a surface of karsting (Fig. 13). This same pattern, i.e., mass extinction preceding evidence for exposure, characterizes other biomere boundaries in inner platform facies in western North America (Robert Thomas, pers. comm., 1997) and in the Appalachians (Loch and Taylor 1995; Taylor et al. 1999). Collections from the 6-7 meters of strata below the Clinetop Member at MEC and in the Glenwood Canyon section confirm the presence of the *Illaenurus* Zone in the Dotsero Formation. Strata above the Clinetop in those sections yielded species characteristic of the overlying *Saukia* Zone, specifically, the *Saukeilla serotina* Subzone. At present, the trilobite data indicate that the underlying *Saukiella junia* Subzone is very thin or absent.

REFERENCES CITED

Allen, G.P., 1992, Sedimentary processes and facies in the Gironde estuary: a recent model for macrotidal estuarine systems, in Smith, D.G., Reinson, G.E., Zaitlin, B.A., and Rahmani, R.A., eds., Canadian Society of Petroleum Geologists Memoir 16, p. 29-40.

Allen, G.P. and Posamentier, H.W., 1993, Sequence stratigraphy and facies model of an incised valley fill: the Gironde Estuary, France: Journal of Sedimentary Petrology, v. 63, p. 378-391.

Allen, J.R.L., 1980, Sand waves: A model of origin and internal structure: Sedimentary Geology, v. 26, p. 281-328.

Anderson, T.B., 1970, Cambrian and Ordovician stratigraphy of the southern Mosquito Range, Colorado: The Mountain Geologist, v. 7, p. 51-64.

Bass, N.W. and Northrop, S.A., 1953, Dotsero and Manitou formations, White River Plateau, Colorado, with special reference to Clinetop Algal Limestone Member of Dotsero Formation: American Association of Petroleum Geologists Bulletin, v. 37, p. 889-912.

Belknap, D.F. and Kraft, J.C., 1981, Preservation potential of transgressive coastal lithosomes on the U.S. Atlantic shelf: Marine Geology, v. 42, p. 429-442.

Belknap, D.F. and Kraft, J.C., 1985, Influence of antecedent geology on stratigraphic preservation potential and evolution of Delaware's barrier system: Marine Geology, v. 63, p. 235-262.

Berg R.R., and Ross, R.J., Jr., 1959, Trilobites from the Peerless and Manitou formations, Colorado: Journal of Paleontology, v. 33, p. 106-119.

Brainerd, A.E., Baldwin, H.L., Jr., and Keyte, I.A., 1933, Pre-Pennsylvanian stratigraphy of Front Range in Colorado: American Association of Petroleum Geologists Bulletin, v. 17, p. 375-396.

Brasier, M.D., 1980, The Lower Cambrian transgression and glauconite-phosphate facies in western Europe: Journal Geological Society of London, v. 137, p. 695-703.

Campbell, J.A., 1972, Lower Paleozoic systems, White River Plateau: in De Voto, R.H., ed., Paleozoic stratigraphy and structural evolution of Colorado: Quarterly of the Colorado School of Mines, v. 67, no. 4, p. 37-62.

Campbell, J.A., 1976, Upper Cambrian stromatolitic biostrome, Clinetop Member of the Dotsero Formation, western Colorado: Geological Society of America Bulletin, v. 87, p. 1331-1335.

Davis, R.A., Jr, Klay, J., and Jewell, P., IV, 1993, Sedimentology and stratigraphy of tidal sand ridges southwest Florida inner shelf: Journal of Sedimentary Petrology, v. 63, p. 91-104.

Erdtmann, B.-D., Early Ordovician eustatic cycles and their bearing on punctuations in early nematophorid (planctic) graptolite evolution: Lecture Notes in Earth Sciences, v. 8, p. 139-152.

Ethington, R.L. and Clark, D.L., 1971, Lower Ordovician conodonts in North America, in W. C. Sweet and S. M. Bergström, ed., Symposium on Conodont Biostratigraphy: Geological Society of America Memoir, 127, p. 63-82.

Ethington, R.L. and Clark, D.L., 1981, Lower and Middle Ordovician con-

odonts from the Ibex area, western Millard County, Utah: Brigham Young University Studies, v. 28, pt. 2, 155 p.

Fenster, M.S., Fitzgerald, D.M., Bohlen, W.F., Lewis, R.S., and Baldwin, C.T., 1990, Stability of giant sand waves in eastern Long Island Sound, U.S.A.: Marine Geology, v. 91, p. 207-225.

Frey, R.W. and Seilacher, A., 1980, Uniformity in marine invertebrate ichnology: Lethaia, v. 13, p. 183-207.

Gerhard, L.C., 1972, Canadian depositional environments and paleotectonics, central Colorado: in De Voto, R.H., ed., Paleozoic stratigraphy and structural evolution of Colorado: Quarterly of the Colorado School of Mines, v. 67, no. 4, p. 1-36.

Gerhard, L.C., 1974, Redescription and new nomenclature of Manitou Formation, Colorado: American Association of Petroleum Geologists Bulletin, v. 58, p. 1397-1406.

Hine, A.C., 1977, Lily Bank, Bahamas: History of an active oolite sand shoal: Journal of Sedimentary Petrology, v. 47, p. 1554-1581.

Houbolt, J.J.H.C., 1982, A comparison of recent shallow marine tidal sand ridges with Miocene sand ridges in Belgium, in Scrutton, R.A. and Talwani, M., eds., The Ocean Floor: New York, John Wiley and Sons, Ltd., p. 69-80.

Johnson, J.H., 1934, Paleozoic formations of the Mosquito Range, Colorado: U.S. Geological Survey Professional Paper 185-B, 43 p.

Johnson, J.H., 1944, Paleozoic stratigraphy of the Sawatch Range, Colorado: Geological Society of America Bulletin, v. 55, p. 303-378.

Johnson, J.H., 1945, A resumé of the Paleozoic stratigraphy of Colorado: Colorado School of Mines Quarterly, v. 40, 109 p.

Landing, E., 1993, Cambrian–Ordovician boundary in the Taconic Allochthon, eastern New York, and its interregional correlation: Journal of Paleontology, v. 67, p. 1-19.

Lewis, J.H., 1965, Petrology and diagenesis of Upper Cambrian rocks of central and western Colorado: Unpublished PhD thesis, University of Colorado, Boulder, Colorado, 184 p.

Loch, J.D., and Taylor, J.F., 1995, High-resolution biostratigraphy in the Upper Cambrian Ore Hill Member of the Gatesburg Formation, south-central Pennsylvania, in Mann, K.O. and Lane, R.L. (eds.), Graphic Correlation, Society for Sedimentary Geology Special Publication 53, p. 131-137.

Lochman-Balk, C., 1956, The Cambrian of the Rocky Mountains and southwest deserts of the United States and adjoining Senora Province, Mexico, in J. Rodgers, ed., El Sistemo Camrico, su paleogeografia y el problema de su base: Internat. Geological Congress, 20th Mexico, v. 2, p. 529-661.

Lochman-Balk, C., 1971, The Cambrian of the craton of the United States, in C. H. Holland, ed., Cambrian of the New World, Lower Paleozoic Rocks of the World: New York, Wiley Interscience, 1, p. 79-167.

Longacre, S.A., 1970, Trilobites of the Upper Cambrian Ptychaspid Biomere, Wilberns Formation, central Texas: Paleontological Society Memoir 4, Journal of Paleontology, v. 44, Supplement, 70 p.

Ludvigson, R., Pratt, B.R. and Westrop, S.R., 1986, The myth of a eustatic sea level drop near the base of the Ibexian Series: New York State Museum Bulletin, v. 462, p. 65-70.

Middleton, G.V. and Southard, J.B., 1984, Mechanics of sediment movement (2nd ed): Society of Economic Paleontologists and Mineralogists, Short Course No. 3, Providence, Rhode Island, 401 p.

Miller, J.F., 1984, Cambrian and earliest Ordovician conodont evolution, biofacies and provincialism, in D. L. Clark, ed., Conodont Biofacies and Provincialism: Geological Society of America Special Paper 196, p. 43-68.

Miller, J.F., 1992, The Lange Ranch Eustatic Event: A regressive-transgressive couplet near the base of the Ordovician System, in Webby, D. B. and Laurie, J. R., eds., Global Perspectives on Ordovician Geology: Rotterdam, Netherlands, A.A. Balkema Publishers, Proceedings of the Sixth International Symposium on the Ordovician System, p. 395-407.

Myrow, P., 1998, Transgressive Stratigraphy and Depositional Framework of

Cambrian Tidal Sandwave Deposits, Peerless Formation, Central Colorado, in Alexander, C., Davis, R., and Henry, J., eds., Clastic Tidal Deposition, SEPM Special Publication.

Myrow, P.M., 1995, Thalassinoides and the enigma of early Paleozoic open-framework burrow systems: Palaios, v. 10, P. 58-74.

Myrow, P.M., Ethington, R.L., and Miller, J.F., 1995, Cambro-Ordovician proximal shelf deposits of Colorado: Short Papers for the Seventh International Symposium on the Ordovician System, Ordovician Odyssey, p. 375-379.

Nicoll, R.S., Laurie, J.R., and Shergold, J.H., 1992, Preliminary correlation of latest Cambrian to Early Ordovician sea level events in Australia and Scandinavia, in B. D. Webby and J. R. Laurie, (eds.), Global Perspectives on Ordovician Geology: Rotterdam, Netherlands, Balkema Publishers, p. 381-394.

Odin, G.S. and Fullagar, P.D., 1988, Geological significance of the glaucony facies, in Odin, G.S., ed., Green Marine Clays: Developments in Sedimentology, Elsevier, Amsterdam, pp. 295-332.

Palmer, A.R., 1965, Biomere–a new kind of biostratigraphic unit: Journal of Paleontology, v. 39, p. 149-153.

Palmer, A.R., 1981, Subdivision of the Sauk Sequence, in M.E. Taylor, ed., Short Papers for the Second International Symposium on the Cambrian System: United States Geological Survey Open-File Report 81-743, p. 160-162.

Palmer, A.R., 1984, The biomere problem: evolution of an idea: Journal of Paleontology, v. 58, p. 599-611.

Pemberton, S. G. and Frey , R. W., 1985, The Glossifungites ichnofacies: modern examples from the Georgia coast, U. S. A., in Curran, H.A., ed., Biogenic Structures: Their Use in Interpreting Depositional Environments: Society of Economic Paleontologists and Mineralogists, Special Publication No. 35, p. 237-259.

Ross, J.R., Jr., and Tweto, O., 1980, Lower Paleozoic sediments and tectonics in Colorado, in, Kent, H.C. and Porter, K.W., eds., Colorado Geology: Rocky Mountain Association of Geologists — 1980 Symposium, p. 47-56.

Sepkoski, J.J., 1982, Flat-pebble conglomerates, storm deposits, and the Cambrian bottom fauna, in Einsele, G. and Seilacher, A., ed., Cyclic and Event Stratigraphy: Springer-Verlag, p. 375-385.

Shapiro, R.S., 1998, Upper Cambrian–lowermost Ordovician stratigraphy and microbialites of the Great Basin, U.S.A.: Unpublished Ph.D. thesis, University of California, Santa Barbara, 435 p.

Stevens, D.N., 1961, Cambrian and Lower Ordovician stratigraphy of central Colorado, in Rocky Mountain Association Geologists Guidebook, 12th Annual Field Conference, p. 7-15.

Stitt, J.H., 1975, Adaptive radiation, trilobite paleoecology, and extinction, Ptychaspid Biomere, Late Cambrian of Oklahoma: Fossils and Strata 4, p. 381-390.

Taylor, J.F., in review, Biomeres and stages: distinctly different but equally valid units in the Upper Cambrian of Laurentia: manuscript in review for publication in Geology, 12 ms. pages, 4 figures.

Taylor, J.F., 1997, Upper Cambrian biomeres and stages, two distinctly different and equally vital stratigraphic units: 2nd International Trilobite Conference, St. Catherines, Ontario, Abstracts Volume, p. 47.

Taylor, J.F., Loch, J.D., and Perfetta, P.R., 1999, Trilobite faunas from Upper Cambrian reefs in the central Appalachians: Journal of Paleontology, v. 32, p. 326-336.

Taylor, J.F., Repetski, J.E. and Orndorff, R.C., 1992, The Stonehenge Transgression: A rapid submergence of the central Appalachian platform in the Early Ordovician, in B. D. Webby and J. R. Laurie, eds., Global Perspectives on Ordovician Geology: Rotterdam, Netherlands, Balkema Publishers, Proceedings of the Sixth International Symposium on the Ordovician System, p. 409-418.

Tweto, O., 1949, Stratigraphy of the Pando area, Eagle County, Colorado: Colorado Science Society Proceedings, v. 15, no. 4, p. 147-235.

Geological Society of America
Field Guide 1
1999

Field guide for the Heart Mountain detachment and associated structures, northeast Absaroka Range, Wyoming

David H. Malone
Department of Geography-Geology, Illinois State University, Normal, Illinois 61790-4400, United States
Thomas A. Hauge
Exxon Production Research Company, P.O. Box 2189, Houston, Texas 77252-2189, United States
Edward C. Beutner
Department of Geosciences, Franklin and Marshall College, Lancaster, Pennsylvania 17604, United States

ABSTRACT

For more than a century, the Heart Mountain Detachment has been an important natural laboratory that has contributed to the education of thousands of students representing most of the colleges and universities of the nation. The purpose of this field trip is to provide a forum in which the important features of the Heart Mountain Detachment (HMD) can be observed and the various explanations for the origin of the structure can be discussed. The foremost questions to be addressed will likely include: What factors (e.g., paleotopographic slope; direction and magnitude of bedding dip; seismicity; eruptive processes, presence and pressure of fluids) triggered the formation of the HMD? What factors facilitated movement across the gently dipping detachment surface? Did the allochthon consist of numerous detached blocks of Paleozoic rocks or was the allochthon continuous and consist of predominantly Eocene volcanic rocks? Were the detached rocks emplaced gradually or catastrophically? What data can be used to constrain the rates of emplacement?

This field guide begins with an overview of the regional setting of the HMD. Detailed site descriptions of the various detachments and their associated structures are integrated into the road log (Figure 1). Most outcrops that are viewed from a distance on this trip are accessible for detailed examination via hikes of 1 to 2 hours. Time constraints preclude visiting more than a few of these exposures for detailed examination during this field trip, but others meriting such examination are identified for future reference. The staging area for this trip is the Double Diamond X Guest Ranch in the upper South Fork Shoshone River valley. The ranch is owned by Russ and Patsy Frazier. We warmly thank Russ and Patsy for their friendship and hospitality.

THE HEART MOUNTAIN DETACHMENT

The HMD is a rootless, low-angle normal fault that accommodated transport of upper-plate rocks for distances of as much as 50 km or more (Figure 1). Transport was largely southeastward, from the northeast flank of the northern Absaroka Mount-ains, a Laramide volcanic center and basement uplift, toward and into the western margin of the Laramide Bighorn Basin. The detachment is preserved over an area of at least 3400 km². The upper plate of the detachment was emplaced during the middle Eocene, during the late stages of the Laramide orogeny. Heart Mountain faulting involved rocks ranging in age from Ordovician to middle Eocene (Figure 2), mostly Paleozoic cratonic strata and andesitic Eocene volcanic rocks. The reader is referred to Pierce (1973) and Hauge (1993) for descriptions of the general features of the HMD. Other field guides to the area include Tucker (1982) and Hauge (1992).

Malone, D. H., Hauge, T. A., and Beutner, E. C., 1999, Field guide for the Heart Mountain detachment and associated structures, northeast Absaroka Range, Wyoming, *in* Lageson, D. R., Lester, A. P., and Trudgill, B. D., eds., Colorado and Adjacent Areas: Boulder, Colorado, Geological Society of America Field Guide 1.

Based on its relationship to Eocene sedimentary and volcanic rocks, the Heart Mountain allochthon was emplaced during the late stages of the Laramide orogeny. Allochthonous Paleozoic rocks at Heart Mountain and McCulloch Peak overlie nonmarine lower Eocene (Wasatchian) Willwood strata, indicating that emplacement of the allochthon postdated most of the Laramide fill of the Bighorn Basin. Thus, most of the offset across the basement-involved fault zone that defines the boundary between the Bighorn Basin and adjacent uplifts to the west had taken place by the time of Heart Mountain faulting. Wise (1983) has argued that the north-trending portion of this fault zone postdated northwest-trending structures along the Bear-tooth range front, and he inferred that the northwest-trending boundary between the Absaroka Mountains and the Beartooth Mountains was also probably an earlier Laramide structure. Thus, development of the presently observed basement framework of Laramide structures in the detachment area was essentially complete when Heart Mountain faulting occurred, and little subsequent tectonic deformation has affected the region. Further constraints on the age of Heart Mountain faulting are provided by relationships west of Buffalo Bill Reservoir, where allochthonous Paleozoic rocks overlie middle Eocene strata (Bridgerian Aycross Formation), indicating emplacement of the allochthonous rocks in this area as middle Eocene or younger (Torres and Gingerich, 1983).

A middle Eocene upper age limit for Heart Mountain faulting is also indicated by the Bridgerian age of volcanic rocks overlying the allochthonous Paleozoic rocks at this locality (Torres and Gingerich, 1983). These Bridgerian rocks are assigned to the Wapiti Formation and Trout Peak Trachyandesite (Pierce and Nelson, 1968). The Wapiti Formation and Trout Peak Trachyandesite have been interpreted as postdating Heart Mountain faulting (Pierce, 1987a), which would indicate that faulting was wholly Bridgerian (pre-Wapiti) in age. Alternatively, rocks assigned to the Wapiti Formation have been interpreted as allochthonous (Hauge, 1985), and the Trout Peak Trachyandesite, which locally comprises the hanging wall of the breakaway fault, has been interpreted as involved in the final phases of Heart Mountain faulting (Hauge, 1990). This alternative interpretation suggests that Heart Mountain faulting was complete slightly later in the Bridgerian (after Trout Peak Trachyandesite time). By either interpretation, Heart Mountain faulting may have been wholly Bridgerian (47.5 to 49.5 Ma; Torres and Gingerich, 1983), though earlier minor movements (Pierce, 1973) are not precluded by the data. Days One and Two (morning) of this field trip provide an introduction to the general features of the HMD, with emphasis on the relationships between Eocene volcanic rocks and Paleozoic sedimentary rocks above and along the HMD.

Two fundamentally different models describing the geometry, kinematic pattern, and emplacement of the upper plate of the HMD have been proposed. For many years, the upper plate was viewed to have been emplaced catastrophically as numerous independent slide blocks (Figure 3a; Bucher 1947; Pierce 1957, 1973, 1987a), and as a result of

this detachment faulting, a tectonically denuded surface was formed. Immediately after faulting had ceased, massive outpourings of Wapiti Formation volcanic rocks were deposited on the detached blocks as well as on the tectonically denuded surface (Pierce 1973, 1987). The most compelling line of evidence for this interpretation is the complete lack of erosion on the exposed fault plane, indicating that the time interval between slide block emplacement and the deposition of the Wapiti Formation must have been very short.

During the 1980s, a different model for the emplacement of the HMD allochthons was advanced. In this view, the upper plate is interpreted to have been a continuous allochthon rather than a series of individual slide blocks (Figure 3b; Hauge 1985, 1990, 1993). Volcanic rocks overlying the detachment, originally viewed as in depositional contact, were reinterpreted as allochthonous, and as comprising much of the upper plate (Hauge 1990). Thus, the continuous allochthon model requires no tectonic denudation or catastrophic emplacement of numerous slide blocks, and the model eliminates the mechanical enigma that tectonic denudation poses. Other continuous allochthon models that do not require tectonic denudation, but do require catastrophic emplacement rates have been advanced by Sales (1983) and Beutner and Craven (1996).

ABSAROKA VOLCANISM

The Eocene volcanic succession within the Absaroka Range has been formally named the Absaroka Volcanic Supergroup (AVS) (Smedes and Prostka, 1972). The rocks of the AVS extend over an area of approximately 18,000 km² most of which is underlain by a Laramide structural basin (Absaroka Basin); the volcanic rocks overlie rock units which range in age from Archean to Eocene. Deeply incised valleys provide excellent natural cross sections, and they display a volcanic stratigraphic succession in excess of 1875 m in thickness. The rocks of the AVS are unconformably overlain to the west by Quaternary volcanic rocks of the Yellowstone Volcanic Plateau. The inferred depositional setting of the AVS is a series of high stratovolcanoes with coalescing alluvial aprons (Sundell, 1993), rather like the Cascade Range or Andes Mountains of today. Volcanic centers define two northwest-trending belts, the locations of which are probably controlled by weakness zones within the Precambrian basement (Chadwick, 1970). Volcanism was initiated in southern Montana during late early Eocene time (53 Ma) (Chadwick, 1970), and continued throughout middle Eocene time, with the locus migrating southeastward until culminating in the southeastern Absaroka Range during late Eocene time (38 Ma; Sundell, 1985).

Reworked epiclastic volcanic rocks, including volcanic breccias, sandstones, conglomerates, siltstones, and claystones are the predominant rock types. Primary volcanic rocks (lava flows, flow breccias, pyroclastic breccias, and tuffs) increase in abundance near the intrusive centers. Three groups comprise the AVS: (1) the lower(?) and middle Eocene Washburn Group, (2) the middle Eocene Sunlight Group, and (3) the middle and

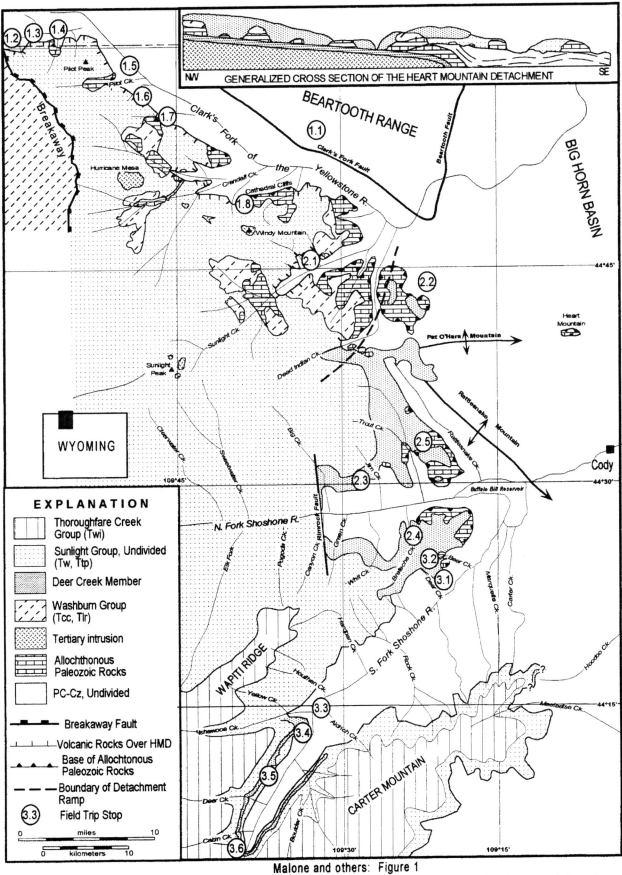

Figure 1. Geologic map of the HMD area (modified from Malone, 1995). The field trip stops discussed in the text are indicated.

Within the figure, the following labels appear:

GENERALIZED CROSS SECTION OF THE HEART MOUNTAIN DETACHMENT

NW / SE

BEARTOOTH RANGE

BIG HORN BASIN

Breakaway

Pilot Peak

Pilot Ck.

Clark's Fork of the Yellowstone R.

Clark's Fork Fault

Beartooth Fault

Hurricane Mesa

Crandall Ck.

Cathedral Cliffs

Windy Mountain

Heart Mountain

Pat O'Hara Mountain

Sunlight Ck.

Sunlight Peak

Dead Indian Ck.

Rattlesnake Mountain

Rattlesnake Ck.

Cody

Clearwater Ck.

Sweetwater Ck.

Big Ck.

Trout Ck.

Jim Ck.

Buffalo Bill Reservoir

WYOMING

109°45'

44°45'

44°30'

N. Fork Shoshone R.

Elk Fork

Pagoda Ck.

Canyon Ck. Rimrock Fault

Whit Ck.

Green Ck.

Breteche Ck.

Deer Ck.

Merquate Ck.

Carter Ck.

Hoodoo Ck.

WAPITI RIDGE

Houlihan Ck.

Yellow Ck.

Hardpan Ck.

S. Fork Shoshone R.

Rock Ck.

Meeteetse Ck.

44°15'

Ishawooa Ck.

Aldrich Ck.

CARTER MOUNTAIN

Deer Ck.

Boulder Ck.

Cabin Ck.

109°30'

109°15'

EXPLANATION

- Thoroughfare Creek Group (Twi)
- Sunlight Group, Undivided (Tw, Ttp)
- Deer Creek Member
- Washburn Group (Tcc, Tlr)
- Tertiary intrusion
- Allochthonous Paleozoic Rocks
- PC-Cz, Undivided

— Breakaway Fault

— Volcanic Rocks Over HMD

— Base of Allochtonous Paleozoic Rocks

— — Boundary of Detachment Ramp

(3.3) Field Trip Stop

0   miles   10

0   kilometers   10

Malone and others: Figure 1

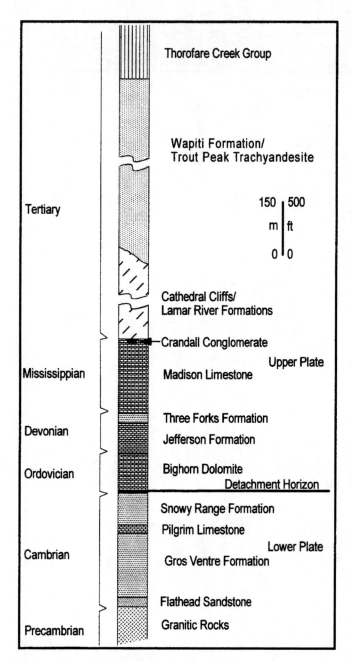

Figure 2. Generalized stratigraphic section showing the formations associated with the HMD, modified from Pierce (1973). Use of the terms Wapiti Formation, Cathedral Cliffs Formation, and Lamar River Formation in the detachment area has proven problematic (see Hauge, 1985, 1990, 1993). Crandall Conglomerate, of Eocene (?) age (Pierce and Nelson, 1971), overlies units as old as the Snowy Range Formation.

upper(?) Eocene Thorofare Creek Group (Smedes and Prostka, 1972, Sundell, 1993). Subdivisions of each of these groups will be observed during this trip. The Washburn Group (Wasatchian-Bridgerian in age) includes the Cathedral Cliffs and Lamar River Formations. These units occur primarily in northern part of the Absaroka Range. The Sunlight Group (Bridgerian in age) includes the Wapiti Formation and the Trout Peak Trachyan-

desite in the north, and the Aycross Formation in the south. The Thorofare Creek Group consists of the Langford, Two Ocean, Wiggins, and Teepee Trail formations. The Langford and Two Ocean formations crop out in the northern Absaroka Range, and the Wiggins and Teepee formations occur through the southerly reaches of the range.

The Deer Creek Member of the Wapiti Formation consists of blocks (individually as large as several km^2 in area) of vent-medial-facies lava flows, breccias, and sandstones within a thin, heterogeneous matrix of boulder- to sand-sized volcaniclastic material. It is interpreted as the deposit of a large debris avalanche, formed by the collapse of a large stratovolcano within the Absaroka Range during the early middle Eocene. The areal extent and volume of the proximal facies of the Deer Creek Member are ~450 km^2 and ~100 km^3, respectively. The unit was first described by Malone (1995) and it was later described in detail by Malone (1996). The Deer Creek member was assigned to the Wapiti Formation because it is within the type section of that unit as defined by Nelson and Pierce (1968). Further work by Malone (1997) indicates that a distal facies of the unit also probably occurs throughout the upper South Fork Shoshone River valley. In its proximal area, the unit extends from the southern end of Sheep Mountain 32 km northward to Dead Indian Creek. The easternmost extent tops Rattlesnake and Pat O'Hara Mountains; to the west, the unit is buried by younger volcanic rocks. A transport distance of 40 km from the center of the inferred source area is indicated. The average thickness of the debris-avalanche deposit is estimated to be ~220 m. Its thickness ranges from zero at its stratigraphic pinch-out to >470 m at the west of Sheep Mountain. The Deer Creek Member is bounded below by an erosional surface that displays both large- and small-scale paleotopography. Overall, the basal surface slopes an average of 3°-4° to the southeast, with >1000 m of total relief on this surface demonstrated. The top of the unit is an erosional surface as well. Overall, the upper surface is relatively flat and slopes gently 1°-2° to the southeast. The age of the Deer Creek Member is closely determined. It falls into the same 2 million year window as Heart Mountain faulting.

DAY 1: PROXIMAL AREAS OF THE HMD

Introduction

Day One begins in Cody, Wyoming. Proceed north from Cody on Highway WYO 120 for 17 miles, then northwest on WYO 296 for 47 miles, then northeast 7 miles on US 212 to the first stop, at the Pilot-Index Overlook on Highway US 212. From there, proceed south and west along Highway US 212 as far west as the Soda Butte Creek picnic area in Yellowstone National Park; then backtrack east along US 212, go southeast along WYO 296, and south on WYO 120 to Cody. Ten USGS geological quadrangle maps (see references) cover much of the detachment area and provide a useful reference. Mileages in the road log are approximate. Interval mileage is italic.

Mileage description

	0.0	Cody, Wyoming. Proceed north on WYO 120.
17.0	17.0	Junction with WYO 296 (Chief Joseph Scenic Highway). Proceed west on WYO 296.
47.0	64.0	Junction with US 212. Proceed east on US 212.
7.0	71.0	Turn left into parking lot of Pilot-Index overlook.

STOP 1-1: OVERVIEW OF THE BEDDING PARALLEL COMPONENT OF THE HMD

This vantage point provides a panoramic view of much of the detachment area. Here we will discuss general features associated with the detachment, which can be summarized as follows (condensed from Hauge, 1993). The Heart Mountain fault can be subdivided into three components. The first is the

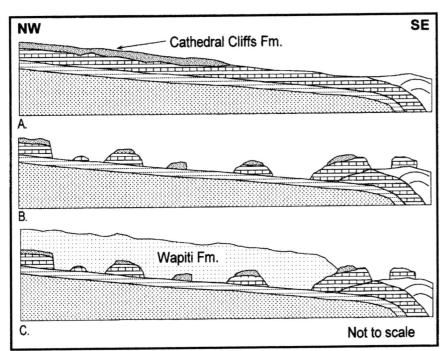

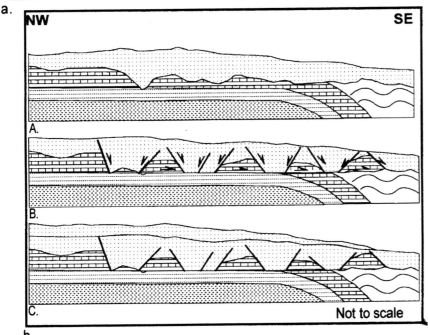

Figure 3. Two models of Heart Mountain faulting. a: The tectonic denudation model, modified from Pierce (1957). (A) Before faulting occurred, volcanic rocks (Cathedral Cliffs Formation) locally overlay Paleozoic strata. (B) When the detachment formed, Paleozoic strata and overlying volcanic rocks detached along a basal Ordovician bedding plane and were emplaced at catastrophic rates as numerous detached blocks along the bedding-plane detachment, up a transgressive ramp, and across the Eocene land surface. (C) Immediately after faulting, catastrophic volcanism (Wapiti Formation) blanketed the disrupted terrane. b: The continuous allochthon model, modified from Hauge (1985). (A) Before faulting occurred, volcanic rocks 1 km or more thick overlay deeply eroded Paleozoic strata and younger strata to the southeast. (B) When the detachment formed, Paleozoic strata and volcanic rocks above the detachment underwent lateral translation and extension as a continuous allochthon, and structurally high (largely volcanic) rocks were downfaulted, tilted, and translated. Displacement was noncatastrophic [1 cm per year?] and coeval with volcanism (feeders out of plane of the section). (C) After faulting ceased, volcanism continued.

breakaway fault, which forms the western boundary of the detachment area and is located near the eastern boundary of Yellowstone National Park. The breakaway fault extends upward from the detachment and ends upward within Eocene volcanic rock, demonstrating that the HMD is a rootless structure. The breakaway fault is the subject of Stop 1-2 of this field trip, and it is described in detail in the text for that stop. The second component of the HMD, which is visible from our present vantage point and is the subject of Stops 1-3 through 1-7 and Stop 2-1 of this field trip, is the portion of the basal detachment that parallels bedding in the homoclinal Paleozoic section of the northeastern Absaroka Mountains.

The bedding-parallel component of the detachment lies along a footwall bedding plane located about 2 m above the base of the Ordovician section. This stratigraphic position is remarkable from the perspective of rock mechanics, as discussed by Pierce (1973) and Hauge (1993), because the detachment lies along a bedding plane within "strong" dolomite rather than within the thick, "weak" Cambrian shales that underlie it by only a few meters. The detachment does not cut below this stratigraphic horizon (see Hauge, 1983, for minor exceptions to this), and upper-plate units are no older than Ordovician, indicating that the detachment is rootless. The bedding-parallel component of the detachment is bounded to the west by the breakaway, to the northeast by erosion, to the southeast by a footwall ramp, and to the southwest, speculatively, by the Black Mountain fault. The bedding-parallel component of the detachment presently dips an average of 3 to 5 degrees to the south-southwest (Pierce, 1985). The autochthon in the area of the bedding-parallel component of the detachment exhibits little deformation either related to or subsequent to faulting. In contrast, the allochthon was strongly extended during Heart Mountain faulting. Allochthonous Paleozoic rocks include 400 m thick untilted sections with little section missing along the detachment; sections tilted up to 30 degrees or more with strata truncating downward at the detachment; and local exposures where Ordovician and Devonian strata are omitted along the detachment, without significant angular discordance between allochthon and autochthon. Volcanic rocks lie between and upon the masses of allochthonous Paleozoic strata, but disagreements exist as to the involvement of these rocks in faulting. Some workers have argued that in most areas the contact between volcanic rocks and the detachment is a "half fault" (Pierce, 1980), the volcanic rocks having been deposited upon the detachment (Pierce, 1987a), or emplaced upon it as a debris-avalanche (Malone, 1994), after it had been tectonically denuded. Others inferred the contact is in many areas (Prostka, 1978) or everywhere tectonic (Hauge, 1982,1985, 1990; Templeton and others, 1995; Beutner and Craven, 1996). This, along with the issue of emplacement rate of of rocks overlying the detachment, is the essence of the current debate about the nature of Heart Mountain faulting.

The third component of the detachment, is a footwall ramp that, in general terms, cuts up section to the southeast, from the

bedding-parallel component of the detachment near the base of the Ordovician up to the middle Eocene. Whereas the footwall of the bedding-parallel component of the detachment is a structurally simple homocline, the footwall of the detachment ramp is the structurally complex transition from the Absaroka Mountains to the Bighorn Basin, and the configuration of the detachment ramp reflects this footwall structure. Pierce (1957, 1960, 1985) subdivided the detachment ramp into (1) a "shear thrust" (1957) or "transgressive fault" (1960), where it climbs abruptly in present-day elevation to the top of Dead Indian Hill, and (2) an "erosion thrust" (1957; after Hewett, 1920) or "fault on former land surface" (1960) from Dead Indian Hill eastward. The 3-to-5-km wide "transgressive fault" presently dips about 10 degrees westward; the "fault on former land surface" as much as 48 km wide, presently dips an average of 1 degree (and locally up to 4 degrees) eastward.

Return to vehicles. Turn around, retrace route west along US 212.

7.0	78.0	Junction with WYO 296. Continue west along US 212.
14.2	92.2 = 0.0	Cooke City, Montana. Continue west along US 212.
3.0	3.0	Silver Gate, Montana.
1.1	4.1	Northeast Entrance to Yellowstone National Park.
1.2	5.3	Turn left into Soda Butte Creek picnic area.
0.3	5.6	Park vehicles at last picnic area. Walk up hill to east for a view of the breakaway fault.

STOP 1-2: VIEW OF THE BREAKAWAY FAULT

From our vantage point the breakaway fault, which forms the western boundary of the detachment area, is exposed across 650 m of relief (Figure 4). At this locality it dips 70 degrees to the east and cuts steeply down through Eocene volcanic rocks and Mississippian, Devonian, and Ordovician sedimentary rocks in its footwall, ending near the base of the Ordovician at the detachment. The hanging wall consists of Eocene volcanic rocks. In other areas, due to limited vertical exposure and difficult access, the trace and cross-sectional geometry of the breakaway are less well known. Pierce (1960, 1980, 1987b) interpreted the breakaway as having been tectonically denuded and interpreted the volcanic rocks overlying it as Wapiti Formation that was deposited on the denuded surface of the breakaway. He described the breakaway fault as a "half-fault," for only one side [the footwall side] is a fault surface; the other side [the hanging-wall side] is a surface of deposition (Pierce, 1980, p. 276). In contrast, the hanging-wall volcanic rocks have been interpreted as in part (Prostka and others, 1975) or wholly (Hauge, 1982, 1985, 1990) allochthonous.

Turn vehicles around and retrace route to Silver Gate.

0.3	5.9	Turn right from picnic area onto US 212.
1.2	7.1	Exit Yellowstone National Park.
1.1	8.2	Pull off road to left (north) near gas station.

STOP 1-3: VIEW OF THE DETACHMENT FROM SILVER GATE, MONTANA (OPTIONAL)

For a distance of 5 km east of the breakaway exposure just described, the detachment is commonly well exposed and is typically overlain by volcanic rocks with a preserved vertical thickness on the order of 500 m. Locally, masses of allochthonous Mississippian and Devonian rocks up to a few tens of meters thick overlie the detachment and underlie the volcanic rocks. The exposure visible from this stop, 2 km east of the breakaway and south of the town of Silver Gate, Montana, was singled out in several publications (Pierce, 1979, 1980, 1987a, 1987b; Pierce and Nelson, 1986) as a showcase example of relationships demonstrating that volcanic rocks lie in situ on the detachment (as well as on allochthonous carbonate rocks) and, hence, that tectonic denudation of the detachment is proven. However, according to Hauge (1990), the relationships cited by these authors are also compatible with tectonic emplacement of the volcanic rocks, and other relationships in this area require that at least the basal 200 m of volcanic rocks in this area are allochthonous. The critical features described at this locality are: 1) the contact between the volcanic rocks and the allochthonous Paleozoic rocks, which is interpreted as depositional (Pierce, 1987a) and as a faulted unconformity (Hauge, 1990); 2) faults within the upper-plate Paleozoic rocks, which either do not (Pierce, 1987a, 1987b) or do (Hauge, 1990) offset the contact between the Paleozoic upper plate and the overlying volcanic rocks; 3) faults within volcanic rocks in the area immediately

west-southwest of the exposure of allochthonous Paleozoic rocks, recognized by Hauge (1990); volcanic flow units dipping 30-45 degrees southeast, recognized by Pierce and others (1973) and Hauge (1990); 4) volcanic rocks underlain along the detachment by striated microbreccia a few hundred meters farther west, recognized by Hauge (1985); 5) clastic dikes of fault breccia, interpreted as having been emplaced after Heart Mountain faulting (Pierce, 1987a) or during faulting (Hauge, 1990). This locality is recommended for detailed examination, with the conflicting interpretations of the relationships in mind. Access, described by Pierce (1987b), is via a climb beginning on the south side of Silver Gate.

Proceed east along US 212.

3.0	11.2	Cooke City
0.7	11.9	Park at junction of US 212 and forest service road to Daisy Pass. Walk across US 212 to south side of road.

STOP 1-4: VIEW OF REPUBLIC MOUNTAIN

This stop affords a view of allochthonous Paleozoic rocks at Republic Mountain and the volcanic rocks that overlie them (Figure 5). The Paleozoic rocks are Ordovician, Devonian, and Mississippian strata that are internally faulted and tilted to small angular discordance with the detachment (Elliott, 1979). These rocks were first recognized as allochthonous by Pierce (1960).

Figure 4. View to south of breakaway of the HMD. Viewpoint is about 1.6 km north of and 600 m higher than the northeast entrance to Yellowstone National Park. The breakaway (b) dips steeply (about 70°) eastward and truncates downward at the detachment (d). Rocks visible in the footwall of the detachment are Pilgrim Limestone (cliff near base of photo), 25 m thick, and overlying shale and limestone of the Snowy Range Formation, about 75 m thick. About 2 m of Ordovician Bighorn Dolomite, not shown, immediately underlies the detachment. The footwall of the breakaway consists of about 90 m of Bighorn Dolomite (O), 90 m of Devonian Three Forks and Jefferson formations (D), 75 m of Mississippian Madison Limestone (M), and about 400 m of Eocene volcanic rocks, partly snow-covered, of the Lamar River and Cathedral Cliffs formations (Tv1). The hanging wall consists of a basal 180 m of Eocene volcanic rocks (Tv2) variably assigned to the Lamar River Formation (Prostka and others, 1975), Lamar River and Cathedral Cliffs formations (Pierce and others, 1973), and the Wapiti Formation (Pierce, 1980), overlain by about 450 m of Eocene volcanic rocks (Tv3) assigned to the Wapiti Formation by these workers. The topographically highest volcanic rocks (Tv4), with apparent subhorizontal bedding, are Wapiti Formation and Trout Peak Trachyandesite (Pierce and others, 1973; Prostka and others, 1975); these strata overlap the breakaway out of view to the southeast (Figure 6 of Pierce, 1980) and therefore postdate Heart Mountain faulting. Note apparent southwestward dips of Tv2, evident above Tv2 annotation. Disagreement exists as to whether the upper-plate rocks were tectonically emplaced (Hauge, 1990) or were deposited on the detachment and breakaway (Pierce, 1987a) (see text).

The volcanic rocks are dominantly massive andesitic volcaniclastic rocks, in which primary stratification is difficult to discern. These volcanic rocks have been variably mapped as Wapiti Formation that postdates Heart Mountain faulting (Pierce, 1978) and as Lamar River and Cathedral Cliffs formations (Nelson and others, 1980; Elliott, 1979). Hauge (1983) described a fault contact between the Eocene volcanic and Paleozoic sedimentary rocks of the upper plate. From our vantage point 30 degree southward dips of the volcanic rocks can be discerned. We will discuss whether these dips are consistent with the volcanic rocks being in situ Wapiti Formation (Wapiti vents are south of this area), or are better explained as tectonic dips.

Proceed eastward on US 212.

6.9   18.8   Park at entrance to Fox Creek campground.

STOP 1-5: VIEW TO WEST OF PILOT PEAK, INDEX PEAK, AND THE UNDERLYING DETACHMENT (OPTIONAL)

This vantage point provides a distant view of a superb exposure of the detachment, Cambrian rocks of its footwall (intruded at this locality by an Eocene latite porphyry sill—Pierce and others, 1973), and volcanic rocks immediately overlying the detachment. The detachment follows a bedding horizon 2 m above the Cambrian-Ordovician contact (Pierce, 1968), and from this view the parallelism of the detachment and footwall strata is evident. The 13-16 degrees southwest dip of the detachment in this area (Pierce and others, 1973) is atypical of most of the bedding-parallel component of the detachment, dips of less than 1 degree, to the south or southwest, being more typical. The volcanic rocks overlying the exposure of the detachment visible from this stop are interpreted as Wapiti Formation on most published maps (Pierce and others, 1973; Pierce, 1978; Nelson and others, 1980). The Wapiti Formation is defined as postdating Heart Mountain faulting (Nelson and Pierce, 1968). Pierce (1968) interpreted the detachment at this locality as having been tectonically denuded and interpreted the volcanic rocks to be in depositional contact with the detachment. Based on observations of the volcanic rocks along the detachment immediately to the north and south of this exposure (the exposure seen from this stop is too steep to be safely accessible), Hauge (1983, 1985) interpreted these volcanic rocks as allochthonous.

From this perspective the characteristic sharp, planar nature of the detachment is particularly impressive. This and other exposures of the detachment show no direct evidence of subaerial exposure; nowhere is it incised by erosion that postdated faulting but predated the overlying volcanic rocks, and nowhere is it overlain by fluvial deposits. These relationships led Pierce (1957, 1973) to infer that the period of time after tectonic denudation of the detachment and deposition of the Wapiti volcanic rocks was very brief. These relationships were cited by Hauge (1985) as supportive of the argument that tectonic denudation did not occur, and the Heart Mountain upper plate was a continuous allochthon

rather than numerous detached blocks. Upper plate rocks that underlie Pilot Peak include Paleozoic rocks visible in the drainages of Fox and Pilot Creeks, volcanic rocks variably mapped as Wapiti Formation (in situ) or Lamar River and Cathedral Cliffs Formations (allochthonous), and Trout Peak Trachyandesite (Pierce and others, 1973). A fault contact between upper-plate Paleozoic and Eocene rocks well exposed in the drainage of Pilot Creek is described in Hauge (1985).

Proceed eastward on US 212.

3.4   22.2   Turn right into Crazy Creek Campground. Park in the loop at the end of the campground road and walk a few meters south.

STOP 1-6: VIEW OF ROCKS OVERLYING THE HMD, FROM PILOT AND INDEX PEAKS TO THE NORTHWEST, TO THE ONEMILE CREEK AREA TO THE SOUTHEAST

Several features and localities of interest are visible from this vantage point. The first (N60-75W) is many of the features seen at Stop 1-5, seen here in a more distant view. The second is a view up Pilot Creek (N75-80W). On the north side of Pilot Creek, a major unconformity, best viewed in early morning sunlight, is visible. This unconformity separates undeformed, subhorizontal volcanic strata from underlying volcanic rocks that, from this view, show no apparent stratification. The volcanic rocks beneath this unconformity, which are mapped as Lamar River and Cathedral Cliffs Formations by Pierce and others (1973) and as mostly Wapiti Formation by Pierce (1978), are characterized by dips as steep as 36 degrees (Pierce and others, 1973; Hauge, 1983), and are variably interpreted as in situ (Pierce and others, 1973: Pierce and Nelson's interpretation) or allochtonous (Pierce and others, 1973; Prostka's interpretation; Hauge, 1983). The volcanic rocks above this unconformity postdate faulting. From the perspective of the continuous allochthon model of Heart Mountain faulting, this view suggests the thickness of the continuous allochthon that was preserved when faulting ceased and the allochthon was overlain by younger volcanic rocks. This preserved thickness, which is less than the thickness of the active allochthon by some unknown amount, is roughly 600 m.

Also visible from this stop are Jim Smith Peak (S70W) and the spot where Jim Smith Creek crosses the detachment (S75W). Like the area south of Silver Gate viewed at Stop 1-3, the excellent detachment exposure at Jim Smith Creek has been a focus of disagreement in the literature. Pierce and others (1991) interpreted the volcanic rocks overlying the detachment on the east wall of Jim Smith Creek as in depositional contact with the detachment, and Pierce and others (1991) interpreted the volcanic rocks on the west wall of the creek as allochthonous (see also Pierce and others, 1973). Hauge (1985) interpreted the volcanic rocks immediately above the detachment on both sides of the creek as allochthonous. Hauge's (1985) inter-

Figure 5. View to west of Republic Mountain, from U.S. Route 212 east of Cooke City, Montana. The allochthon consists of Ordovician, Devonian, and Mississippian sedimentary rocks (MO) overlain by Eocene volcanic rocks (Tv). Note southward dip of volcanic strata, which are Lamar River and Cathedral Cliffs formations of Pierce and others (1973) and Wapiti Formation of Pierce (1978). Detachment (d) parallels footwall strata (C). Prominent cliff in the footwall is Pilgrim Limestone. Total relief shown is about 750 m.

pretation was based on mesoscopic features (Figures 6 and 7) supplemented by thin-section examination. Pierce and others (1991) provide descriptions of thin sections of the detachment horizon that they interpret as incompatible with tectonic emplacement of the volcanic rocks. I (Hauge) view the features described by Pierce and others as compatible with tectonic emplacement of the volcanic rocks. This locality is accessible via a Forest Service road: turn south off of US 212 just west of Crazy Creek, cross the bridge over the Clarks Fork, drive past the B Four Ranch 1.5 mi (this is a Forest Service road; public access is permitted), and hike south about 1.5 km. Time and weather permitting, we will visit this locality today. Throughout the area of exposure of the bedding-plane component of the detachment, numerous normal faults, most probably with small offset, are present in volcanic rocks overlying the detachment. Hauge (1983, 1985) measured the orientations of these faults and their slickenlines, and Buetner and Craven (1996) interpret a vertical contraction axis and an approximately horizontal extension axis oriented N59W-S59E. This is compatible with the interpretation that they were emplaced as part of a continuous expanding allochthon (Beutner and Craven, 1996).

Proceed eastward on US 212.

2.8   25.0 = 0.0   Turn right onto WY 296.
6.4   6.4   Turn west on Park Co. Road XUX. Go 3.5 mi. and park near the bridge over Squaw Cr., which is just past the Squaw Cr. trailhead.

STOP 1-7: SQUAW CREEK AREA

This stop will examine the critical relationship, alluded to earlier, between the bedding plane portion of the Heart Mountain detachment fault and overlying volcanogenic rocks. Malone argues that the volcanics were, at least in large part, emplaced by megaslandsliding onto the denuded fault surface immediately following dispersal of the carbonate blocks. Hauge argues that the volcanics originally overlay the Paleozoic carbonates and moved down upon normal faults to fill gaps between carbonate blocks

created during slow (~typical geologic strain rates) collapse and extension of the continuous allochthon. Beutner agrees with the kinematics of Hauge but maintains that movement was catastrophic and aided by a gas-fluidized cushion (now represented by microbreccia) at the base of the allochthon. To observe most exposures of volcanics on the fault requires a climb of 1000' to 1500' through brush. The outcrops we will visit are not large but demonstrate the relations well and are easily reached by trail. *(NOTE: THIS IS GRIZZLY COUNTRY AND WE HAVE SEEN THEM ALONG THIS TRAIL. WE STRONGLY RECOMMEND CARRYING BEAR PEPPER SPRAY AND AGAINST GOING INTO THIS AREA ALONE.)*

Go 6.4 miles east on the Chief Joseph Highway from its intersection with U.S. 212 (0.1 mile E. of the bridge over the Clark's Fork) and go up the Squaw Cr. trail across a clearcut and into the woods. Massive outcrops of Crandall Conglomerate are uphill to the south and can be examined with a short, brisk climb or in float. In the first small drainage crossed, Crandall Conglomerate composed of angular fragments (channel margin talus?) can be seen in the bed of the ravine. When the Squaw Cr. trail turns south at a sign just after crossing a small stream, leave it and continue west on the trail along the bench on top of the Pilgrim Limestone. Cross another small stram and then Squaw Cr., where the trail is washed out. Immediately downstream from the washout is an outcrop of Pilgrim with an attitude of N60°W; 31°NE. This outcrop is probably on the NE flank of the Blacktail anticline, which must plunge out between here and the Pilgrim cliffs 1 km to the NW. Immediately after crossing Squaw Cr., turn left (W) where the trail forks and go upstream <1/8 mile, crossing an open slope, to the first drainage from the right (NW). Go up this small stream. The fault is well exposed in the first ravine from the right (NE), where it is marked by reddish microbreccia ~20 cm thick underlying shattered volcaniclastics and overlying comminuted Cambrian shale. If you want samples, *please* do not sample the microbreccia in place—there are sufficient float blocks available. The fault is poorly exposed (grey microbreccia up to 0.5 m thick) in the next small drainage; it appears to be offset by a small fault between these locations. It is well exposed further up in the main stream, with pale grey-green microbreccia 3-5 cm thick containing superb tool marks (asymmetric grooves) (Figure 8) on its base indicating movement toward S50°-55°E. The foot-wall in this area is sheared shale and fractured limestone of the Snowy Range Formation. The stratigraphic level of the fault is lower and the footwall is more intensely deformed than usual here, presumably because the fault is cutting through the preexisting Blacktail anticline. The microbreccia at the base of the volcanics is, as usual, a massive, cohesive rock with no internal shear planes or macroscopic orientation to its fabric. It contains volcanic and carbonate clasts in a microcrystalline carbonate-rich matrix.

Return to Highway 212 and proceed eastward.
1.4 7.8   Cross Crandall Creek
1.8 9.6   Pull off highway to the right into Scenic View
                    pull-out.

STOP 1-8: CATHEDRAL CLIFFS AREA

Several features of interest are visible in this panoramic view. Hunter Peak 5 km to the northwest (N60W), is underlain by allochthonous Paleozoic rocks that traveled perhaps 6 to 15 km across the detachment. South of Hunter Peak are the drainage of Crandall Creek (bearing due W to S80W) and the high country of

Hurricane Mesa 11 km due west. Windy Mountain is the highest peak visible to the south. Both Hurricane Mesa and Windy Mountain are underlain by Absaroka volcanic and intrusive rocks. The carbonate exposure immediately to the south is Cathedral Cliffs. At this stop we will discuss the relationship between Absaroka volcanic rocks and upper-plate Paleozoic rocks in two areas: at the west end of Cathedral Cliffs (near Corral Creek), and along the length of the face of Cathedral Cliffs.

The first topic of this stop is the poorly exposed contact between the upper-plate Paleozoic sedimentary rocks and Eocene volcanic rocks that abut them to the west along Corral Creek. This area is best viewed from the road that leads to the K-Z Ranch from the west; we will stop at this better vantage point if time permits. The volcanic rocks of this area were mapped as Wapiti Formation, in depositional contact with the detachment and Paleozoic upper plate, by Pierce and Nelson (1971) and Pierce (1978). There has been disagreement in the literature, however, as to whether volcanic rocks in this area were deposited on the detachment and against the allochthonous Paleozoic rocks (Pierce, 1987a) or were tectonically emplaced (Hauge, 1985, 1990) (Figure 9).

Cathedral Cliffs (Figure 10) is a 5 km wide exposure of allochthonous Paleozoic sedimentary rocks, with Eocene volcanic rocks along much of the skyline. The Paleozoic rocks appear remarkably little deformed, despite a probable transport distance of 5 to 15 km. In most of this exposure, the base of the allochthonous Paleozoic rocks consists of shattered Bighorn Dolomite, with

Figure 6. Volcanic rocks along the bedding-plane detachment immediately west of Jim Smith Creek, showing basal shatter zone up to 10 m thick, steeply dipping truncated strata (right of center), and internal faulting of volcanic rocks, The detachment and underlying autochthonous strata are also visible. View is westward. From Hauge, 1985.

Figure 7. View obliquely downward of detachment at Jim Smith Creek. Pen and pencil in lower half of photo lie parallel to two sets of striae on fault gouge that veneers the autochthon. Pencil at upper right parallels striae on sheared surface ~10 cm above, and subparallel to, the detachment.

little section omitted along the detachment. The allochthonous Paleozoic rocks are bounded to the west by volcanic rocks, as was described immediately above. They are bounded to the east by volcanic rocks that were mapped by Pierce and Nelson (1971) and Pierce (1978) as Wapiti Formation (in depositional contact with the detachment) and were interpreted by Hauge (1983, 1985) as allochthonous. An asymmetric graben, about 200 m wide at its exposed base, is visible near the center of the exposure of allochthonous Paleozoic rocks (Pierce and Nelson, 1971). Hauge (1990) interpreted this small graben as representative of an early phase of extension of the allochthon. He envisioned downfaulting of Paleozoic and Eocene rocks within grabens such as these, with continued extension leading to the downfaulting of broad expanses of Eocene volcanic rocks to the detachment horizon, such as are seen east and west of Cathedral Cliffs (Hauge, 1990). The involvement of volcanic rocks in the formation of the small graben visible

at this stop is indicated by the fault contacts between Eocene and Paleozoic rocks (Pierce and Nelson, 1971; Hauge, 1983) and within volcanic rocks (Hauge, 1983) at this locality. From this perspective, the volcanic rocks on the skyline seem to overlie the allochthonous Paleozoic rocks. Instead, the volcanic rocks are in fault contact with the Paleozoic rocks across a steeply south-dipping fault. From this vantage, this south-dipping contact can be seen best in the area of the graben and, farther west, where Paleozoic rocks form the highest cliffs. Numerous Eocene latitic to basaltic dikes intrude the upper plate rocks (both Paleozoic and Eocene) at this and numerous other localities. These dikes, which die out downward at or within a few m of detachment, are restricted to the upper plate. Although many hundreds have been mapped within the allochthon (e.g., Pierce and Nelson, 1971; Nelson and others, 1980), only a few have been identified within the autochthon (Hauge, 1983, 1985). Pierce (1987a) interpreted these

Figure 8. Groove casts on the base of the microbreccia from the exposure on a tributary to Squaw Creek. Many of the "tools" (small limestone fragments) which produced the grooves are at the southeast end of the grooves, indicating movement toward the S50°-55°E.

dikes as laterally intruded and younger than Heart Mountain faulting. Voight (1974a, 1974b) and Erskine and Kudo (1991) interpreted these dikes as allochthonous, truncated at their bases by Heart Mountain faulting. Hauge (1985, 1990) inferred that they were intruded laterally and are allochthonous, and further he argued that the volume they occupy was created by extension of the allochthon that was accommodated by slip along the detachment. Intrusion of many of these dikes occurred during the final stages of Heart Mountain faulting (Hauge, 1990).

Proceed eastward on WYO 296.

24.2	33.8	Dead Indian Pass.
14.3	48.1	Junction with WYO 120. Turn right (south) toward Cody.
17.0	65.1	Cody, Wyoming. End of Day One.

DAY TWO: DISTAL AREAS OF THE HMD; DEER CREEK MEMBER

Introduction

Day Two begins at the Double Diamond X Ranch. Proceed northeast on the South Fork Road to Cody, Wyoming. The caravan will proceed north from Cody on Highway WYO 120 for 17 miles, then northwest on WYO 296 for 24 miles to the Sunlight Basin Road. We will proceed westward into Sunlight Basin to the exposure of the HMD at the base of White Mountain. We will then retrace our route to Cody, stopping at Dead Indian Pass to view and discuss the footwall-ramp component ("transgressive phase" and "fault-on-former-land-surface phase" of Pierce, 1973) of the HMD, including Heart Mountain itself. From Cody, we will proceed westward into the Shoshone River drainage, where we will further discuss the "fault on former land surface," emphasizing discussion of the proximal

facies of the Deer Creek Member of the Wapiti Formation. Day Two ends at the Double Diamond X ranch for dinner and lodging. USGS geological quadrangle maps (see references) cover much of the area and provide a useful reference. Mileages in the road log are approximate. See the 1990 Wyoming Geological Association Guidebook for a detailed road log.

Mileage description

	0.0	Cody, Wyoming. Proceed north on WYO 120.
17.0	17.0	Junction with WYO 296 (Chief Joseph Scenic Highway). Proceed west on WYO 296.
24.0	41.0=0.0	Junction with Sunlight Basin Road. Go west on Sunlight Basin Road. Note: mileages within Sunlight Basin are taken from topographic maps and have not been field checked.
5.3	5.3	Turn right into Wyoming Game and Fish public access fishing area. Proceed ENE approx. 1.4 miles, crossing Sunlight Creek and passing through barbed-wire gates, to the base of White Mountain.
1.4	6.7	White Mountain

STOP 2-1: WHITE MOUNTAIN

Since the first recognition that the white marbles and monzonitic stock of White Mountain are allochthonous on the HMD (Nelson, 1969), this excellent exposure of the HMD has received considerable attention, particularly in light of the possible role of a sill of volcanic gas in triggering and sustaining movement on the detachment (e.g., Hughes 1970a, 1970b, 1970c, 1973; Pierce and Nelson, 1970; Nelson and others, 1972, 1973; Beutner and Craven, 1996). Beutner and Craven (1996) describe the following mesoscopic features at White Mountain: 1) footwall exposures of several meters of Cambrian Snowy Range Formation, overlain depositionally by 2 meters of Bighorn Dolomite; 2) a sharp planar contact between the footwall and 2 to 3.5 m of well-indurated, massive fault microbreccia; 3) an irregular contact between the microbreccia and the overlying dolomitic marble of the allochthon; 4) an abrupt downward decrease in metamorphism at the base of the allochthon, and a less dramatic decrease at the base of the microbreccia; 5) cataclasite dikes up to 1 m thick within the allochthon. According to Beutner and Craven, thin sections of the detachment microbreccia reveal volcanic glass grains with primary shapes and accreted grains equivalent to accretionary and armored lapilli (Figure 11). They interpret the fault-rock constituents and fabric to imply that injection of volcanic glass along the detachment horizon triggered detachment and catastrophic emplacement of the Heart Mountain allochthon. They also interpret a chain of intrusive plugs that trends NW from White Mountain as a kinematic indicator, akin to a hot-spot trace in the moving allochthon.

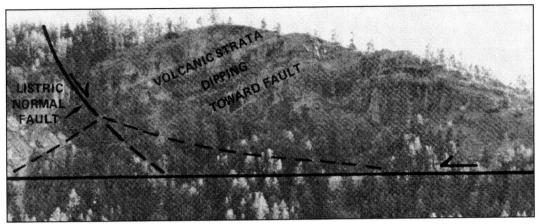

Figure 9. Photograph of relationships along the detachment west of Cathedral Cliffs.

Figure 10. View of Cathedral Cliffs. Dash-C is Heart Mountain footwall rocks, mostly Cambrian; MO is allochthonous Ordovician-Mississippian cratonic strata; Tv is Tertiary volcanic rocks; d is HMD.

These observations and inferences, in conjunction with their inferences regarding the kinematics of emplacement of the volcanic rocks associated with Heart Mountain faulting, led them to conclude that Heart Mountain faulting comprised catastrophic emplacement of a continuous allochthon. Pierce (1965, 1978) and Pierce and Nelson (1971) interpreted these volcanic rocks as younger than Heart Mountain faulting, but Hauge (1983, 1985) and Pierce (1985) interpreted them as allochthonous. Hauge (1985) described a normal fault within the volcanic rocks that is exposed across 200 m of topographic relief. This is one of the best accessible exposures of a fault within volcanic rocks in the upper plate.

Retrace route through Sunlight Basin to WYO 296; turn right onto 296 and proceed eastward up the switch backs to Dead Indian Pass.

0.0		White Mountain.
6.7	6.7	Junction with WYO 296 (Chief Joseph Scenic Highway). Proceed east on WYO 296.
10.0	16.7	Pull off Wyoming Highway 296 to right (west) side of highway at the overlook at Dead Indian Pass.

STOP 2-2: DISCUSSION OF DETACHMENT RAMP

Dead Indian Pass provides a spectacular view of the Beartooth Plateau to the north, the Clarks Fork fault that bounds its southern flank in this area, and the Absaroka Mountains to the west and southwest. The prominent valley immediately to the west in the Absaroka Mountains is Sunlight Basin, visited at Stop 2-1; the distinctive peak at N80W along the north side of Sunlight Basin is White Mountain. The bedding-parallel component of the HMD underlies much of the country to the west, where Eocene volcanic rocks overlie Paleozoic strata. Our discussion will focus on the transition from the bedding-parallel component of the HMD to its transgressive ramp component, which is visible from this vantage point near the foot of Dead Indian Hill. The detachment ramp is described in the text of Stop 1-1. A perspective on the scale of the HMD can be appreciated at this stop. The breakaway, which we visited at Stop 1-2, is roughly 50 km northwest of our present location, Heart Mountain is about 25 km east-southeast of our present location, and McCulloch Peak, the most distal klippe of the allochthon, is 50 km to the east-southeast. The Paleozoic strata of Heart Mountain and McCulloch Peak are thought to have originated above the bedding-plane component of the detachment, requiring a minimum displacement of 50 km along the detachment (Pierce, 1957). This displacement magnitude is not representative of the entire allochthon; displacements as small as a few kilometers are suggested for allochthonous Paleozoic rocks near the breakaway. Given the remarkable scale and rootless nature of the HMD, it remains a most enigmatic feature, despite attention in the literature that spans a century (Hauge, 1993). Whether some form of the tectonic denudation model, the continuous allochthon model, or some as yet unformulated model will best explain Heart Mountain faulting, it is clear at present that the enigma is far from resolved.

Proceed east on WYO 296

14.0	30.7	Junction with WYO 120. Proceed south on WYO 120.
17.0	47.7=0.0	Junction of WYO 120 with US14-16-20 (Sheridan Rd.) in Cody, Wyoming. Turn right and follow US14-16-20 westward, through the Rattlesnake/Cedar Mountain anticline, past Buffalo Bill Dam, toward Yellowstone National Park.
6.0	6.0	Buffalo Bill Dam.
10.0	16.0	Pull off onto north side of highway for view of the east side of the Trout Creek valley.
1.5	17.5	Turn right (north) on Jim Creek Road. Proceed northward to the Four Bear Ranch.
2.0	19.5	Park at the end of the public road for an overview to the north of the stratigraphy of the Wapiti Formation at Jim Mountain.

STOP 2-3: OVERVIEW OF THE WAPITI FORMATION AT JIM MOUNTAIN

The Wapiti Formation was initially defined by Pierce and Nelson (1968) as the 1500 m thick exposures of dark-colored breccias, conglomerates, sandstones, and lava flows in the North Fork Shoshone River valley. These rocks were previously included in the upper parts of the Early Basic Breccia of Hague and others (1899). The Wapiti Formation is overlain by massive lava flows of the Trout Peak Trachyandesite. Throughout the deposition of the Wapiti Formation and the Trout Peak Tra-

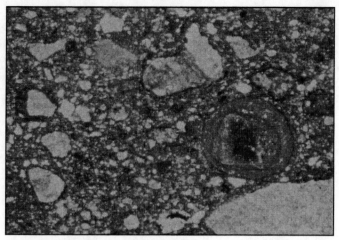

Figure 11. Photomicrograph of accreted grains in microbreccia from White Mountain. Accreted grains are common in (recognized so far at eight localities from Silvergate to Steamboat), and perhaps unique to, the microbreccia along the HMD. The larger grain has inner and outer layers of accreted, concentrically arrayed microbreccia, each contained by a dark film, suggesting a two-stage growth history. Numerous smaller grains with accreted margins are also present.

chyandesite, the principal center of intrusive/extrusive activity was the Sunlight Peak vent-complex (Pierce and Nelson, 1968), as indicated by the distribution of vent-facies rocks and the presence of a spectacular radial dike swarm. No Early Middle Eocene volcanic vents have been described south of the Sunlight Peak area. To the south and east, basin conditions existed then, and a thick sequence of medial-distal-facies volcanic rocks accumulated. Malone (1996) subdivided the Wapiti Formation into several formal and informal lithostratigraphic units (Figure 12). Several of these units were defined previously and used by earlier workers in laterally adjacent regions. All of these units are time transgressive (with the exception of the Deer Creek Member, which is considered to be isochronous), and recognition is based on rock type, color, and overall texture. The basal unit of the Wapiti Formation at this locality is the Deer Creek Member. The Deer Creek Member consists here of several blocks (>500 m in diameter) of medial-proximal facies sandstone, breccia, and conglomerate. Strata within these blocks range from steeply inclined to subhorizontal. Several smaller blocks of Madison Limestone occur at various structural positions between the volcanic blocks (Figure 13).

The Deer Creek Member is overlain by a 0-170 m succession of light-colored, distal-facies, epiclastic rocks (upper stratified member of Malone, 1996) The upper stratified member is very similar to the lower stratified member of the Wapiti Formation that occurs throughout the upper South Fork Shoshone River valley. Many of the conglomerates at the base of the upper stratified member bear quartzite clasts which range from pebble-to cobble-size, and which are typically well-rounded. The provenance of the quartzite clasts is likely the Proterozoic Belt Supergroup strata of the Laramide-age Targhee uplift in Idaho. The upper stratified member is overlain by a >250 m succession of dark-colored, medial-proximal-facies breccias, conglomerates, and sandstones. This succession has been informally named the lower breccias member (Malone, 1996). The lower breccias member is wedge-shaped and thins gradually to the south. Several thin yellow and green tuff beds occur sporadically throughout the unit, recording intermittent explosive eruptions from a distant, and probably northerly source. Medial-proximal-facies breccias occur throughout the North Fork Shoshone Valley west of the Wapiti area. Further work is necessary to determine the stratigraphic relations between these two areas. The Jim Mountain Member overlies the lower breccias member and was formally named by Nelson and Pierce (1968) to include the >300 m thick succession of trachyandesite lavas within the Wapiti Formation exposed in the steep upper slopes of Jim Mountain. The massive lava flows are in marked contrast with the overlying and underlying breccia units. Individual flows range in thickness from 10-60 m. The Jim Mountain Member gradually pinches out to the west in the upper North Fork Shoshone River valley. Where absent, the position occupied by the Jim Mountain Member is marked by an unconformity (Nelson and others, 1980). This unconformity is a relatively persistent marker horizon, and allows recognition of the dominantly proximal-facies upper breccias member above from the medial-facies lower breccias member below. Malone (1996) defined the upper breccias member of the Wapiti Formation as the dark-colored, massive, vent- and proximal-facies epiclastic and pyroclastic breccias that lie above the Jim Mountain Member and below the Trout Peak Trachyandesite. Time permitting, we will proceed to the Jim Creek Trail Head and hike to the base of the exposure for a closer inspection of this locality.

Turn caravan around and proceed back down the Jim Creek Road to the junction with US 14-16-20. Numerous outcrops of light gray debris-avalanche matrix occur on either side of the road.

2.0	21.5	Junction with US 14-16-20. Turn right (west). Proceed west to the bridge over the Shoshone River.
0.5	22.0	Turn left (sharp) on to gravel road and proceed east. The grassy foothills to the east are underlain by the Willwood Formation. The bedrock in this area is largely overlain by Quaternary landslide deposits and colluvium.
3.2	25.2	Turn right on the Breteche Creek road. This is a private road. Permission must be secured from Mr. Bo Polk before entering. We will make several short overview stops along this road. Time permitting, we will make a short traverse to the base of a debris avalanche block to observe the contact with the underlying matrix.
3.0	28.2	End of the Breteche Creek Road.

STOP 2-4: BRETECHE CREEK AREA

In the Breteche Creek area, the Deer Creek Member is poorly exposed and probably consists of a number of smaller blocks whose margins are poorly defined. In many cases, bedding within the blocks is difficult to discern. Smaller blocks, <150 m in diameter, exhibit a wider variety of bedding orientations, a result of the breaking up of larger blocks and greater relative movement between blocks during transport. A striking feature of the Deer Creek Member of this area is the presence of fragments (as much as 150 m in diameter) of older units suspended at structurally high positions within the unit (Figure 14). Most fragments are of the Mississippian Madison Formation, and they were probably derived from the tops and sides of allochthonous upper remnants of the Heart Mountain fault during the emplacement of the debris avalanche. Other fragments are of the lower Eocene Willwood Formation. The origin of the Willwood Formation fragments is less straightforward. Two possibilities exist that could account for these fragments (1) during generation of the debris avalanche, a sector of the volcanic edifice, as well as some of the subvolcanic basement could have been mobilized; and (2) the Willwood Formation fragments could have been entrained from the land surface during emplacement. The former explanation is preferred because it is difficult to visualize the debris avalanche entraining large fragments composed dominantly of incompetent mudstone and having these fragments remain coherent during continued transport.

Turn caravan around and proceed southward to the ranch road entrance.

3.0	31.2	Junction with gravel road. Turn right (east). Proceed east. To the south is the SheepMountain allochthon of the HMD.
3.43	4.6	Junction with N-S road at west end of Buffalo Bill State Park. Turn right (north).
0.2	34.8	Cross Shoshone River.
0.4	35.2	Junction with US 14-16-20. Turn right (east).
4.2	39.4	Junction of US 14-16-20 with the Mooncrest Ranch road. Turn left (north) on Mooncrest Ranch road. This road runs north for twelve miles along Rattlesnake Creek. This road is a private road. Permission must be granted by Mr. Dick Geving, ranch foreman. Proceed northward. Rattlesnake Mountain is to the east; Logan Mountain is to the west.
6.8	46.2	Junction with Logan Mountain Road. Turn left (west). Follow the road to where the gravel ends (approximately 2.5 miles).

STOP 2-5: OVERVIEW OF THE DEER CREEK MEMBER AND HMD FROM THE TOP OF LOGAN MOUNTAIN

This stop provides an excellent vantage of northernmost (closest to source) mapped extent of the Deer Creek Member. The

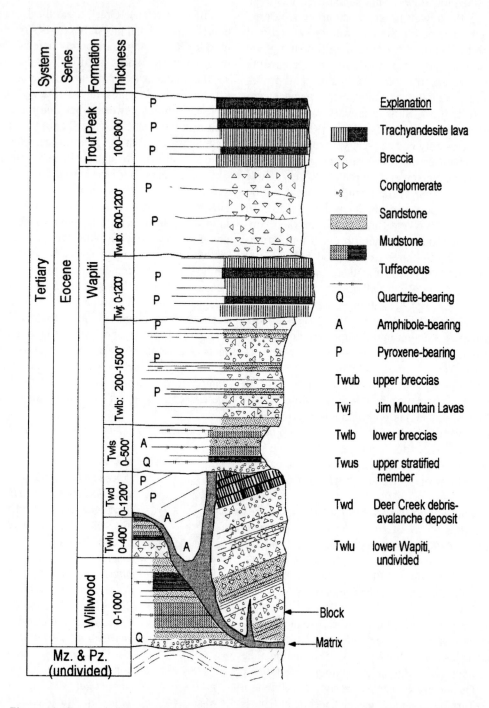

Figure 12. Composite stratigraphic column of Eocene rocks exposed in the lower North Fork Shoshone River Valley (from Malone, 1995).

divide between the Clarks Fork and Shoshone River drainage is the ridge to the north. To the east is the Rattlesnake Mountain anticline. In the lower elevations to the north and west is the Deer Creek Member. The high peaks to the west are Trout and Dead Indian peaks. Deer Creek Member blocks are commonly elongate or lenticular in shape and they make up about 80-90% of the total volume and area of the Deer Creek Member. The largest block identified thus far is about 12 km long and 5 km wide (Figure 15). It occupies most of the area between Rattlesnake and Trout creeks (lower elevations to the west). Blocks decrease in size and increase in number and structural complexity to the southeast. The rock type, facies, and depositional environments of the individual units within blocks are similar to other vent-medial-facies units exposed throughout the northeastern Absaroka Range.

Key marker beds within blocks have not been identified, but a crude coarsening upward stratigraphy has been recognized in some of the larger blocks with vent-facies rocks increasing in abundance toward the tops. The basal parts of blocks are composed primarily of well-stratified, medial-facies epiclastic breccias and sandstones. These units are overlain by massive laharic breccias, flow breccias and trachyandesite lava flows. This coarsening upward relationship is indicative of the pre-collapse stratigraphic configuration of the volcanic edifice, and it reflects a rather mature stage of development. The proportion of vent-facies rocks within blocks is higher in this area and they are the dominant rocks exposed between Robber's Roost and Rattlesnake creeks. This prevalence of vent-facies rocks indicates a proximity to the source area. Throughout this area, the interior

Figure 13. Chaotic deformation within the Deer Creek Member of the Wapiti Formation. Photograph of the complex internal structure of the Deer Creek Member between Jim and Dunn Creeks. View to the northwest. The debris-avalanche matrix occurs at the base of the unit (Twdm). Several debris-avalanche blocks of different attitudes are separated by dismembered block material (matrix?). Several small (< 50 m in diameter) "floating" blocks of Madison Limestone occur within these zones.

Figure 14. Fragment of Willwood Formation sandstone and mudstone east of Breteche Creek (from Malone, 1996) View to the east. This block is more than 190 m in diameter and occurs about 125 m above the base of the Deer Creek Member. The existence of Willwood Formation blocks within the Deer Creek Member is enigmatic. These blocks were probably entrained from topographic irregularities along the Eocene land surface during emplacement of the Deer Creek Member.

structures of individual blocks are relatively simple with only a slight backward tilt toward the inferred source area. Minor local changes in attitude within the larger blocks are due either to local variations in initial dip or to broad folds and unmappable faults. Bedding within blocks becomes highly deformed toward the base and lateral margins, with folds, faults, and fractures common. Some of these larger blocks are intruded by swarms of trachyandesite dikes, most of which trend roughly north-south. These dikes are typically perpendicular to bedding. Some dikes are truncated along bedding-parallel shear planes, and all are truncated along block-matrix contacts. This relation indicates that the intrusion of the dike swarm almost surely predates the emplacement of the debris avalanche. In this area, both the upper plate allochthons and the Deer Creek Member occur within a broad N-S trending paleovalley. At the time of emplacement, Rattlesnake Mountain stood as a prominent topographic barrier to the dispersal of the allochthons. Exposures of the Deer Creek Member do not occur east of this locality. Time permitting, the contact between the Logan Mountain Block and the Deer Creek Member can be examined in detail. Turn caravan around and proceed back to US 14-16-20. Return to the Double Diamond X Ranch via Cody and the South Fork Road.

DAY 3: DEER CREEK MEMBER

Introduction

Day 3, like Day 2 begins at the Double Diamond X Ranch. We will spend today observing the volcanic stratigraphy and detached extensional structures in the South Fork Shoshone River valley. The morning stops will include overviews and detailed examinations of the proximal facies of the Deer Creek Member in the Sheep Mountain area. The afternoon stops will focus on the stratigraphy of the upper South Fork Shoshone River valley and the distal facies of the Deer Creek Member. The mileage between stops is approximate.

Mileage description

	0.0	Double Diamond X Ranch on the South Fork Road.
3.5	3.5	Bridge over the South Fork Shoshone River.
3.3	6.8	TE Ranch Road
4.4	11.2	Bridge over Rock Creek
4.7	15.9	Carter Mountain Road
2.2	18.1	Park at sign on the north side of the highway that discusses the origin of Castle Rock.

STOP 3-1: OVERVIEW OF THE TYPE AREA OF THE DEER CREEK MEMBER

This stop provides an excellent vantage of the type area of the Deer Creek Member (about 5 km to the north). At this locality the relief on the Eocene land surface, two mountain-size blocks, and the matrix beneath and between these blocks can be observed (Figures 16 and 17). The steep dark brown exposures of the eastern block consists of proximal-medial facies breccias, sandstones and conglomerates that dip about 30° to the north. Beneath this block the Deer Creek Member matrix forms a 10 m thick veneer above the Willwood Formation. The matrix at this locality has an imbricate fabric. Petrified wood is locally common between these imbrications and indicate a forest was probably overrun during emplacement. The basal contact of the Deer Creek Member here preserves the early middle Eocene

Figure 15. View to the north of the large backward rotated block along Trout Creek. The stratification within this block dips about 20-30° to the northwest. The thicker layers at the top of the block, where stratification is more apparent, are mostly trachyandesite lava flows. The lower units are mostly breccias and sandstones, and they reflect an internal coarsening upward stratigraphy within individual blocks.

topography. The grassy covered area between the large blocks consists of a chaotic assortment of matrix, small volcanic blocks, small blocks of the Willwood Formation, and small blocks of Mississippian limestone. Pierce and Nelson (1969) map a fault contact between the Wapiti Formation and adjacent Willwood Formation along Deer Creek. This interpretation is based on an abrupt change in stratigraphic level of the underlying Willwood Formation west of the inferred fault zone. Malone (1994) interprets the contact to be depositional, and attributes the change in stratigraphic position to the deposition of the younger Deer Creek Member of the Wapiti Formation on the east side of a steep middle Eocene hill that is composed of the Willwood Formation. A detailed examination of this locality is highly recommended if time permits. The exposure can be reached by several horse trails from the Castle Rock Ranch.

Continue northeast on the South Fork Road.

2.8	20.9	Junction with the lower South Fork road. Turn left (north).
0.2	21.1	Cross the South Fork Shoshone River.
1.5	22.6	Junction with Castle Rock Ranch road. Turn left (west). Sheep Mountain allochthon is to the north.
1.4	24.0	Junction with Hidden Valley Ranch road. Turn right (northwest). Proceed northwest to the Hidden Valley Ranch at the end of the lane.
1.1	25.1	Park cars and ask permission for access. The Sheep Mountain allochthon is to the east, two smaller allochthons occur to the west. To the north is a large volcanic block of the Deer Creek Member.

STOP 3-2: BEAR CREEK AREA

This locality provides an excellent opportunity to view the structural relationship between the Deer Creek Member and HMD and to observe the small-scale structure of the Deer Creek Member Matrix. Traverse on horse trail along the south side of Bear Creek for about 2.5 km. The trail slowly climbs up hill away from the creek. Several smaller volcanic and limestone blocks (~10 m in diameter) occur along the trail about ~120 m above the creek. View to the north of the contact between a Deer Creek Member block and an upper plate allochthon of the HMD (Figure 18). At this locality, the dip of volcanic strata is gently to the northwest. The Paleozoic strata of the upper plate allochthon also dip gently to the northwest. A 10-20 m wide zone of gray matrix occurs between these two blocks. The high-angle truncation of volcanic strata along the contact indicates that these volcanic rocks are allochthonous. Proceed northward down to Bear Creek and up the other side along the contact to view the Deer Creek Member matrix (the climb is about 150 m).

The debris-avalanche matrix is a highly sheared and pervasively brecciated zone that is found beneath and between blocks. The recognition of matrix is the key to mapping the distribution of the unit and to understanding its origin. The matrix is typically light gray and bears a strong resemblance to ready-mix concrete. Most of the matrix consists of sand-sized particles, which consist in decreasing abundance, of lithic fragments, hornblende, pyroxene, glass, feldspar, and quartz. Pebble- to boulder-sized fragments are composed dominantly of light gray and pink hornblende andesite, and lesser dark gray trachyandesite. Matrix usually breaks through, rather than around, enclosed clasts, yield-

Figure 16. Panoramic view to the north of the type area of the Deer Creek Member of the Wapiti Formation from the South Fork Shoshone River Valley, about 3 km away (from Malone, 1995). The light-colored, grassy foothills are underlain by the Willwood Formation (Twl) and Cody Shale (Kc). The heavy dashed line is the early middle Eocene land surface with more than 320 m of relief. In this scene, two blocks (>1 km in diameter) are visible (Twdb). The block to the right (east) consists of about 250 m of interbedded breccias, sandstones, and conglomerates, and dips about 25° to the north. The two blocks are bounded by a poorly exposed, lighter-colored interval of matrix. Matrix (Twdm) also occurs beneath each block but is too thin to be resolved from this distance.

Figure 17. View to the east of the type area of the Deer Creek Member of the Wapiti Formation (from Malone, 1996). Shown here are two blocks separated by a zone of matrix up to 25 m wide. The block to the left (north) dips about 35° to the northwest, and the block to the right (south) dips about 10° to the north. This locality displays an imbricate relationship between Deer Creek Member blocks. The width of photo is about 320 m.

ing smooth and rounded outcrops. Matrix was formed by the progressive disaggregation of block material during transport. Thus, matrix is entirely tectonic in origin, and no primary (depositional) layering is present. Matrix as defined here also includes all volcanic fragments <25 m in diameter. Broken and shattered clasts (ranging from pebble to boulder size) are the most common structural feature within the matrix. Clasts of competent, dark-colored trachyandesite lava display a wide variety of fracture patterns, ranging from well-developed conjugate shear sets to pervasive shattering. Most of the shattering and fracturing of individual clasts does not penetrate the surrounding finer grained material, and the orientations of fractures within adjacent clasts are variable. In some places, minor displacement along the fractures has produced visible offset. In extreme cases, fractures within clasts were intruded by a slurry-like injection of sand-sized material. Less common, but also significant, are strained clasts. Most strained clasts are composed of incompetent, light-colored, hornblende andesite. As seen in thin section, the boundaries of strained clasts are difficult to discern, and intrusive-appearing contacts between these clasts and the surrounding material are common. The progressive deformation of these incompetent clasts led eventually to their complete destruction and homogenization into the surrounding material. Planar and curviplanar clastic dikes are found throughout the matrix. Individual dikes range in width from a few centimeters to a meter, and are as much as ten meters in length. Most clastic dikes are light gray and composed dominantly of fine sand- to small-pebble-sized material. At a few localities, multiple episodes of clastic dikes are present and complex cross-cutting relations among dikes are common, indicating that the matrix was dynamic, and that high pore-fluid pressures existed within the matrix throughout the emplacement of the unit. Some of these clastic dikes

intrude overlying blocks. These clastic dikes were likely injected late, and they indicate that some blocks were strong enough to support the development of extension fractures during the late stages of emplacement. Return to vehicles via the horse trail along Bear Creek. Retrace route back along the South Fork Road to the TE Ranch road.

| *41.4* | 0.0 | Junction of the South Fork Road with the TE Ranch Road. |
| *1.6* | 1.6 | Park at "Entering Shoshone National Forest" Sign. |

STOP 3-3: OVERVIEW OF THE UPPER SOUTH FORK SHOSHONE RIVER VALLEY

More than 1550 m of layered volcanic rocks are exposed in the upper South Fork Shoshone River valley. In this area, rocks from the Wapiti Formation, the Trout Peak Trachyandesite, and the Wiggins Formation have all been identified (Malone, 1997; Figures 19 and 20). The Willwood Formation is locally exposed at the base of the volcanic succession in the Ishawooa Hills and Slide Mountain areas. The following discussion is modified from Malone (1997). A geologic map and manuscript describing the stratigraphy and structure of this area is in preparation. The Deer Creek Member in this area is extremely poorly sorted and consists of particles ranging from silt- to boulder-size. No internal stratification within the unit has been recognized, indicating that the entire unit was likely emplaced during a single depositional event. The unit is matrix-supported and resembles a very large laharic breccia. Most clasts consist of well- to sub-rounded, dark gray trachyandesite, red pyroxene andesite porphyry; and light gray hornblende andesite; the dark gray trachyandesite is the most abundant.

Figure 18. View to the north of the contact between the Deer Creek Member and allochthonous Paleozoic block.

Several large, irregular inclusions of trachyandesite lava and epiclastic breccia occur throughout the deposit, but they are concentrated in the middle. Some of these inclusions are comprised of light colored, well-stratified epiclastic sandstones and breccias, but this variety is rare. These inclusions range in size from about 3 m to more than 155 m in diameter. Layering within these inclusions is often apparent from a distance, but upon closer inspection it is difficult to discern. Where present, this layering appears to be highly contorted, and it is truncated along the margins of the inclusions. Contacts of these inclusions range from sharp to gradational. Where contacts are gradational and the volcanic material within and adjacent to the inclusion are texturally similar, the contact is only recognized by a change in color. Where contacts between the inclusions and adjacent material are sharp, the contacts are locally striated—most commonly where the inclusions are comprised of trachyandesite lava. These striations are typically on the inclusion rather than on the surrounding material. In most cases (a total of five striated inclusions were observed), the striations are unidirectional and subhorizontal and indicate laminar rather than turbulent flow.

Decker (1990) interpreted these inclusions as the result of the liquefaction and dismemberment of the overlying lava flows, flow breccias, and epiclastic breccias. In the present study, no disruption of the overlying lava flows was observed, and all liquefaction domains are probably confined to the subjacent well-stratified succession. More likely, most of these inclusions are probably disaggregated debris-avalanche blocks that were suspended and supported buoyantly by the strength of the surrounding matrix material when the predominant transport mechanism evolved from slide to flow. The origin of the lighter colored inclusions is less obvious, as no light-colored strata within Deer Creek Member has been recognized elsewhere. It is possible that these inclusions were incorporated from the subjacent units during emplacement. The Deer Creek Member here

averages about 600 ft in thickness, and ranges from a low of 90 m between Sheephead and Hunter creeks to a high of 450 m between Legg and Schoolhouse creeks. The top of the unit is mostly flat, and the variation in thickness is mainly a function of relief on the lower surface. In general, the unit is sheet-like in geometry, but it thins gradually to the south and west. In most cases, the relief on the basal surface is less dramatic. Throughout most of the area, the lower contact occurs between 2240 m and 2305 m in elevation. These values are similar to the higher elevations observed along the base of the Deer Creek Member in the lower North and South Fork Shoshone River valleys.

It is likely that a distal-facies of the Deer Creek Member occurs throughout the upper South Fork Shoshone River valley. Evidence for this assertion include: 1) the presently known southeastern extent of the Deer Creek Member is only 18 km from the Ishawooa Hills area; the only stratigraphic pinch-out of the proximal facies of the Deer Creek Member occurs near Jordan Creek at an elevation of about 2460 m; this pinch-out is attributed to a paleotopographic high, and the unit could well occur to the southwest of this paleohill at lower elevations; 2) in both areas, the rocks display a similar color, texture, scale, and degree of deformation; 3) in both areas, the units overlie a major early middle Eocene unconformity; 4) kinematic data indicate a southeastward transport direction. This is consistent with a collapse directed outward from the inferred source area. The structural differences between the two areas can be explained by a difference in the emplacement mechanisms. In the lower North and South Fork Shoshone River valleys, sliding was the dominant emplacement mechanism, and blocks remained relatively intact and undeformed. With further transport, blocks disaggregated, and the dominant emplacement mechanism evolved from slide to flow, with the resultant deposit having characteristics resembling a very large, laharic debris flow deposit. The total distance from the inferred source area at Sunlight Peak to the extreme southern exposure of the unit in this area is at least 58 km. This increases

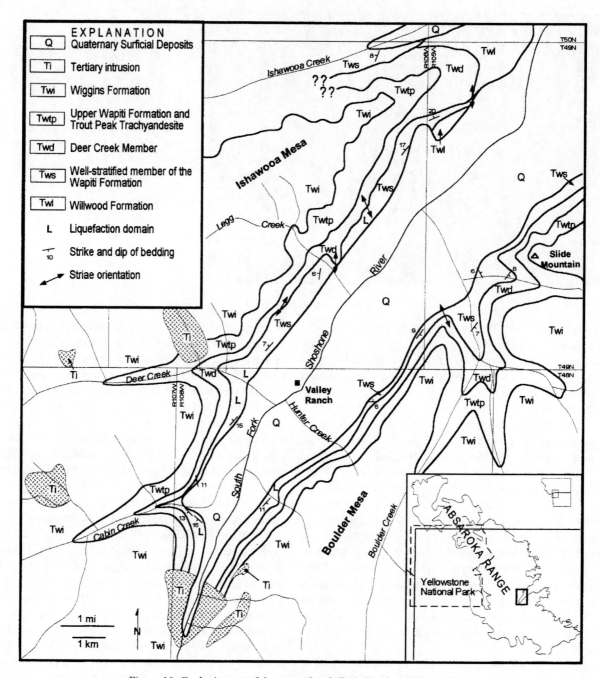

Figure 19. Geologic map of the upper South Fork Shoshone River valley.

the previously known transport distance by some 18 km. The actual maximum transport distance is probably greater, as the southernmost part is concealed by younger volcanic rocks. Another implication of the assignment of these rocks to the Deer Creek Member is that a clear distinction can be made between the respective distribution areas of the latter and the upper plate of the HMD. The closest upper plate remnant is in the Sheep Mountain area which is more than 32 km to the northeast, and no Paleozoic limestone fragments of any size were identified within the Deer Creek Member in this area. This demonstrates that they are areally distinct and is another reason to suggest that the Deer Creek Member and Heart Mountain allochthon were emplaced sequentially rather than coevally. The emplacement of the Deer Creek Member could have contributed to the liquefaction-related deformation in the underlying well-stratified epiclastic units, and several domains of chaotically dismembered bedding have been identified (Decker, 1990) at Deer Creek Canyon, between Deer and Cabin Creeks, between Sheepeater and Hunter Creeks, and Southwest of Legg Creek. The deformation in these domains has been attributed to in-situ liquefaction caused by excess pore fluid

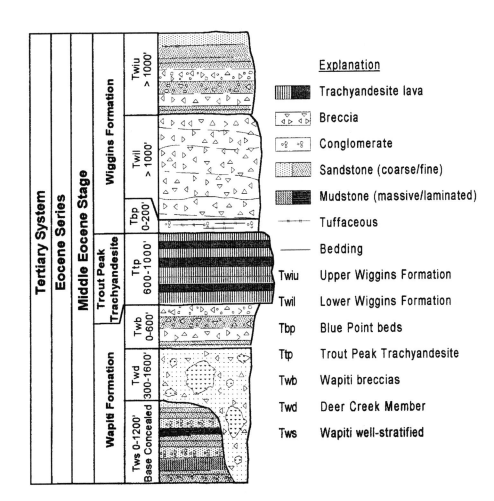

Explanation

	Trachyandesite lava
	Breccia
	Conglomerate
	Sandstone (coarse/fine)
	Mudstone (massive/laminated)
	Tuffaceous
	Bedding
Twiu	Upper Wiggins Formation
Twil	Lower Wiggins Formation
Tbp	Blue Point beds
Ttp	Trout Peak Trachyandesite
Twb	Wapiti breccias
Twd	Deer Creek Member
Tws	Wapiti well-stratified

Figure 20. Stratigraphic column of the Eocene volcanic rocks exposed in the upper South Fork Shoshone River valley.

pressures resulting from cyclic loading during large earthquakes. It is possible that the liquefaction-related deformation may have been induced by impulsive loading caused by the emplacement of the debris avalanche rather than repetitive loading during earthquakes. Unfortunately, there is no way to evaluate or quantify the relative importance of either stimulus.

Proceed southwest on the South Fork Road.

| 1.9 | 3.5 | Bridge over the South Fork Shoshone River. |
| 1.8 | 5.3 | Park at the red U.S. Forest Service gate and proceed north across the sage covered foot hills to the base of the cliff. |

STOP 3-4: DISTAL FACIES OF THE DEER CREEK MEMBER IN THE ISHAWOOA HILLS AREA

This location provides an opportunity to closely observe an outcrop of the distal facies of the Deer Creek Member. Access to the Deer Creek Member is easiest here because the lower contact is a relatively low elevation (paleovalley). The lower 30 m of the exposure consists of northward-dipping epiclastic sandstone, breccia, and conglomerate. The base of the Deer Creek Member is well-exposed here and changes stratigraphic

position rapidly. The underlying well-stratified succession is commonly extensively fractured just beneath the contact. Several large, irregular inclusions of trachyandesite lava can be observed here within the Deer Creek Member. The contacts of these inclusions are striated. These striations are subhorizontal or gently inclined and trend nearly north-south.

Return to the vehicles and proceed south along the South Fork Road.

1.5	6.8	Double Diamond X Ranch on right. Slide Mountain to the east.
0.8	7.6	Ishawooa Mesa Trail Head
1.9	9.5	Bridge over Legg Creek.

STOP 3-5: OVERVIEW OF DRAMATIC PALEO-TOPOGRAPHY NORTHEAST OF LEGG CREEK

Pull vehicles off the road near the bridge over Legg Creek. View to the north of a spectacular paleovalley filled by the Deer Creek Member. During emplacement, the Deer Creek Member filled and topped steep gorges that were carved in the underlying well-stratified succession (Figure 20). The west contact is nearly vertical for about 300 m of paleorelief. The east contact is less steep; the paleoslope on this side of the ancient canyon is about 20 degrees. The origin and preservation of such a feature is enigmatic; it probably would not have been preserved if it had not been immediately and fully buried by the deposition of the Deer Creek Member. This relationship is not recognized on the opposite cliff on the south side of the South Fork valley.

Along the west side of this paleovalley, there is evidence of drag of the well-stratified units near the contact. In addition, sub-horizontal striations were observed in two places along the contact. Both of these structures indicate that the feature is not merely an unconformity but rather both an unconformity and a fault at the same time. This relationship is remarkably similar to what is observed along the breakaway of the HMD, and a brief comparison of the features of each is warranted. 1) Although at its southernmost mapped point, the breakaway is buried by younger Wapiti Formation rocks and is more than (48 km) away, the Legg Creek locality is along strike with the breakaway fault. 2) Both the breakaway and this structure are half-faults (Pierce, 1979), the lower surface is a fault surface while the upper surface is a site of deposition. 3) Both the breakaway and this structure display a similar relief across a short lateral distance. 4) Massive, dark-colored breccias immediately buried both the breakaway and this feature. It is likely that the Deer Creek Member is present within the hanging wall of both localities. 5) The footwall of each is the western side and consists of light colored, well-stratified medial-distal-facies volcanic rocks. 6) Both structures are similar (probably identical) in age. Despite these similarities, it is unlikely that this structure in the upper South Fork valley is a southern extension of the breakaway of the HMD. Other large paleovalleys exist in the Ishawooa Hills area and along Little Boulder Creek. The paleorelief in these areas is less spectacular, but it is still in excess of 250 m. Unfortunately, since exposures are limited to the canyon walls, it is impossible to trace these paleovalleys laterally.

Proceed southwest along the South Fork Road.
4.5   14.0   End of the South Fork Road.

STOP 3.6: OVERVIEW OF DISTAL FACIES OF THE DEER CREEK MEMBER AT CABIN CREEK

The Deer Creek member is beautifully exposed in the steep cliff face just north of and 170 m above Cabin Creek. At this locality, the unit is approximately 100 m in thickness. Several large inclusions of bedded distal facies rocks occur near the base of the exposure. Two light-colored dacite dikes intrude the units. The discontinuous geometry of these dikes indicate that the Deer Creek Member was not fully lithified as the dikes intruded. These dikes are related to a series of small plutons in the area that intruded during the deposition of the overlying Wiggins Formation.

REFERENCES CITED

Beutner, E.C. and Craven, A.E., 1996, Volcanic fluidization and the Heart Mountain detachment, WY: Geology, v. 24, p. 595-598.

Bucher, W. H., 1947, Heart Mountain problem, in Guidebook, Wyoming Geological Association Field Conference for 1947, Bighorn Basin, p. 189-197.

Chadwick, R.A., 1970, Belts of eruptive centers in the Absaroka-Gallatin Volcanic Province, Wyoming and Montana: Geological Society of America Bulletin, v. 81, p. 267-273.

Decker, P.L., 1990, Style and mechanics of liquefaction-related deformation, lower Absaroka Volcanic Supergroup (Eocene), Wyoming: Geological Society of America Special Paper #240, 71 pp.

Elliott, J. E., 1979, Geologic map of the southwestern part of the Cooke City quadrangle, Montana and Wyoming: United States Geological Survey Miscellaneous Investigations Series, Map I-1084, 1:24,000.

Erskine, D. W., and Kudo, A. M., 1991, Evidence from a Shoshonite dike in support of the continuous allochthon model, Heart Mountain fault, northwestern Wyoming [Abstract]: EOS., v. 72, no. 44, p. 490.

Hague, A., Iddings, J.P., Weed, W.H., Walcott, C.D., Girty, G.H., Stanton, T.W., and Knowleton, F.G., 1899, Descriptive geology, petrography, and paleontology, Part II of Geology of the Yellowstone National Park; U. S. Geological Survey Monograph #32, 893 pp.

Hauge, T. A., 1982, The HMD fault, northwest Wyoming: Involvement of Absaroka volcanic rock: Wyoming Geological Association, 33rd Annual Field Conference, Guidebook, p. 175-179.

Hauge, T. A., 1983, Geometry and kinematics of the HMD fault, northwestern Wyoming and Montana [Ph. D. thesis]: Los Angeles, California, University of Southern California, 265 p.

Hauge, T. A., 1985, Gravity-spreading origin of the Heart Mountain allochthon, northwestern Wyoming: Geological Society of America Bulletin, v. 96, p. 1440-1456.

Hauge, T. A., 1990, Continuous-allochthon model of Heart Mountain faulting: Geological Society of America Bulletin, v. 102, p. 1174-1188.

Hauge, T. A., 1992, HMD, Northwestern Wyoming, in Elliott, J. E., ed, Guidebook for the Red Lodge - Beartooth Mountains - Stillwater Area: Northwest Geology v. 20-21 (Tobacco Root Geological Society, Seventeenth Annual Field Conference), p. 21-46.

Hauge, T. A., 1993, The HMD, northwestern Wyoming: 100 years of controversy: in Snoke, A. W., Steadtman, J. R., and Roberts, S. M., eds., Geology of Wyoming (Blackstone - Love Volume), Wyoming Geological Survey Memoir No. 5, p. 530-571.

Hewett, D. F., 1920, The Heart Mountain overthrust, Wyoming: Journal of Geology, v. 28, p. 536-557.

Hughes, C. J., 1970a, Role of cohesive strength in the mechanics of overthrust faulting and of landsliding: Discussion: Geological Society of America Bulletin v. 81, p. 607-608.

Hughes, C. J., 1970b, The HMD fault - a volcanic phenomenon?: Journal of Geology, v. 78, no. 1, p. 107-116.

Hughes, C. J., 1970c, The HMD fault - a volcanic phenomenon? A Reply: Journal of Geology, v. 78, p. 629-630.

Hughes, C. J., 1973, Igneous activity, metamorphism, and Heart Mountain faulting at White Mountain, northwestern Wyoming: Discussion: Geological Society of America Bulletin, v. 84, p. 3109-3110.

Malone, D.H., 1994, A Debris-Avalanche Origin for Absaroka Volcanic Rocks Overlying the HMD, Northwest Wyoming. Unpublished Ph.D. dissertation, The University of Wisconsin, 292 pp.

Malone, D.H., 1995, A very large debris-avalanche deposit within the Eocene volcanic succession of the Northeasten Absaroka Range, Wyoming: Geology, v. 23, no.7, p.661-664.

Malone, D. H., 1996, Revised Stratigraphy of Eocene Volcanic Rocks in the Lower North and South Fork Shoshone River Valleys, Wyoming: Wyoming Geological Association Annual Field Conference Guidebook, vol. 47, p. 109-138.

Malone, D.H., 1997, Recognition of a Distal Facies Greatly Extends the Domain of the Deer Creek Debris-Avalanche Deposit (Eocene), Absaroka Range, Wyoming: Wyoming Geological Association Annual Field Conference Guidebook, vol. 48, p. 1-9.

Nelson, W. H., and Pierce, W. G., 1968, Wapiti Formation and Trout Peak trachyandesite, northwestern Wyoming: United States Geological Survey Bulletin, 1254-H, p. H1-H11.

Nelson, W. H., Pierce, W. G., Parsons, W. H., and Brophy, G. P., 1972, Igneous activity, metamorphism, and Heart Mountain faulting at White Mountain, northwestern Wyoming: Geological Society of America Bulletin, v. 83, no. 9, p. 2607-2620.

Nelson, W. H., Pierce, W. G., and Brophy, G. P., 1973, Igneous activity, metamorphism, and Heart Mountain faulting at White Mountain, northwestern Wyoming: Reply: Geological Society of America Bulletin, v. 84, p. 3111-3112.

Nelson, W. H., Prostka, H. J., and Williams, F. E., 1980, Geology and mineral resources of the North Absaroka Wilderness and vicinity, Park County, Wyoming: United States Geological Survey Bulletin 1447, with sections on Mineralization of the Cooke City mining district by James E. Elliott and Aeromagnetic survey by Donald L. Peterson, 101 p.

Pierce, W. G., 1957, Heart Mountain and South Fork detachment thrusts of Wyoming: American Association of Petroleum Geologists Bulletin, v. 41, no. 4, p. 591-626.

Pierce, W. G., 1960, The "break-away" point of the HMD fault in northwestern Wyoming: in Geological Survey Research 1960: United States Geological Survey Professional Paper 400-B, p. B236-B237.

Pierce, W. G., 1965, Geologic map of the Deep Lake Quadrangle, Park County, Wyoming: U.S. Geological Survey Geologic Quadrangle Map GQ-478, 1:62,500.

Pierce, W. G., 1968, Tectonic denudation as exemplified by the Heart Mountain fault, Wyoming, in Orogenic Belts: 23rd International Geological Congress, Prague, Czechoslovakia, 1968, Report, Section 3, Proceedings: p. 191-197.

Pierce, W. G., 1973, Principal features of the Heart Mountain fault, and the mechanism problem, in DeJong, K. A., and Scholten, R., editors, Gravity and tectonics: New York, John Wiley and Sons, p. 457-471.Pierce, W. G., 1978, Geologic map of the Cody 1° x 2° Quadrangle, northwestern Wyoming: United States Geological Survey Miscellaneous Field Studies Map MF-963, 1:62,500.

Pierce, W. G., 1979, Clastic dikes of Heart Mountain fault breccia, northwestern Wyoming, and their significance: United States Geological Survey Professional Paper 1133, p. 1-25.

Pierce, W. G., 1980, The Heart Mountain break-away fault, northwestern Wyoming: Geological Society of America Bulletin, Part 1, v. 91, p. 272-281.

Pierce, W. G., 1985, Map showing present configuration of Heart Mountain fault and related features, Wyoming and Montana: Geological Survey of Wyoming, Map Series MS-15, 1:125,000.

Pierce, W. G., 1987a, The case for tectonic denudation by the Heart Mountain fault - a response: Geological Society of America Bulletin, v. 99, p. 552-568.

Pierce, W. G., 1987b, HMD fault and clastic dikes of fault breccia, and Heart Mountain break-away fault, Wyoming and Montana, in Beus, S. S., editor, Geological Society of America Centennial Field Guide, Rocky Mountain Section, p. 147-154.

Pierce, W. G., and Nelson, W. H., 1968, Geologic map of the Pat O'Hara Mountain Quadrangle, Park County, Wyoming: U.S. Geol. Survey Geological Quadrangle Map GQ-755, 1:62,500.

Pierce, W. G., and Nelson, W. H., 1969, Geologic Map of the Wapiti Quadrangle, Park County, Wyoming: U.S. Geol. Survey Geological Quadrangle Map GQ-778, 1:62,500.

Pierce, W. G., and Nelson, W. H., 1970, The HMD fault -a volcanic phenomenon? A discussion: Journal of Geology, v. 78, p. 116-122.

Pierce, W. G., and Nelson, W. H., 1971, Geologic map of the Beartooth Butte quadrangle, Park County, Wyoming: United States Geological Survey Geological Quadrangle Map GQ-935, 1:62,500.

Pierce, W. G., and Nelson, W. H., 1986, Some features indicating tectonic denudation by the Heart Mountain fault: Guidebook, 1986 Montana Geological Society -- Yellowstone-Bighorn Research Association Field

Conference, p. 155-164.

Pierce, W. G., Nelson, W. H., and Prostka, H. J., 1973, Geologic Map of the Pilot Peak Quadrangle, Park County, Wyoming, and Park County, Montana: U.S. Geological Survey Miscellaneous Geologic Investigations Map I-816, 1:62,500.

Pierce, W. G., Nelson, W. H., Tokarski, A. K., and Piekarska, E., 1991, Heart Mountain, Wyoming, detachment lineations -- are they in microbreccia or in volcanic tuff?; Geological Society of America Bulletin, v. 103, p. 1133-1145.

Prostka, H. J., 1978, Heart Mountain fault and Absaroka volcanism, Wyoming and Montana, U.S.A., in Voight, B., editor, Rockslides and Avalanches, I, Natural Phenomena: Amsterdam, Oxford, New York, Elsevier Scientific Publishing Company, p. 423-437.

Prostka, H. J., Ruppel, E. T., and Christiansen, R. L., 1975, Geologic map of the Abiathar Peak Quadrangle, Yellowstone National Park, Wyoming and Montana: U.S. Geological Survey Geologic Quadrangle Map GQ-1244, 1:62,500.

Sales, J.K., 1983, Heart Mountain; blocks in a giant volcanic rock glacier: Wyoming Geological Association, 34th Annual Field Conference Guidebook, p. 117-165.

Smedes, H. W., and Prostka, H. J., 1972, Stratigraphic framework of the Absaroka volcanic supergroup in the Yellowstone National Park region: U. S. Geol. Survey Prof. Paper 729-C, 33 p.

Sundell, K.A., 1985, The Castle Rocks Chaos; a gigantic Eocene landslide-debris flow within the southeastern Absaroka Range, Wyoming: Unpublished Ph.D. Dissertation, University of California, 283 pp.

Sundell, K.A., 1993, The Absaroka volcanic province, in Snoke A.W., Steidtman, J.R., and Roberts, S.M., eds, Geology of Wyoming: Geological Survey Memoir #5, p. 572-603.

Templeton, A. S., Sweeny, J. Jr., Manske, H., Tilghman, J. F., Calhoun, S. C., Violich, A., and Chamberlain, C. Paige, 1995, Fluids and the Heart Mountain fault revisited: Geology, v. 23, p. 929-932.

Torres, V., and Gingerich, P. D., 1983, Summary of Eocene stratigraphy at the base of Jim Mountain, North Fork of the Shoshone River, northwestern Wyoming: Wyoming Geological Association 34th Annual Field Conference Guidebook, p. 205-208.

Tucker, T. E., 1982, Dead Indian Hill to Yellowstone National Park, Northeast Entrance: Wyoming Geological Association, 33rd Annual Field Conference, Guidebook, p. 380-386.

Voight, B., 1974a, Architecture and mechanics of the Heart Mountain and South Fork rockslides, in Voight, B. and Voight, M. A., editors, Rock Mechanics: The American Northwest, 3rd Congress International Society of Rock Mechanics Expedition Guidebook: Special Publication, Experiment Station, College of Earth and Mineral Sciences, The Pennsylvania State University, p. 26-36.

Voight, B., 1974b, Roadlog: Wapiti-Heart Mountain Area - Canyon, in Voight, B. and Voight, M. A., editors, Rock Mechanics: The American Northwest, 3rd Congress International Society of Rock Mechanics Expedition Guidebook: Special Publication, Experiment Station, College of Earth and Mineral Sciences, The Pennsylvania State University, p. 112-124.

Wise, D. U., 1983, Overprinting of Laramide structural grains in the Clarks Fork canyon area and eastern Beartooth Mountains of Wyoming: Wyoming Geological Association 34th Annual Field Conference Guidebook, p. 77-87.

Printed in U.S.A.